Social History of Africa

COLONIAL CONSCRIPTS

Social History of Africa
Series Editors:
Allen Isaacman and Luise White

Belinda Bozzoli
(with the assistance of Mmantho Nkotsoe)
Women of Phokeng:
Consciousness, Life Strategy, and Migrancy
in South Africa, 1900–1983

Myron Echenberg
Colonial Conscripts:
The Tirailleurs Sénégalais *in French West Africa,*
1857–1960

Kristin Mann and Richard Roberts (editors)
Law in Colonial Africa

COLONIAL CONSCRIPTS

The *Tirailleurs Sénégalais*
in French West Africa,
1857–1960

Myron Echenberg
McGill University

HEINEMANN
Portsmouth, NH

JAMES CURREY
London

Heinemann
A division of Reed Elsevier Inc.
361 Hanover Street
Portsmouth, NH 03801
Offices and agents throughout the world

James Currey Ltd
54b Thornhill Square, Islington
London N1 1BE

© 1991 by Myron Echenberg. All rights reserved. No part of this book may be reproduced in any form or by electronic or mechanical means, including information storage and retrieval systems, without permission in writing from the publisher, except by a reviewer, who may quote brief passages in a review.

First published 1991.

Credits for reprinted material are on p. xix.

Library of Congress Cataloging-in-Publication Data
Echenberg, Myron J.
  Colonial conscripts : the *Tirailleurs Sénégalais* in French West Africa, 1857–1960 / Myron Echenberg.
    p. cm. — (Social history of Africa series)
  Includes bibliographical references and index.
  ISBN 0-435-08048-2. — ISBN 0-435-08052-0 (pbk.)
  1. France. Armeé. *Tirailleurs sénégalais*—History.   2. France. Armée—Colonial troops—History—19th century.   3. France. Armée—Colonial troops—History—20th century.   4. Africa, French-speaking West—Race relations.   5. Africa, French-speaking West—History, Military.   I. Title.   II. Series.
UA703.T725E28 1991
355.3'59'0944—dc20   90-43157

British Library Cataloguing in Publication Data
Echenberg, Myron
  Colonial conscripts: the Tirailleurs Sénégalais in French West Africa, 1857–1960.   (Social history of Africa).
  1. West Africa. Warfare, history
  I. Title II. Series
  355.020966

  ISBN 0-85255-651-9
  ISBN 0-85255-601-2 pbk

Cover design by Jenny Greenleaf.
Text design and prepress work by G&H Soho Ltd.
Printed in the United States of America.
Docutech PB 2002

To my mother, Anne Linds Echenberg,
and the memory of my father,
Leon Echenberg

CONTENTS

	Illustrations	ix
	Tables	xi
	Acknowledgments	xiii
	Caveats, Spelling Conventions, and Abbreviations	xv
	Credits	xix
1	Four *Tirailleur* Armies	1
2	Slaves into Soldiers	7
3	The *Tirailleurs Sénégalais* and the First World War	25
4	Military Conscription in Theory and Practice	47
5	Consequences of Conscription: Military Migration	70
6	"*Morts pour la France*": The *Tirailleurs Sénégalais* and the Second World War	87
7	The New Army, 1945–1960	105
8	French West African Veterans, 1945–1960	127
9	Veterans and French West African Politics, 1945–1960	146
10	Epilogue: Collective Memory and the African Military	164
	Appendix: Questionnaire for Ex-*Tirailleurs Sénégalais*	171
	Notes	175
	A Note on Sources	213
	Sources	217
	Index	229

ILLUSTRATIONS

		Page
Map	Ethnic groups in French West Africa	16–17

Figures

3.1	Size of the *Tirailleurs Sénégalais*, 1890–1953	26
3.2	Annual levies for the *Tirailleurs Sénégalais*, 1908–1954	27
4.1	Quota averages to produce an annual levy of 10,000 men, 1920–1947	53–56
4.2	Quota averages to produce an annual levy of 10,000 men, 1920–1947, by regions	57
4.3	Conscription averages for 1926 by colony (including *cercle* of Bougouni, Soudan)	60
5.1	Average percentages of absentees per 100 men on draft lists, 1923–1946	71
5.2	Percentage of absentees per 100 men on draft lists, 1923–1946	72
5.3	Colonies of origin for the 1,600 men of the first *Tirailleurs Sénégalais* Regiment (Saint-Louis, Senegal), in percentages, 1926	77
5.4	Distribution of veterans by residence, and of soldiers by colony of origin, in percentages, 1952	83
8.1	Dakar-based veterans' lobbies, 1949	129

Photographs

Division of the spoils, Conquest period	11
Tirailleurs and officers with mounted artillery, French West Africa in the late nineteenth century	13

	Page
Tirailleurs marching in Paris on Bastille Day, 1914	30
Mangin and the *Tirailleurs Sénégalais* in Morocco, 1913	31
Colors of the *1er Régiment des Tirailleurs Sénégalais*, First World War	39
Captain Abdel-Kader Mademba	40
Mobile draft board in French West Africa, interwar period	52
The first episode of *"Mamadou s'en va-t'en guerre,"* 1939	90
A *Tirailleur* helpless during the Wehrmacht offensive, May–June 1940	95
Tirailleurs Sénégalais reach Tripoli with Leclerc, January 1943	98
Ecole des Enfants de Troupe, French West Africa, circa 1950	119
EFORTOM classes, circa 1955	124
Tirailleurs Sénégalais veterans with de Gaulle, 1958	142
General de Gaulle decorating a *Tirailleur* sergeant, circa 1944	143
Memorial at Airaines to the Fifty-third *Régiment d'Infanterie Coloniale Mixte Sénégalais*	167
Memorial at Airaines to Captain Charles N'Tchoréré	168

TABLES

		Page
2.1	Growth of the *Tirailleurs Sénégalais*, 1820–1914	7
2.2	Origins and prices of slaves purchased for the *Tirailleurs Sénégalais*	9
2.3	Enlistment bonuses for *Tirailleurs*, 1853–1904	12
3.1	Composition of the French Conquest Army of Morocco, 1908–1913	28
4.1	Male-to-female sex ratios based on the survey of 1922–1923 by colony (and by region and ethnic group for Upper Volta)	49
4.2	Conscription figures for 1926 by colony (including *cercle* of Bougouni, Soudan)	59
4.3	Recruitment of second-portion workers and first-portion soldiers in French Soudan, 1928–1946	62
5.1	General mortality per 1,000 men in the French Colonial Army, 1920–1938	81
7.1	A comparison of recruitment in FWA between 1928 and 1948	109
7.2	A comparison of absentee rates between 1923–1946 and 1948	111
7.3	Comparison of skills in French language of first and second portions of 1936 and 1946 cohorts	114
7.4	African and European officers in *T. S.* units, 1908–1954	118
7.5	African cadets attending EMPAs in 1953–1954	120
7.6	African officers at EFORTOM, 1958–1965	123
8.1	Veterans in FWA, 1952 and 1959	128
9.1	*Office National*'s budget allocated by territory, 1948–1949	155
9.2	Results of the Liger campaign as of July 1950	160

ACKNOWLEDGMENTS

My interest in the subject of African soldiers in French military service began in the 1960s when I first visited the former colonies of French West Africa as a young graduate student then researching another subject. Like most outsiders, I required interpreters to translate oral information into French and to help me find my way about the towns and villages of what was then Upper Volta and is now Burkina Faso. When I sought such assistance, I found that most of those who came forward were African ex-servicemen, men who had learned their French in the *Tirailleurs Sénégalais*. One who helped me greatly and who became my friend was Dramane Sarambé, a Maraka-speaking small businessman and farmer of Tougan. Dramane had been unfortunate enough to have done his military service in the labor brigades of the infamous "second portion" of the military draft (see Chapter 5), and his vivid description of his and others' sufferings on the *Office du Niger* project aroused my compassion and historical curiosity.

Another of my interpreters was Goulli Zo, a Samo-speaking ex-sergeant also from Tougan. He had served over twelve years in the *Tirailleurs Sénégalais*, a period of almost constant belligerency which included the Allied liberation of France, a tour of duty in French Indochina, and a stretch in Algeria. Goulli was a man with a foot in two worlds. On the one hand he knew and valued the history and cultural traditions of Samo society before the French conquest; and on the other he preserved great pride in the accomplishments of the French Colonial Army of which he was a part. Here is how Goulli Zo viewed the capacity of North Vietnam to challenge—let alone defeat—the strongest superpower in the world:

> We fought as best we could in Indochina but in the end the VietMinh were too strong for us. They were fighting to be free and we were fighting because we were told to do so by our officers. There is a difference. The Americans think they will do better because they are stronger than us. I do not think they will win.*

As the years passed, my thoughts about the life of Goulli Zo and other Africans in French uniform, and about their ambiguous but strong bonds which tied them to France, finally germinated into a research project. For their inspiration I thank both Dramane Sarambé and Goulli Zo. I have also learned much from interviews with other former soldiers in West Africa, and their names are listed in a section of the Sources.

Among others in Burkina Faso to whom I am grateful I would like to mention the Traoré brothers—Moussa, Moustapha, and Aliou. Together with their families,

---

*Interview with Goulli Zo, Tougan, Burkina Faso, 7 April 1967.

the Traorés shared with me their understanding of their country's past, and always made me feel welcome at their homes in Tougan, Koudougou, and Ouagadougou. In Senegal, Richard Barry and Emma and Dominique Zidouemba graciously extended their hospitality to me.

Funds for overseas research in West Africa and France have been generously provided by the Social Sciences and Humanities Research Council of Canada, while a grant from the Research Grants Committee, Faculty of Graduate Studies and Research, McGill University, helped in the preparation of statistical material.

Especially helpful in guiding me through their collections were chief archivists Jean-François Maurel at Aix-en-Provence, Saliou M'Baye at Dakar, and head librarian Mme. Marie-Ange Durand of the *Centre Militaire d'Information et de Documentation sur l'Outre-Mer* (CMIDOM) in Versailles. I thank also the industrious staff of the Inter-Library Loan Department, McLennan Library, McGill University.

Bruno Zidouemba in Dakar, Moustapha Traoré in Ouagadougou, and Jim Giblin and Katherine Gombay in Montreal all served as able research assistants at various times. My special gratitude extends to Pat Manning, who believed the subject was important and who urged me to consolidate my thoughts and earlier writings into book form. He along with the following colleagues read all or parts of the manuscript: Fred Cooper, Jean Filipovich, Marc Michel, Allen Isaacman, and Luise White. The last two have graciously included this work in their new series, Social History of Africa, with Heinemann.

Another long list of people—Robin Cohen, Jean Copans, Dennis Cordell, Fred Dreisziger, the late Joel Gregory, Peter Gutkind, Bogumil Jewsiewicki, Paul Lovejoy, Victor Piché, and Richard Rathbone—were editors who read parts of my work in earlier versions. For responding at different stages of this study to my varying argument, I would also like to thank Christiane Akpo, Dan Aronson, Ralph Austen, Abdoulaye Bathily, Bruce Berman, Ray Betts, Pierre Boulle, Bonnie Campbell, Frank Chalk, Tony Clayton, Fred Cooper, Phil Curtin, Mamadou Diouf, Raymond Gervais, Wendell Holbrook, David Killingray, Martin Klein, Ella Laffey, René Lemarchand, Mike Mason, Mohamed Mbodj, Stephen Randall, Larry Stokes, Robert Vogel, and Immanuel Wallerstein. I apologize to all those whom I have named and even more to those whom I have forgotten to name when I have been obstinate in rejecting their counsel. I alone am responsible for errors of fact or logic.

I owe a final debt of gratitude to my wife, Eva, and my children, David and Margo. Although at times I gave them reason to regard my manuscript as an albatross I would never escape, they never doubted either the value of the effort or my capacity to see it through to completion.

CAVEATS, SPELLING CONVENTIONS, AND ABBREVIATIONS

Historians writing in English about modern France and its overseas extensions often offend their French readers by demonstrating a conscious or unintended bias. I hope that I have not struck an anti-French tone in this book, for that is certainly not my intention. If I am critical of French politicians and generals whose decisions affected the lives of so many thousands of Africans, it is as colonial mandarins and not as representatives of French civilization that I have viewed them. Always I have attempted to judge their failings against the standards French democratic norms would set for them, and certainly not against any idealized sense, for instance, that the agents of any other colonial power would have behaved more equitably. My subject is the French Colonial Army, but the historical literature indicates that the British Colonial military perpetuated the same sorts of abuses, and the same compromises with principle. If in the overall imperial scheme of things the British did not conscript soldiers from their African colonies on as grandiose a scale as the French, it was because they recruited more heavily from their possessions in India instead.

Spelling of French West African place names always involves awkward choices in an English text. For the most part, I have used English spellings and have avoided postindependence neologisms. Thus, for present-day Burkina Faso I have preferred Upper Volta to Haute-Volta. But I have made exceptions; the Republic of Côte d'Ivoire adamantly rejects the English translation of its name, and so I have obliged by using the French form throughout. To distinguish it from the former British territory in the Upper Nile Valley in eastern Africa, for modern-day Mali I use French Soudan, or simply Soudan. Also to avoid confusion with other former colonies, I use French Cameroun and French Guinea for the Republic of Cameroun and for Guinée-Conakry, respectively. Senegal appears without accents, but I have preserved the acute accents in such proper names as Léopold Sédar Senghor or Moussa Traoré.

The translation of military designations from French to English presents another set of problems. Throughout this text, I have used the name given to black African soldiers in French service—that is, *"Tirailleurs Sénégalais."* I prefer this choice despite its inaccuracy, and despite the preference former soldiers from territories other than Senegal would undoubtedly have for a more generic term like

"*Tirailleurs Africains*." *Tirailleur* may, of course, be rendered into English as "sharpshooter," "rifleman," or even "scout."[1] In earlier writings I have used the British-style term "Senegalese Rifles" or "African *Tirailleurs*." The latter designation avoids the impression that these regiments were made up primarily or exclusively of men from Senegal; in fact, by the twentieth century, recruitment extended throughout French West Africa. On the other hand, the term not only awkwardly mixes French and English, it tends to obscure an existentialist myth of French colonialism that each and every black African subject who ever visited the metropolis experienced: that for the French all blacks were "Senegalese" whatever their ethnic or regional origins.

It may also be useful at this point to situate the *Tirailleurs Sénégalais* within the French army as a whole. Black African troops were organized in their own regiments as a significant part of the French Colonial Army, or *La Coloniale*, and were commanded mainly by French officers but with a small minority of junior officers who were themselves African. Though the *Tirailleurs Sénégalais* were a major component of the Colonial Army, other largely segregated units included the Malagasy *Tirailleurs*, the Algerian *Tirailleurs*, the Moroccan *Tirailleurs*, and several other more specialized units.[2] The French *Armée d'Afrique*, or Army of Africa, on the other hand, while part of the Colonial Army, consisted of men of European origin recruited from among the so-called *colons*, or settlers living throughout French North Africa. The Army of Africa, in particular, should definitely not be confused with the *Tirailleurs Sénégalais*, since African subjects were not allowed in this exclusive force. No inventory of French forces in Africa should omit the ubiquitous Foreign Legion, most decidedly also a part of the Colonial Army, though consisting by definition of men who were neither French citizens nor subjects. In contrast to the Colonial Army, the French Metropolitan Army refers to the territorial regiments of metropolitan France, as well as to special units officered by men who held commissions as officers of the Metropolitan Army. In times of crisis and disaster— as in 1940 and the years immediately following—these distinctions grew blurred. For most of the twentieth century, however, very clear distinctions existed between the Metropolitan and Colonial armies and among their respective officers.

These cleavages were never exclusively along racial lines, though race was clearly a factor. Most of the rank and file in the Metropolitan Army was white, and in the Colonial Army, nonwhite. However, the Foreign Legion and the Army of Africa were exceptions in the colonial Army, and French nonwhite citizens from the French Antilles and Réunion, for example, enlisted as a rule in French territorial regiments of the Metropolitan Army. So too would a nonwhite Frenchman born and

---

[1] The last option is provided by Anthony Clayton, although he also has chosen to remain with *Tirailleur Sénégalais* in his *France, Soldiers and Africa* (London: Brassey's, 1988).

[2] The point needs stressing for specialists in sub-Saharan Africa. North African troops, especially those from Morocco, contributed significantly to France's war machine in the twentieth century. See Charles-Robert Agéron, *Les Algériens Musulmans et la France*, vol. 2 (Paris: Presses Universitaires de France, 1968), 1139–89; and Clayton, *France, Soldiers and Africa*, 244–306. Clayton is also the best guide through the labyrinth of French army terms for the various colonial units.

raised, for instance, in Toulouse. For the citizens of the Four communes of Senegal, unit affiliations were more complex. Unless they were officers, these privileged French West Africans would not normally serve in the *Tirailleurs Sénégalais*, but would instead be seconded either to North African or Metropolitan units.

ABBREVIATIONS

AAAMCC	*Association Amicale des Anciens Militaires Coloniaux de Carrière*
AACVGAOF	*Association des Anciens Combattants et Victimes de Guerre de l'Afrique Occidentale Française* (President: Papa Seck Douta)
AEF	*Afrique Equatoriale Française*, translated as French Equatorial Africa, or FEA.
AGAGBAOF	*Association Générale des Amputés et Grands Blessés* (President: Paye Dame Baye; and later, Momar Baye)
ANSHA	*Archives Nationales, Service Historique de l'Armée*
ANS	*Archives Nationales du Sénégal*
ANSOM	*Archives Nationales, Section Outre-Mer*
AOF	*Afrique Occidentale Française*, translated as French West Africa, or FWA
BEPC	*Brevet d'Etudes du Premier Cycle*
BDS	*Bloc Démocratique Sénégalais*
CEP	*Certificat d'Etudes Primaires*
CFA franc	*Communauté Financière Africaine* franc
CIC	Commander in chief
CMIDOM	*Centre Militaire d'Information et de Documentation sur l'Outre-Mer*
DIC	*Division d'Infanterie Coloniale*
EETs	*Ecoles des Enfants de Troupe*
EFORTOM	*Ecole de Formation des Officiers Réguliers des Territoires d'Outre-Mer*
EMPA	*Ecole Militaire de Préparation Africaine*
ESMIA	*Ecole Spéciale Militaire Interarmes*
FEA	French Equatorial Africa
FNPG	*Fédération National des Prisonniers de Guerre* (President: Papa Guèye Fall)
FWA	French West Africa
GCS	*Général Commandant Supérieur pour l'AOF*, translated as Commander in Chief, French West Africa
IOM	*Indépendants d'Outre-Mer*
MPS	*Mouvement Populaire Sénégalais*
Office	*Office National des Anciens Combattants*

PCF	*Parti Communiste Française*
POWs	Prisoners of war
RDA	*Rassemblement Démocratique Africain*
RICMS	*Régiments d'Infanterie Coloniale Mixtes Sénégalais*
RPF	*Rassemblement du Peuple Français*
RTS	*Régiments des Tirailleurs Sénégalais*
SFIO	*Section Française de l'Intérnationale Ouvrière*
US	*Union Soudanaise*

CREDITS

Permission is gratefully acknowledged to publish in revised form the following articles or chapters:

The University of Wisconsin Press for Chapter 2, from Paul E. Lovejoy (ed.), *Africans in Bondage: Studies in Slavery and the Slave Trade* (Madison, 1986), pp. 311-33; the *Canadian Journal of African Studies* for parts of Chapters 4 and 5, which appeared as "Paying the Blood Tax: Military Conscription in French West Africa, 1914-1929," 9 (1975), 171-92, and as "Les Migrations militaires en Afrique occidentale française, 1900-1945," 14 (1980), 429-50; Westview Press for part of Chapter 4, from Dennis D. Cordell and Joel W. Gregory (eds.), *African Population and Capitalism: Historical Perspectives* (Boulder and London, 1987), pp. 95-108; and to Cambridge University Press, publishers of the *Journal of African History*, for Chapter 6, which appeared as "*Morts pour la France*": The African Soldier in France during the Second World War," 26 (1985), 363-80.

Photographic Credits

Courtesy of Roger Bruge, page 95; CMIDOM, pages 13, 30, 39, 52, 119, 143; Myron Echenberg, pages 167, 168; *La Gazette du Tirailleur*, page 90; *Revue des Troupes Coloniales*, pages 31, 40, 98, 124 (both); *Souvenir et Devenir*, page 142.

1

Four Tirailleur *Armies*

> *The Black Troops represent African civilization reborn, and constitute our crowning achievement.*
>
> Charles Mangin, *La Force Noire*

This book traces the history of a large and diverse group of West Africans, the men who served in the *Tirailleurs Sénégalais* regiments of the French Colonial Army. Wherever possible, the main focus is on how the rank-and-file soldiers, officers, and veterans who made up the African military responded to their ambiguous and often contradictory position within the colonial social formation. Unfortunately, the nature of the surviving evidence does not permit a complete picture of the experiences of the African military.[1] French military records do, of course, concentrate on military and political questions, but, judging by what has survived, it is disappointing to find that the army showed little interest in how its colonial soldiers lived their lives either while in service or after discharge.

Exceptions do occur. Military statistics occasionally do shed light on the fitness and skill levels of recruits, and death rates in service seemed always to have been a concern of army authorities. Brief glimpses of sex and family life or political affiliation show up through surviving correspondence in the files of military security, which censored the mail of soldiers. Another approach used in this study to compensate obliquely for the paucity of official records is to examine the African military within a framework bounded by such issues as labor, migration, and demography.

This study represents only the beginning of an effort to understand how the African military lived. Further research is required; that focusing upon the lives of former soldiers in the years after 1945 and based upon oral data is especially needed before this last generation of colonial veterans dies out. In the meantime, the African military remains but one of a series of groups about which we still know too little. Equally obscure are the lives of other, nonmilitary, groups in colonial French West Africa (FWA). Longshoremen and railway workers, civil servants and clerks, small businessmen and petty commodity producers, all await fuller scholarly treatment than they have thus far received.[2]

Some readers may be also disappointed to find that certain aspects of the history of the *Tirailleurs Sénégalais* are given less attention than they deserve. Little has been added to what has already been written about the numerous combat experiences of the *Tirailleurs Sénégalais* in two world wars and countless conquest and police operations. Second, the painful experiences of the *Tirailleurs* when called upon to participate in late colonial "counterinsurgency" operations in Indochina and North Africa cannot yet be well treated because archives remain closed for these still recent and painful events.[3]

The *Tirailleurs Sénégalais* have evoked vivid yet contradictory images among adversaries and allies from the time of their founding by Governor Faideherbe of Senegal in 1857 to their phaseout after the coming of independence to French West Africa in 1960. In Germany, the extreme right exploited a racial prejudice shared throughout the Western world against France's decision to use colonial troops in European combat in the First and Second world wars and as occupying forces in the Rhineland after 1918. In *Mein Kampf*, Hitler railed several times against black Africans, whose presence in the French Army he saw as only the latest "pollution," or "negrification" of the once pure Frankish blood of the French.[4] Nazi propaganda in this instance reinforced the existing racial prejudice of prominent members of the German military.[5] Rommel and other important officers in the Second World War considered it an insult to have to fight against such "inferior" races.[6]

Images of the *Tirailleurs Sénégalais* in the English-speaking world are much vaguer. British audiences frequently associate the *Tirailleurs* with Gurkha regiments, with the East African units of the King's African Rifles, and with Nigerian or Gold Coast regiments of the West African Frontier Force, all of them standard examples of colonial mercenary soldiers in the British Empire. Few would remember, however, that the *Tirailleurs Sénégalais* fought beside Canadian and British troops at Vimy Ridge and the Somme in the First World War. Similarly, few Americans today recall that in 1918 black American soldiers were temporarily amalgamated with *Tirailleurs Sénégalais* in the trenches of France.[7]

In France itself, *Tirailleur* images could be romantic, stereotyped, or accurate. An older generation pictured gaudily clad black African regiments parading through the Arc de Triomphe on Bastille Day. Somewhat younger French citizens remembered their childhood and the caricatured portrait of a grinning African soldier on billboards throughout France, the advertising symbol of "Banania," a popular breakfast cereal. Frenchmen born and raised in the colonies often had more substantive memories, as, for example, when the *Tirailleurs Sénégalais* helped protect the lives and property of the French community in riot-torn Casablanca in the mid-1950s. In the main, French memories of the *Tirailleurs Sénégalais* were favorable, even if the events they recalled were not.

In the former French empire a negative image of the *Tirailleurs Sénégalais* prevailed. For many colonial subjects, black African soldiers were a detested symbol, France's "black watchdogs of empire," to use the phrase of Léopold Sédar Senghor, who, while not opposed to the idea of African military service, deplored the French practice of using African troops in a servile manner.[8] Frantz Fanon has described the deep psychological imprint in Madagascar of the black African sergeant as torturer, bogeyman, and general agent of French coercion.[9]

As for African soldiers in French uniform, they developed their own particular variant of the dominant French ideology of *assimilation*, that is, full absorption into the French nation.[10] For civilians in FWA, *assimilation* offered to a tiny minority of highly educated Africans the prospect of full intellectual membership in the most enlightened polity in the world, the French nation. No African intellectual ever symbolized this dominant ideology of assimilation more dramatically than Léopold Sédar Senghor, talented teacher, prize-winning poet, president of the Republic of Senegal for over two decades, and finally, member of the *Académie Française*.[11] On the other hand, the military version of *assimilation* was a form of Jacobinism. It owed more to the barracks of southern France than to the hallowed halls of the Sorbonne, yet was equally effective as an instrument of control. Dispensed to the African military, and especially to those who served as career soldiers in the *Tirailleurs Sénégalais* for periods ranging from fifteen to twenty-five years, this doctrine suggested that colonial soldiers shared both the burdens and benefits of defending the French republic against its enemies. If military service entailed difficult and dangerous assignments far from home, it also offered such substantive and symbolic rewards as steady employment, medical care, pensions, decorations, and other indications of an enhanced social standing.

Albert Memmi has sought to explain the place of African soldiers in colonial ideology and social structure. He sees them, along with interpreters and other minor officials, as newly assimilated yet quintessential intermediaries, caught between colonizer and colonized, who ended by adopting the colonizer's ideology completely. He does not paint an admiring portrait of them:

> They [the recently assimilated] push a colonial mentality to excess, display proud disdain for the colonized and continually show off their borrowed rank, which often belies a vulgar brutality and avidity. Still too impressed by their privileges they savor them and defend them with fear and harshness; and when colonization is imperiled, they provide it with its most dynamic defenders, its shock troops and sometimes its instigators.
>
> The representatives of the authorities, cadres, policemen, etc. recruited from among the colonized, form a category of the colonized which attempts to escape from its political and social condition. But in so doing, by choosing to place themselves in the colonizer's service to protect his interests exclusively they end up by adopting his ideology, even with regard to their own values and their own lives.[12]

Born a Jew in Tunisia under French rule, Memmi felt well placed to understand the intermediary because of his own predicament as a minority which, while colonized, sought to distance itself from the majority of Muslim Tunisians. Yet, despite its insights, his explanation is insufficient to account for an essential contradiction in the African soldiers' experience. It does not recognize that African soldiers, although a fundamental part of the coercive state apparatus of rule, were themselves coerced into service. Rather than being a caricature of colonialism, African soldiers were, perhaps more than other groups, a mirror of colonialism and a reflection of its most basic contradictions. The hierarchical and paternalist institution to which they belonged was a metaphor for colonialism itself. Unlike the situation in which a French citizen in the metropole found himself when he left the hierarchy of military life to return to the privileges of civil society, the African ex-soldier and veteran entered into civilian colonial society where the rules were less clear but still unequal. Soldiers in service received, in turn, conflicting messages. Those who rose through the ranks to become African officers faced, it is true, a life of opportunity and identification with their rulers, but always at a disadvantage and at a price.

Situating African soldiers within the colonial social formation in FWA is a difficult analytical task. Teodor Shanin, following Leon Trotsky's analysis of the Russian army in revolution, has argued persuasively that an army is a copy of the society it serves, with elements in it from all sectors.[13] Soldiers may become, like students, "class-like non-classes," or "temporary classes," and may act as an avant-garde in a revolutionary situation. But for the *Tirailleurs Sénégalais* in French West Africa, conditions were not revolutionary and the society which the *Tirailleurs Sénégalais* served most directly was located in the metropole. The end of the Second World War was one such privileged moment when African veterans became a "temporary class," but the crisis was soon over. FWA itself was a society in transition to capitalism, with household and petty commodity production remaining widespread through the entire period covered by this study.

A more permanent class standing can be assigned to the small but significant number of soldiers who made a career of military service, especially those who became officers. Such men formed part of what Nicos Poulantzas defined as a "new petty bourgeoisie," a specific class made up of salaried, nonproductive workers, not of the proletariat but capable of alliance with them or with the bourgeoisie.[14] These new elements included a long list of people who offered services rather than goods: managers, clerks, minor bureaucrats, lawyers and other professionals, and intellectuals of all sorts. The "old" petty bourgeoisie would have included specifically those groups mentioned by Karl Marx: small manufacturers, shopkeepers, peasants, and artisans.[15]

In Africa, especially after the Second World War, both old and new components of this large grouping grew exponentially with urbanization and economic diversification. Professions and occupations as diverse as porters and teamsters, photographers and mechanics, launderers and prostitutes, newspapermen and politicians, senior policemen and African military officers could all be included in this large grouping. At present, too much vital information about many of these trades is missing to permit the establishment of boundaries among them, let alone to delineate their alliances and ideologies. While African officers in particular were an emerging profession in FWA by the 1950s, not enough is yet known about their life experiences to be able to incorporate them within a prefabricated grouping such as the "petty bourgeoisie."

The institution in which they served is often as incompletely understood as the men within its ranks. The *Tirailleurs Sénégalais* were unique in the colonial experience of the Western powers in Africa. Only France brought about an intense militarization of its African colonies. Only France instituted universal male conscription in peace as well as in war from 1912 until 1960. To be sure, some functional similarities did exist with the colonial units of the British and the Belgians, for instance. All colonial systems have understood the practical advantages of recruiting part of their coercive police and occupation force from among the local population. Similarly, other colonial powers used indigenous troops to defend their colonies from external attack, as illustrated dramatically by the German colonial forces in Tanganyika under Colonel Paul von Lettow-Vorbeck during the First World War.[16] But what distinguished France from other powers was its determination to use the *Tirailleurs Sénégalais* extensively as an expeditionary force in every corner of the French empire, whether for purposes of conquest, occupation, or later, counterinsurgency. From these uses came still another, the defense of the mother country. France was the only colonial power to bring Africans by the thousands to the trenches of northeastern Europe in the First World War, and to form a key element in the continental defense of France in the late 1930s. While it is true that Great Britain used colonial troops from the nonwhite tropics as well as from the European dominions overseas in both world wars, the British studiously avoided assigning nonwhite troops either for home defense, to oppose a European enemy, or to occupy enemy territory.[17] In short, France did what other colonial powers dared not do: arm and train large numbers of potentially rebellious colonial subjects in what was euphemistically called the art of modern warfare but what could also be described as modern methods of exercising military will against their European overlords.

The *Tirailleurs Sénégalais* underwent dramatic change over the course of its one-century lifespan. Most *Tirailleurs* in 1857 were still being issued muzzle-loaders, while some of their successors a century later included *Tirailleurs Sénégalais* paratroopers who had jumped into the inferno of Dien Bien Phu in 1954, or into Port Said during the ill-fated and ill-advised Anglo-French operation at the Suez Canal in 1956. To illustrate the process of change, this study traces the history of the *Tirailleurs Sénégalais* chronologically through what might be termed four different *Tirailleur* armies.

Chapter 2 examines the Conquest Army, which lasted from 1857 to 1905, when the French West African Federation came into being under civilian authority. Basically a mercenary force, it consisted of slaves and others drawn from the lowest social ranks, together with an important minority of well-born Africans as military intermediaries standing between the French officers who commanded and the African troops who followed orders.

After 1905, when France's colonial possessions had become largely secured south of the Sahara, the *Tirailleurs Sénégalais* entered a transitional phase as an occupation army until 1919. Chapter 3 focuses on the mythology of the *Tirailleurs Sénégalais* during this period as well as on the political circumstances leading to the adoption of permanent universal male conscription in peacetime as a result of the Conscription Law of 1919.

Chapters 4, 5, and 6 examine the Conscript Army, occupying the third phase of the *Tirailleurs Sénégalais*, from 1919 to the end of the Second World War in 1945. While Chapter 4 concentrates on the theory and practice of military conscription in FWA, Chapter 5 examines the consequences of the conscription system, a vast movement of people in a veritable military migration. The involvement of the *Tirailleurs Sénégalais* in the momentous events of the Second World War forms the subject for Chapter 6.

The fourth and final phase of the *Tirailleurs Sénégalais*, characterized by at least the rhetoric of professionalization, lasted from 1946 until the coming of independence to FWA in 1960, and is examined in the remaining chapters. Chapter 7 explores the new military conditions for active servicemen in the changed circumstances of postwar French colonialism. Chapters 8 and 9 examine the struggles of ex-servicemen for justice in a French community still wedded to the ideology of assimilation. The first of these chapters details the economic and social, and the second, the political, dimensions to the veterans' experience.

2

Slaves into Soldiers

> *What serious concerns about his future prospects can the African have when he feeds himself with a banana and uses the rays of the sun for clothing? Given the carefree life he leads, the only regular profession which we have brought, the military career, seems attractive to this man. Seduced by the triple attractions of glory, profit and the uniform, he decides to sign on as a volunteer.*
>
> Hippolyte-Victor Marceau, *Le Tirailleur soudanais*

While Governor Louis Faidherbe of Senegal created the first permanent units of black African soldiers under French rule in West Africa and named them the *Tirailleurs Sénégalais* in 1857, in fact their roots are much older.[1] They can be traced back to the first years of company rule in seventeenth-century Senegambia, when British and French military recruiters took on local Africans as soldiers and sailors in order to augment the European units which formed the core of the small company detachments.[2]

During the course of the nineteenth century the *Tirailleurs Sénégalais* underwent several transformations, as Table 2.1 indicates. In the small French enclave of

TABLE 2.1 Growth of the *Tirailleurs Sénégalais*, 1820–1914

Year	Size	Year	Size	Year	Size
1820	23	1862	900	1900	8,400
1823	125	1867	1,000	1902	8,639
1827	200	1872	625	1904	9,000
1831	400	1882	1,200	1911	11,980
1839	150	1886	1,600	1912	12,920
1848	250	1888	2,000	1913	14,790
1852	350	1891	2,400	1914	17,356
1857	500	1893	5,087	1920s	48,000
		1895	5,987		

Sources: Maurice Abadie, *La Défense des colonies* (Paris: Lavauzelle, 1937), 210–16; ANS, 2G12 7, and 4D143 100.

Senegal at the beginning of the post-Napoleonic era in 1820, black African soldiers constituted only a fraction of the 383-man *Bataillon d'Afrique*. By 1823, the first all-African company of 125 men was formed. There then followed very slow growth to midcentury, a rise during Faidherbe's tenure in the 1850s, and a decrease thereafter, until a rather sharp increase occurred in the conquest decades that ended the century. A dramatic growth then took place during the years of military occupation from 1890–1904, and especially after the introduction of civilian rule after 1904, until, by the beginning of 1914, the *Tirailleurs Sénégalais* had come to include almost 18,000 men organized in six regiments.

While political decisions lay behind each of these shifts, it is also clear that recruitment methods and policies contributed dramatically to these changes. A brief examination of the *Tirailleurs Sénégalais* in its three phases before the First World War is in order.

Early Recruitment Methods

Early recruitment occurred in the same random way that obtained for local labor generally.[3] Much of it involved slaves, though the free mulatto sons of European and African unions sometimes found employment as well. The Wolof word for sailor, *laptot*, in time came to be generic in Senegambia for a military employee of the French, but the term had higher status when it was associated with local merchant marine employment.[4] Pure military labor, consisting largely of chore and fatigue duties, was restricted to slaves and others of servile origin.

The first departure from this practice occurred in 1819 when the Ministry of Marine ordered the recruitment of "companies of color," in which men could enlist in return for a signing-on bonus or premium.[5] These men were attached to the French units of Senegal. In fact, this premium was paid to their masters, and the men were then indentured to military service for periods ranging from twelve to fourteen years. Instructions even specified that purchase prices should not exceed 300 francs.[6] This purchase system, called *rachat*, or "repurchase," was to remain in place for most of the nineteenth century.[7] Table 2.2 indicates the varying prices paid for indentured men and the regions where they were obtained. As in the slave trade generally, the market rather than government edict determined prices paid and sources tapped for such soldiers.

The steady demand for African soldiers can be explained in part by death rates for the European troops among whom they served. Thus, in 1815, within six months of the dispatch of seventy-five French military cadres to Senegal, some fifty-seven had died.[8] Colossal European death rates were partly a function of the poor quality of the soldiers as well as the hostile disease environment, where malaria in particular proved a terrible scourge. For instance, the four hundred soldiers sent out to Senegal in 1819 on penal duty were "composed of drunkards and marauders, disturbed and profoundly vicious men," wrote Colonel Schmaltz, military commander in Senegal at that time. Such men not only "indulged themselves in all sorts of excess," they were also serving in Senegal without fixed terms.[9] These men believed, perhaps correctly, that they would die in Senegal unless they could be

TABLE 2.2 Origins and prices of slaves purchased for the *Tirailleurs Sénégalais*

Date	Region	Price	Details
1820	Gorée	400 francs	30 garrison soldiers for duty at Gorée
1820	Galam	150 Fr	
1828	Bissagos	350 Fr	
1831	Bakel	325 Fr	Bought by Galam Company
1833	Bakel	335 Fr or less	Galam Company paid less and turned profit
1839	Bissao and Cacheo	300 Fr	Bought by le Sieur Marbeau
1853	Casamance	200 Fr	

Sources: Claude Faure, "La Garrison européenne du Sénégal (1779–1858)," *Revue d'histoire des colonies* 8 (1920): 5–108; ANS, 2B 5, 1B 176, 2B 15, and 2B 18.

sent home for medical reasons, and many tried deliberately to make themselves ill enough to be repatriated and still survive.

Soon the need for Africans to replace such unreliable troops was augmented by yet another novelty, the exporting of West African soldiers to fight wars of conquest and occupation in other corners of the expanding French empire. In 1827 some two hundred Wolof soldiers were sent to Madagascar, followed in 1831 by the dispatching of 220 troops to Guyana.[10]

The practice of buying slaves to satisfy expedient needs for military labor was not without its difficulties. Apart from its cost, the *rachat* system was an embarrassment and a contradiction as Western Europe moved closer to abolitionist principles.[11] The Marbeau affair in the early 1840s was particularly scandalous, since it implicated the highest levels of government. The governor of Senegal licensed Marbeau, a French trader, to purchase slaves for the military in Portuguese regions of Guinea. The affair came to light when the British Anti-Slavery Squadron captured him en route to Bissao and confiscated his ship.[12]

The emancipation of slaves as a result of the 1848 revolution further complicated French recruitment of African soldiers by means of *rachat*. The Emancipation Act explicitly prohibited the practice, and the French military in Senegal bravely stated its intention to rely upon volunteer soldiers from among the free population of Saint-Louis, if necessary through a formal conscription law modeled after the existing practice in France.[13]

The immediate impact of emancipation was a recruiting crisis. In the three years following 1848, only three volunteers came forward, hardly enough to meet the two hundred minimum required to maintain garrison strength.[14] Free Africans of military age, in their twenties and thirties, were said to find military labor degrading, especially as the army was the preserve of slaves and ex-slaves and the pay poor in comparison with wages for *laptots*, for example.[15] An interministerial commission assigned in 1851 to study the future of the colony concluded that the tried method of *rachat* would have to be preserved and used alongside the volunteer system, one requiring fourteen years of service and the other seven.[16] It was clear enough that the *rachat* system would have to provide the bulk of the recruits when, by 1854, the quota of African soldiers was raised to the then unprecedented figure of 750 men, 200 of whom were to be dispatched abroad to Mayotte, Madagascar, Martinique, and Guyana. To raise such numbers, the French had to draw on slave purchases in Sedhiou in the Casamance and even Gabon. To avoid any repetition of incidents with the British, ministerial orders called for naval officers on warships to bring back the indentured soldiers.[17]

When Louis Faidherbe became governor of Senegal in 1854 with ambitious plans to expand French territorial control in the Senegal valley and beyond, his first task was to address the crisis in military recruitment. Determined to conduct not one but two military campaigns a year against opponents upriver, Faidherbe needed more soldiers, and, like his predecessors, he realized that the disease factor made the dispatch of European troops too costly.[18] Unlike previous governors, however, Faidherbe believed that African soldiers could become effective combat soldiers, not simply indentured military laborers (*captifs-hommes-de-corvée* was the evocative French phrase used).[19] He therefore began a lengthy campaign to transform the status of the African soldier from indentured military laborer to fully fledged regular combat soldier.[20]

Faidherbe's efforts were rewarded when Napoleon III signed the decree of 21 July 1857 creating the *Tirailleurs Sénégalais*.[21] The decree brought local troops in Senegal to battalion strength by doubling the existing companies from two to four and by segregating the Africans into their own units, with their own distinctive uniforms. Apart from the formal separation of units, terms of service were to be virtually identical to those of European units. Rules and regulations of regular navy infantry regiments were to apply with regard to allowances and retirement pensions, and the battalion was to be headed by a major, with officers and non-commissioned officers drawn from regular navy infantry regiments. One significant difference highlights the rationale for the entire reform. The pay of *Tirailleur* soldiers was to be excluded from normal navy stipends and instead was to be determined locally by the governor of Senegal and his officials. Faidherbe had shrewdly and accurately calculated that this economy would appeal to planners and politicians in Paris.

Faidherbe relied on a combination of measures to attract Africans to fill the expanded ranks.[22] The new uniform was colorful and designed to lure young men. Degrading physical labor was to be severely limited; the men were to be infantry soldiers, their days taken up with training for this task. Most importantly, the regular stipend was to be raised from previous levels, and incentive bonuses were to be paid not only as lump sums to first-time recruits as had been the system under *rachat* (the money going to the former master) but also to veterans who were encouraged to re-enlist for up to four-year terms. Without completely excluding *rachat*, Faidherbe hoped in this fashion to move gradually to a volunteer professional army that attracted recruits through a combination of material and ideological incentives. While we cannot be sure which incentives proved the most effective, in combination Faidherbe's reforms certainly achieved their goal. By the beginning of 1858 he had his four companies at full strength; moreover, only 160 of the 500 men were holdovers from earlier days when *rachat* was virtually the exclusive technique used.[23]

Division of the spoils, Conquest period

One new incentive was implicit in the greatly expanded *Tirailleurs Sénégalais*, the promise of booty. Faidherbe personally took a rigorous moral line against taking spoils of war, but he was not able to eradicate this practice.[24] On the contrary, the prospects of being on the winning side and of taking booty were obvious to Africans who had the taste and skills for the military vocation. This phenomenon became especially apparent during the expansion of the *Tirailleurs Sénégalais* during the conquest decades at the end of the century. Nevertheless, *rachat* did not disappear immediately; on the contrary it remained the prime method of recruitment despite Faidherbe's distaste for it. After his departure, the number of *Tirailleurs* shrank for a time; it then remained constant for the next twenty-five years, during which time *rachat* was practiced regularly.

Recruitment in the Conquest Era, 1886–1905

Throughout Africa in the second half of the nineteenth century, slavery was in crisis. Colonial antislavery policies gained strength as the various European states began to formalize their presence on the African continent. Yet the end of slavery did not rapidly transform relations between former masters and slaves.[25] Not only did other forms of unfree labor such as pawning and forced labor persist, colonial states proved more inclined to protect the rights of owners over those of slaves. African slaves and former slaves in this period faced extreme uncertainty in their relations with masters and with the colonial state. For example, escape might have proved easier, but it was also easier to become victimized by any new authority figure whether local or foreign. It is against this background that French military recruitment in the last decades of the century should be understood.

In a formal sense, the *rachat* system had ceased by 1882, as the French no longer purchased slaves on the open market for the *Tirailleurs Sénégalais*.[26] In its place, three new forms of recruitment enabled the *Tirailleurs Sénégalais* to grow to the proportions necessary to carry out the conquest of West Africa. Two of these methods, the payment of an enlistment bonus to a slave's master and the incorporation of prisoners-of-war and porters, were marked by at least as much coercion as the discredited technique of *rachat*. The third method, the exercise of a career choice by what might reasonably be called the first free mercenaries of the *Tirailleurs Sénégalais*, represented something new.

In 1868 Faidherbe's successor, Pinet-Laprade, inaugurated the first of these new practices, the transfer of enlistment bonuses to slave masters as a means of terminating their ownership over their human property.[27] Enlistment bonuses rose from 100 francs for a seven-year contract in 1853 to 300 francs for the same term in 1894, before falling off slightly by 1905 (Table 2.3).

While this modified form of purchase resembled the old *rachat* system, Pinet-Laprade made it clear that it enabled the administration to avoid dealing directly with slave traders while offering the former slave the opportunity of formally ending his dependence on his master, even if the army was to become his new patron:

> The new plan is designed to encourage purchase of soldiers from their masters; our enlistment bonus of 200 francs is in fact the highest price of a fully grown and well built male slave....
>
> I authorize you to advance the necessary funds, not to exceed 200 francs, to slaves who wish to join the service, and to do this only at the moment they are ready to embark for Saint-Louis by steamboat so they can indemnify their masters.[28]

As demand for soldiers outstripped supply, particularly in the active years of conquest in the 1890s, French commanders were given a wide degree of latitude in their recruitment methods, leading to the second method cited above. In the region of West Volta, east of Sikasso, in 1897, for example, the French commander Major Caudrelier, faced with the task of commanding and occupying a huge region only

TABLE 2.3 Enlistment bonuses for *Tirailleurs*, 1853–1904

Year	Bonus	Bonus per Year
1853	100 Fr for 7-year contract	14 Fr
1868	200 Fr for 7-year contract	29 Fr
1880	240 Fr for 6-year contract	40 Fr
1881	300 Fr for 7-year contract	43 Fr
1894	300 Fr for 7-year contract	43 Fr
1905	120 Fr for 4-year contract	30 Fr

Sources: Boutillier, "Captifs en A.O.F.," 529; Faure, "Garrison européenne," 98, 104; ANS, 4D 29, 4D 30.

Tirailleurs and officers with mounted artillery, French West Africa in the late nineteenth century

recently devastated by Samori's forces, ordered each *chef de poste* to recruit soldiers into regular service to bring the *Tirailleurs Sénégalais* to battalion strength.[29] One junior officer took the directive to justify the seizure of a caravan of roughly three hundred porters arriving with supplies. He locked the exits, brought out *Tirailleurs Sénégalais* uniforms, and after two months of drill, had two combat companies of former porters. Several men apparently adjusted well enough to have re-enlisted and were still in service in 1904 when they participated in the occupation of the Côte d'Ivoire.

Royal slaves, who were also incorporated, are difficult to classify. Some of these slaves (called *diomba* in the various Mande states) were commanders of slave soldiers (*sofa*) who had become rich and powerful in the service of West African notables, much in the fashion that the Ottomans confided great power and wealth to their military slaves.[30] Many owned slaves in their own right. One French eyewitness reported that the *diomba* and their *sofa* soldiers who were taken as prisoners steadfastly refused freedom and requested a privileged status with the French.[31] Presumably they hoped to preserve the same standing within the French system as they had achieved earlier. During the conquest years, French practice gave them some reason to expect these favors. Initially at least, *diomba* retained their weapons and continued to keep the wives, children, and even slaves they had acquired from their former masters. Emancipation would have also implied freedom for their dependents.[32]

The third recruitment technique, the incorporation of West African career soldiers, represented the interplay of market forces. Some African army leaders were no doubt attracted by the superior weapons of the French, their winning record, and the promise of sharing in the spoils of war. This seems, for example, to have been the motive of Babato, a Zaberma slave raider in the Voltaic region,[33] and of the defeated remnant of the Maraka forces of Al-Kari of Boussé of the same region.[34] Oral traditions indicate that many of these leaders brought their entire units into French service, and they were enlisted, no questions asked, by the manpower-short French commanders of Upper-Senegal-and-Niger in 1898.[35] Another case, this time with fewer options, was that of officers of Samori Touré. After his capture by the French in 1898, several of his sons and lesser commanders were incorporated into the *Tirailleurs Sénégalais* and sent to Saharan outposts as a test of their loyalty.[36] As leaders turned their coats inside out, many of their slave soldiers followed suit, no doubt with little or no consultation.

By these methods, the *Tirailleurs Sénégalais* tripled in size in the chaotic years of the conquest. Nevertheless, in spite of the predominant coercion involved, a few Africans were beginning to exercise choices, opting for military service in preference to the alternative of tilling the soil or obtaining occasional wage labor. In this process a new phenomenon was emerging. Members of the traditional ruling families of West Africa were beginning to join the predominantly servile *Tirailleurs Sénégalais*, entering the ranks as noncommissioned or even native officers.

French recruitment in the nineteenth century guaranteed not only a predominance of slaves and other men of servile origin but also the reliance on specific ethnic groups. In the early years of recruiting, cost and availability rather than regional or ethnic preferences dominated French practice. By the 1840s, Governor Bouet-Willaumez found it difficult to recruit in western Senegal.[37] Recruits from the upper Senegal valley seemed as vulnerable to the unhealthy climate of the town of Saint-Louis as were Europeans, a circumstance which the governor blamed on malignant sea breezes. Former slaves from the nearby Wolof states of Cayor and Walo, on the other hand, were said to desert military service too easily. Consequently Bouet-Willaumez's regime turned to the Bambara states of the interior as a source of slaves. By midcentury, Bambara recruits were singled out for their military "qualities": "Among others, those who belong to the Bambara race are, it may be said, as good as white soldiers, as they have the advantage over them of being immune to all the hardships of the climate."[38]

In the helter-skelter recruiting of the conquest era, no statistics on regional, let alone ethnic, origins of recruits were compiled. Nonetheless the French preference for Bambara recruits is clear, even if a wide variety of other ethnic groups were also represented in the *Tirailleurs Sénégalais*. One experienced French officer, Captain Marceau, published his opinions in 1911, and claimed without evidence that two-thirds of all recruits were Bambara speakers.[39] The Bambara *tirailleur*, added Marceau, was "an uncouth fellow who possessed all the strong warrior's virtues, but who did not unfortunately show much intelligence"; in other words, he was a compliant, stolid peasant soldier obedient to his new French masters. In contrast, the Wolof soldier was spoiled by his long association with France and had become "a terrible snob" toward other Africans whom he regarded as "savages."

> [The Wolof] is a "Rights of Man" nigger; a citizen without civilization; an elector who disdains our uniforms; he is no longer fit to be a *tirailleur*. As a result, even at the regiment in Saint-Louis, we find few Wolofs.

Marceau believed that the best officers came from the ranks of the Tukolor:

[The Tukolor] is a pure warrior. He's a born soldier who does not unfortunately always take to our military discipline with good grace.

Apart from their subjectivity, what is remarkable about such opinions is that they were so long lasting, in part no doubt because they were self-fulfilling.[40] As late as the 1950s, officers were repeating the homilies of Marceau and of Mangin regarding "warlike races."[41] Mangin, for example, had doubted whether people of the forest regions had the discipline and physical stamina to make good *Tirailleurs*.[42] Mangin's strong influence on military policy, together with the fact that forest zones were the last to be occupied by the French, meant that only after 1919, when universal male conscription was systematically introduced, did forest peoples begin to form a modest minority in the *Tirailleurs Sénégalais*.[43]

As Bambara came to dominate the ranks and noncommissioned grades, the Bambara language became, alongside the "petit-nègre" French that the military insisted upon, the colonial army's vernacular language. New recruits from Bambara and related Mande-speaking regions of the savanna came to find the army a more hospitable institution than did, say, Agni speakers from southern Côte d'Ivoire. Thus patterns established in the formative years were to hold sway decades later.

If the ethnic or regional composition of the early *Tirailleurs Sénégalais* cannot be established precisely, some estimates are possible. Not surprisingly, intensive recruitment correlated with areas of high population density and often with areas where the percentages of slaves were also high. As the data on slavery compiled by French officials in the years from 1903 to 1906 indicate, *cercles* (the equivalent of *départements* in France) with slave populations of at least 50 percent were also well represented in the army.[44] These included such *cercles* as Bakel in Senegal; Kong in Côte d'Ivoire; Odienné, Kankan, and Dinguiray in Guinea; and Bafoulabé, Kita, Bamako, Siguiri, Nioro, and Sikasso in Soudan.[45] Though not an area of heavy slave concentration, the densely populated Mossi plateau in Upper Volta furnished its share of recruits as well, but only once conquest was completed after 1900.

At the other end of the social hierarchy in West Africa, a small but significant elite of soldiers who belonged to the families of defeated African rulers joined the *Tirailleurs Sénégalais*. Approximately twenty of Samori's sons had served in the *Tirailleurs Sénégalais* during the First World War, for example.[46] General Mangin remembered encountering a Lieutenant Touré, who was one of Samori's sons, serving as a member of General Gouraud's escort in the Rhineland in 1919. The irony was evident since, as a young officer, Gouraud had captured Samori in 1898.[47] Bouna, a son of the Wolof resistance figure Ali Bouri N'Diaye, volunteered for service in 1914 and was made an officer in the Legion of Honor soon afterward.[48] A son of Mamadu Lamine, the Senegalese resistance figure, also served in the *Tirailleurs Sénégalais*.[49]

While no single explanation could account for all cases, one clue can be obtained from the initial association of Samori's kinsmen with French forces. When Samori and his immediate family and entourage were captured in 1898, fifteen of the group, including Samori and his son Sarankeni Mori, were exiled to French Equatorial Africa. Four other sons were given lighter sentences. They were sent to Nioro to be incorporated into the cavalry units of the frontier guards and dispersed among the various French frontier posts on the desert side, Timbuktu included,

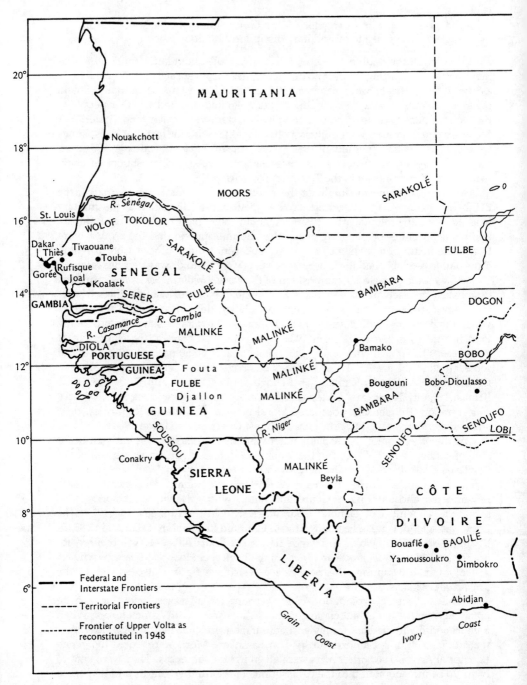

Ethnic groups in French West Africa

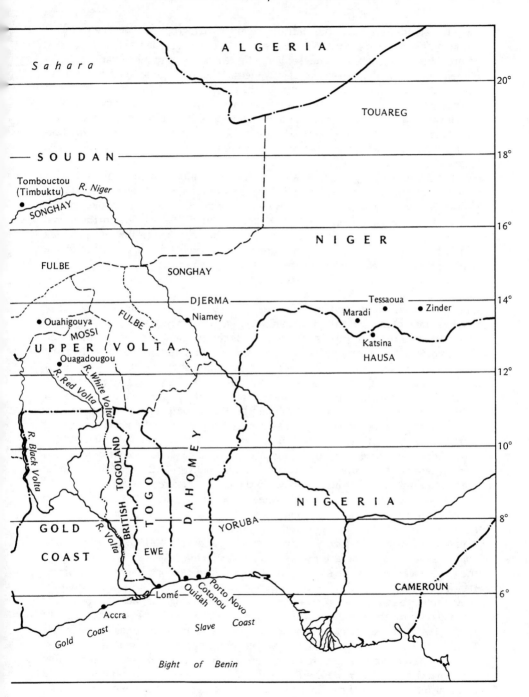

there to defend French conquest territory against Tuareg opposition. The youngest of Samori's sons were placed in French schools for sons of chiefs.⁵⁰ It was not long before these Tourés, professional soldiers in a real sense, and mercenaries in spite of themselves, found themselves formally members of the regular *Tirailleurs*. Cut off from and indeed forbidden to return to their homeland in Guinea in what had become the *cercle* of Kankan, the Touré family went over to French service as an act of self-preservation. As for the grandiose French military officers of the day, the recruitment of defeated African leaders and their entourage symbolized the thoroughness of their victory in a manner that suggested the triumphs of Rome and its legions.⁵¹

The *Tirailleurs Sénégalais* paid a price for such dubious mixtures of recruits. A collection of armed men consisting in large part of royal slaves, war captives, and people largely of low social origins, led by sons of former warlords, created considerable unpopularity among the African population. Slave soldiers in traditional armies had been given to excess and were generally disliked in the Senegambian and Niger valleys.⁵² Now, soldiers were not only *captifs de blancs* (slaves of the whites), they were regarded by many as mercenary renegades.⁵³

The low status of the *Tirailleurs Sénégalais* produced few volunteers from the higher strata of African society, a situation the French tried to rectify on occasion. One short-lived experiment in 1890 was the creation of a separate battalion of mounted men of free birth, to be named the *Chasseurs Sénégalais*.⁵⁴ It was suggested at the time that the prestige of a cavalry unit, its confinement to well-born men, and the promise of booty were sufficient to induce recruitment and that no enlistment bonus was needed, another indication, incidentally, that the bonus was clearly recognized as a form of purchasing a slave soldier. Over the years, French recruiters listed a variety of reasons to explain the reluctance of the population to volunteer for military service. It was noted that Muslim leaders in particular discouraged their followers from military service because it meant exposure to the French and their Christian ways.⁵⁵ Other reasons mentioned were that the men disliked fatigue duties, of which there were too many, that young noncommissioned officers and African cadres were transferred too often, and, finally, that those Africans under French cultural influence simply did not find the military career attractive.⁵⁶

Thirty years later, French authorities began to recognize the importance of economic factors in African decisions to enter military service. Thus, in 1907, the lieutenant governor of Guinea noted that while Muslim leaders in particular continued to be firmly opposed to military service, and held back their children and clients, a more important factor was the greater attractiveness of alternative employment.⁵⁷ African youths, he said, particularly disliked the long voyages and uncertainties of service abroad, whether in Congo or Madagascar, but most importantly, they wanted to stay in the colony and take on better paying jobs; he cited the case of railroad construction in regions such as Kouroussa, where labor was in short supply. In this *cercle*, of the fifteen recruits supplied in October 1907, most were in fact Bambara speakers originating from Ségou, the inference being that they were probably slaves and strangers to the region. In a word, the governor believed that economic changes had made the *Tirailleurs Sénégalais* economically unattractive. "It could also be said that the calling of *Tirailleur* no longer has the same attraction for Blacks as in the days when soldiers returning from campaign brought back slaves and booty."⁵⁸ Remarks such as these contradicted the frequently repeated homily of that time that Africans were unwilling to work for wages.

African Officers in the Early *Tirailleurs Sénégalais*

> *Prince Anabia, the only difference between you and me is that between black and white.*
>
> Attributed to Louis XIV upon recognizing one of his African officers, cited in "La France et les cadres africains et malgaches: 1, les cadres avant l'indépendance."

While the Sun King's remarks above were probably apocryphal, they help illustrate a tradition among well-born Africans as officers in French military service. Various French governments from the *ancien régime* through the nineteenth century used Africans as officers. General Faidherbe's policies as governor of Senegal marked a departure from earlier practices with regard to African officers as well as for enlisted men.⁵⁹ Faidherbe, a believer in the Jacobin ideology of assimilation and of a democratic army, made his thinking clear in a speech he gave while awarding the rank of second lieutenant to one Alioun Sall, a Senegalese soldier who had distinguished himself in the *Tirailleurs Sénégalais*. Just as the army had been the vehicle of advancement for many French officers, Faidherbe himself included, the *Tirailleurs Sénégalais* played a similar role for some Africans. Upward mobility through the army ranks was also a convenient solution to a personal problem. Faidherbe's own illegitimate son, whose mother was a Khassonké concubine, eventually became a so-called native officer in the French army.⁶⁰ In his public address, Faidherbe observed:

> This position has been specially created to reward the numerous examples of devotion displayed by M. Alioun Sall since the beginning of the war. This nomination has, in addition, a more general significance: it demonstrates that, even for loftier positions in our social hierarchy, colour is no longer a reason for exclusion and it is in this sense that the black population should understand it. . . .
>
> The [officer's] career will be open to them under the same conditions as for any one else. Only the most capable will succeed. Those who obstinately prefer ignorance to civilization will remain in the lowly ranks of society, as is the case in all the countries of the world.⁶¹

While Faidherbe's assimilationist views were consistent with the universalist principles of the French Revolution, not all French officers shared his confidence in the ability of Africans officers to succeed in a military career. The French military conqueror Archinard had a poor opinion of the *Tirailleurs Sénégalais* and of their African native officers. He argued that African troops needed five months each year of training in order to spend the rest of the year on campaign. While the troops were in his opinion slow to learn and possessed of poor memories, their leaders suffered from "weakness of character."⁶²

If one source of officers for the *Tirailleurs Sénégalais* was the emerging African bourgeoisie, it cannot be said that the army was a popular option for this group.⁶³

The illegitimate sons of French officials like Faidherbe were exceptions, as was Benjamin Adekempi-Thompson, the son of a merchant from Dahomey. Born in Porto-Novo, he attended military school in France, was named a second lieutenant in the *Tirailleurs Sénégalais* in 1901, but does not seem to have had a long military career.[64] In practice, however, educated young men of the middle class were not attracted to a career which was dominated by an overwhelming number of people of slave origin, leavened perhaps by a minority of native officers representing the old African nobility, but with French officers dominant at junior, intermediate, and senior rank.

As noted above, sons of defeated African rulers and notables were a more important source of native officers than were the offspring of merchants.[65] Faidherbe, in 1856, had actively pursued such recruits as part of a general design to recruit intermediaries when he opened his so-called *école des otages*, or school for hostages, in Saint-Louis. Its name reflected the fact that its first pupils were claimed by the French from the families of African notables to assure their cooperation with the new colonial regime. Later the name was changed to *Ecole des Fils de Chefs et des Interprètes* (or school for sons of chiefs and interpreters) to reflect its primary function, which was to produce intermediaries literate in French. By the time it was closed in 1871, this school had produced among its 103 students 56 "loyal" graduates, that is, men who served the French administration well. Of these, 9 were interpreters, 11 were native chiefs, 2 were African officers in the *Tirailleurs Sénégalais*, and the rest divided among clerks on trading vessels, teaching assistants in the colony schools, and small businessmen.[66] Despite the small yield of officers, this early school signaled an important French need in those early years of rule: to recruit and win the allegiance of a segment of African leadership to help them maintain power. Perhaps as an attempt to remedy the shortage of educated African officers, by the 1880s a small school in Saint-Louis had been opened with the title *L'Ecole des Enfants de Troupe*.[67]

One of the graduates of this school was Macodou M'Baye, perhaps the most outstanding African officer of the Conquest Army. Born in Dagana, Senegal, in 1870, he entered the *Ecole des Enfants de Troupe* at age twelve. At eighteen, he joined the *Tirailleurs Sénégalais*, serving first in garrisons in the Senegal Valley before joining the campaign of General Dodds against Dahomey. There he distinguished himself sufficiently that Dodds promoted him to sergeant and took a paternal interest in his subsequent career. He served in further conquest operations in Guinea, Mauritania, and Côte d'Ivoire, and volunteered for duty in the First World War at the age of forty-four. Promoted to second lieutenant in 1894, and four years later to lieutenant, he remained at that rank until his retirement in 1918 after 30 years of service. His death at Saint-Louis in 1948 earned him an obituary in the SFIO newspaper, *L'AOF*, where he was described as a loyal and brave soldier.[68]

However colorful it may have been, Macodou M'Baye's career illustrated two somber lessons for Africans hoping to become officers in the early *Tirailleurs Sénégalais*. First, even if they came from well-born families, Africans would have to begin with the rank of noncommissioned officer and achieve promotion through exceptional acts of bravery and leadership in combat. Second, their advancement beyond lieutenant was most unlikely. The only African officers to advance beyond that rank were two members of the Sy family of Futa Toro in the Senegal Valley, Captain Mahmadou Racine Sy, and his nephew, Abdel-Kader Mademba.[69]

Daily Life and Conditions of Service

The current historical literature permits only a sketchy portrait of daily life and conditions of service in the early years of the *Tirailleurs Sénégalais*.[70] It is possible, nevertheless, to discuss at least such issues as casualty rates from combat and disease, diet, family life, and soldiers' wages.

A fundamental reason for the expansion of the *Tirailleurs Sénégalais* had been the severity for European soldiers of the West African disease environment. While Africans enjoyed the benefits of inherited and acquired immunities over their European counterparts in the army, it also appears that they took better care of themselves. Faidherbe, when planning his African army, had advised against issuing a daily wine ration to African soldiers. He urged that its value of 30 centimes daily be given in supplemental food rations instead. The African soldiers, whether Muslim or not, were not consumers of wine and would sell their wine rations to French soldiers, who "would be drunk from morning to night and would die of dysentery. Drunkenness here is the cause of death for half the white soldiers. In addition, there would be no discipline."[71]

Of course, disease was a constant danger for Africans as well. One epidemic occurred in 1896, immediately prior to the dispatch of soldiers for Madagascar. Cerebrospinal meningitis attacked thirty-five recruits, of whom five died, while an unspecified number were hospitalized with pneumonia. The head of the *Service de Santé* feared that the epidemic might spread wherever soldiers served, but that proved not to be the case.[72]

As in most armies, in the *Tirailleurs Sénégalais* death from disease usually outpaced mortality from combat operations. Among the soldiers fighting in the conquest of Morocco in 1911–1912, there were 155 deaths from disease compared with 62 killed and 70 wounded in combat. The overall mortality rate for Africans in that year was approximately fifteen per one thousand, which compared favorably with the death rate of seventeen per one thousand for Europeans in the same military theater.[73] Neither overall casualty rate was particularly high.

In the conquest of West Africa itself, the considerable differences in military technology had much to do with combat casualties. In this respect, Africans serving in the *Tirailleurs Sénégalais* were much more fortunate than their counterparts in the various African armies of resistance. In his letters home to France, a young French soldier named Orsat provided graphic testimony to the differing technologies. Here is his description of the Gras repeating rifle, model 1884, which was standard issue by the 1890s:

> You have no idea of the effect of the Gras rifle: a bullet in the head takes off the skull, a bullet in the chest makes a hole the size of a plate in the victim's back, the limbs are crushed and the bones broken in a horrible manner.[74]

On the other hand, African arms caused Orsat little concern:

Fortunately, the wounds from their bullets are not generally very dangerous; they only have bullets of worked iron which have to be shot from up close to produce a bad effect.[75]

If the wounds inflicted in West African campaigns were not usually fatal, they were nevertheless real and cumulative. Orsat tells us that the section of the company he commanded consisted of forty men, no less than thirty-six of whom had been previously wounded in battle.[76] Other references to casualties indicate that certain French combats were worse than others. Frey, discussing various Senegal campaigns, rated the following as worst: Dailmath in 1855 cost 150 men killed or wounded; Guémou in 1859, 136 casualties; Sabouciré in 1878, 67 casualties.[77] The Dahomean campaigns in the early 1890s cost heavy losses of 251 killed and 371 wounded, though these figures may have included irregulars as well as regular forces. At the siege of Sikasso against Babemba, the French lost 2 officers and 42 soldiers killed, and over 100 wounded.[78]

Soldiers' diets differed considerably from what West Africans customarily ate. From an early date, rice, not then a common food in the West African savanna, was the staple daily ration for soldiers. An early mention of rice occurs in 1827 for African troops serving in Madagascar.[79] Service abroad could mean changes. Thus in French Guyana, the staple became either 500 grams of rice or 750 grams of cassava, or 250 grams of cod, salt beef, or fresh meat.[80] During the 1880s, the French were importing rice from Indochina into West Africa to feed their soldiers. To cut costs, the *Tirailleurs Sénégalais* began requisitioning millet from conquered territories, and after 1895, imports of rice dropped off. Total imports of goods to Soudan show a decline from 1,200 tons in 1894–1985 to half that amount in the following year.[81] Millet warehouses were ordered built throughout the occupied regions. An official account in 1903 acknowledged that local millet requisitions were still going on, even though by that date the practice of "informal" requisitioning had been prohibited in favor of a systematic tax organized by civilian administrators.[82] By 1911, the daily ration was once again rice (550 grams), together with 400 grams of fresh meat and small quantities of coffee, sugar, cooking oil, and salt. Firewood was also issued to individual soldiers and their families.[83]

In the early nineteenth century, African soldiers complained about being forced to live in barracks away from their families.[84] By Faidherbe's day, soldiers were allowed to live off base with their wives, to go home for meals, and to come to camp only for daily drill and exercises.[85] In the garrison town of Saint-Louis, for example, home of the First Regiment of the *Tirailleurs Sénégalais*, European officers and men lived in the barracks of the military camp N'Dar-Tonte, while Africans lived off base.[86]

Even on campaign it was clear that Africans lived quite separately from their European officers. One account from Sousou country in Guinea in the late 1880s indicates that African soldiers actually set up their living quarters outside the wall of the French fort.[87] Orsat gives a typically colorful description of the African soldier and his family on campaign:

> Now the campsite is virtually complete; it's funny to see the *Tirailleurs* building their square straw huts, carefully arranged in rows, each a home for their wives, children, slaves, everybody. On marches, the entire ensem-

ble follows the column far in the rear, but arrives at the day's campsite in time to pound the millet and prepare the *Tirailleur's couscous*.[88]

Women clearly played an important if dependent role in the life of an African soldier in the *Tirailleurs Sénégalais*. They tended to the needs of their husbands or masters in numerous ways; as mothers for the soldiers' children, as cooks and accountants performing daily miracles on the meager salaries and rations, and sometimes even as companions in battle. In the harsh conditions of desert combat at Adrar on one occasion, munitions carriers fled in panic and the women took up the munitions and supplied the decimated lines of their husbands with cartridges.[89] They were the hardest workers on campaign. As soon as the bivouac began, they were to be seen collecting firewood, lighting fires, carrying water and meals to the men on guard duty, and of course, cooking.[90] Their role as companions and entertainers was also recognized, through performances of music and dancing and through sexual favors. The French military encouraged family life in the camps, seeing in women and children a force for stability and a situation much preferred to the problems presented by prostitution and camp followers. This attitude was all the more interesting in that it anticipated by four decades the view of French authorities in the 1940s, which argued that "stabilized" Africans with wives and families made for a better work force.[91] Typically, the wife of a soldier was part of the convoy, cooking not only for her husband but for his bachelor friends, in charge of and responsible for camp property. So militarized did women become that they acquired the status and prestige of their husbands, a trait perhaps common in many armies. One old Africa hand among French officers described how a group of indignant army wives, one hundred strong and led by the wife of the African adjutant, started off on a march to Bamako, some 300 kilometers away, "in order to bring 'grievance' to the colonel" over the abusive action of a young officer. The officer, totally contrite, was forced to back down rather than let this happen.[92]

A variety of African civilians soon gravitated around the military camps in West Africa. These camp followers included artisans and even long-distance kola traders and other merchants, all living from the needs of the barracks economy. In general the French found these activities salutary, but worried about "boys" and former cooks with bad records, as well as women of low morals, tending to congregate in these camp towns.[93]

Soldiers' wages did not compare favorably with civilian rates in French West Africa. The first reference to a daily wage, in 1827, set pay at 5 centimes a day for rank-and-file soldiers.[94] Clearly this must have been a singularly unattractive rate because only eighteen months later the base pay had become 25 centimes a day.[95] Not until the dust had settled from conquest, and the spoils of war that accompanied victories had been acquired, did wages increase. In 1910 the base pay for soldiers in the *Tirailleurs Sénégalais* was 60 centimes a day, or 250 francs per annum, roughly half the pay of daily wage laborers and one quarter the wage earned by civilian policemen.[96] Still another potential disadvantage for soldiers was the French army's practice of withholding 25 percent of salary to be paid out as a lump sum upon discharge.[97]

On the other hand, two potential benefits that day laborers did not enjoy were extended to the African military. First, the soldier and his family were exempt from head taxes as long as he was in service.[98] The second perquisite was a pension plan, provided by the decree of 1889, but only to those Africans who served in the *Tirailleurs Sénégalais* for twenty-five years; only in 1904 did a second decree allow a proportional pension for those with fifteen years of service.[99] While such stringent rules limited benefits to a small minority, in 1910 303 soldiers, or 3 percent of the 12,000 *Tirailleurs Sénégalais*, had twelve years or more of seniority and were likely to qualify eventually for a proportional pension.[100]

The *Tirailleurs Sénégalais* served France well during the conquest period. Basically a mercenary force, it consisted primarily of men from the lowest social ranks, but with an important minority of well-born Africans as military intermediaries standing between the French officers who commanded and the African troops who followed orders. Little ideological allegiance, whether to the state or the ethnic group, bound men to their duties. Rather, as in most premodern and early-modern military organizations, patron-client ties bound men to their leaders. Nevertheless, it would be an error to assume that terms and conditions of military service had not changed. For some soldiers, the army had become a salaried career, carrying perquisites not available to civilians, and beginning a process of social differentiation that was to mark government service during the colonial period. Even the common rank-and-file *Tirailleur* by 1904 had gone well beyond his slave soldier predecessor who had watered the gardens and tended goats in the old French forts of the Senegal river and coast fifty years before.

With the creation of the French West African Federation in 1905, France announced an end to the military phase of its rule. For the *Tirailleurs Sénégalais* this marked a transitional phase from an army of conquest to an army of occupation. The next chapter examines the important changes which took place in the *Tirailleurs Sénégalais* from 1905 to 1919.

3

The Tirailleurs Sénégalais *and the First World War*

> *It is not a matter of introducing obligatory military service in West Africa, but of finding there a certain number of volunteers each year of age to become career soldiers who would retire after twelve to fifteen years of service.*
>
> Charles Mangin, La Force Noire

The *Tirailleurs Sénégalais* continued to grow dramatically in the first years of civilian rule after 1905. Figures 3.1 and 3.2 illustrate the size of the *Tirailleurs Sénégalais* from 1890 through to the early 1950s and the annual levies necessary to maintain these forces at strength. The figures reveal clearly that with the exception of the war years, the period of greatest expansion occurred after the beginning of the century and before the outbreak of war in 1914. French officials of that period were forced to learn a basic axiom of military force: that it required more soldiers to occupy conquered territory than it had to win it initially. The problem for military recruiters, however, was that the period following the creation of FWA was one of unprecedented economic expansion, and demand for civilian labor made a career in the *Tirailleurs Sénégalais* forces far less attractive than more remunerative jobs opening up on the railways, in the towns and cities, and in capitalist agriculture.[1]

Signs of a recruiting crisis were evident after 1905. When the time came for African *Tirailleurs* to be sent to Morocco to participate in the French conquest there, the army was forced to offer a special overseas bonus to meet its short-term commitments there.[2] Despite assurances by Charles Mangin, advocate of the Black Army, that West Africa was easily able to supply the military manpower, the shortage of men in 1912 prompted Governor-General William Ponty to take an important and unprecedented step.[3] With Mangin's support, Ponty enacted a law calling for partial conscription. At the time of this legislation no one had anticipated the significance that the First World War would give this new military obligation. Yet, on the basis of this 1912 law, 170,000 black Africans served in the *Tirailleurs Sénégalais* during the war.[4] Even more significantly, when it was replaced by a

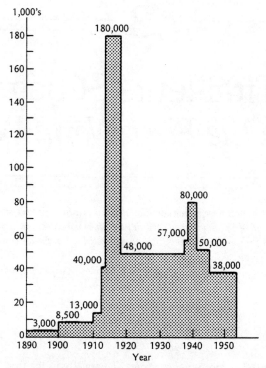

FIGURE 3.1. Size of the *Tirailleurs Sénégalais*, 1890–1953

Source: Compiled from various Annual Recruitment Reports in ANS, séries 2G, 4D, and 5D; and from recruitment data for 1953–1954 in CMIDOM, Versailles.

similar but more elaborate peacetime code in 1919, obligatory universal military service for African males became the single most important aspect of military life in colonial FWA. In effect, a small conquest army of largely slave mercenaries had been transformed into a mass army of conscripts drawn from virtually every level of West African society.

Transition to Partial Conscription

Four elements were at work in producing the rapid growth of the *Tirailleurs Sénégalais* in the early years of this century. First, the conquest itself was by no means over by 1905. The effective occupation of much of Côte d'Ivoire had yet to be achieved, and many pockets of primary resistance flourished.[5] Second, occupation troops were needed both to police and to administer the vast territory of Upper-Senegal-and-Niger, parts of which remained under military administration until after the First World War. Third, garrison troops were installed in permanent strong points throughout FWA. Kindia in Guinea was one such locale, and Dakar, the new

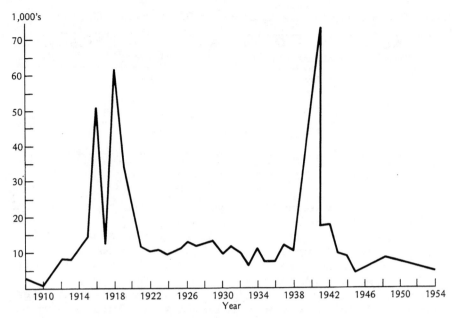

FIGURE 3.2. Annual levies for the *Tirailleurs Sénégalais*, 1908–1954

Source: Compiled from various Annual Recruitment Reports in ANS, séries 2G, 4D, and 5D; and from recruitment data for 1953–1954 in CMIDOM, Versailles.

federal capital, became a military center as well, with a permanent regimental garrison of three thousand men installed there.[6]

Fourth, as French colonial expansion proceeded apace elsewhere in Africa, the decision was taken at the highest levels in Paris to use the *Tirailleurs Sénégalais* as an expeditionary force outside West Africa.[7] The dispatch of African soldiers elsewhere had long been secondary to the primary task of maintaining order at home in West Africa. Now that West Africa was becoming secure, the use of the *Tirailleurs Sénégalais* to help create and then defend the far-flung French empire became a more important commitment. Thus, the *Tirailleurs Sénégalais* played an important and controversial role in the conquest of Morocco.[8] In that last venture of French expansion in Africa, the *Tirailleurs Sénégalais* constituted between 9 and 15 percent of the entire French Conquest Army, with dramatic increases occurring for the years 1912 and 1913 (See Table 3.1).

These increased manpower needs could not be met by the old rough-and-ready techniques of recruitment. The *rachat* method had effectively ended by the 1880s, and wholesale incorporation of defeated soldiers and prisoners became impractical once the major African armies had been defeated and dismantled. Perhaps a trickle of recruits could be procured as owners put their remaining young slaves forward to tap the enlistment bonus. Perhaps also some refugees from the *villages de liberté* (settlements of ex-slaves and fugitives) and from the slave exodus at Banamba and

TABLE 3.1 Composition of the French Conquest Army of Morocco, 1908–1913

Year	Total Forces	Europeans	Algerians and Tunisians	Black Africans	Percentage Black Africans
1908	10,476	4,260	4,872	1,344	13
1909	7,325	3,681	2,532	1,112	15
1910	6,228	3,314	2,166	748	12
1911	17,819	10,094	5,945	1,780	10
1912	41,065	23,882	13,314	3,870	9
1913	61,692	35,867	17,573	8,252	13

Source: Ministère de la Guerre, *Statistique médicale de l'armée métropolitaine et de l'armée coloniale: Statistique du Maroc de 1907 à 1919* (Paris: Imprimerie Nationale, 1922).

elsewhere produced new recruits.[9] The army, however, was not in a strong position to compete with other employers for the available labor.[10]

It was against this background that Charles Mangin entered the scene.[11] This diminutive, feisty colonial officer was born in Metz, Lorraine, in 1866, to a conservative, devoutly Catholic family.[12] Mangin was much affected by the exile of his family from the lost province of Lorraine after 1871, and his later military career was characterized by a strong desire for revenge against Germany. Even more important was his lifelong association with the French Colonial Army in Africa. He began his military career under Archinard in the conquest of the western Sudan, and was a member of Marchand's frustrating Fashoda expedition in 1898–1899.[13] Above all, Mangin led in the propagation if not in the conception of what he called *La Force Noire*, the idea of a large black African army in the service of France.[14]

Mangin's thinking about the utility to France of African soldiers can be traced to a novelist and conservative opponent of the Third Republic with whom he must have been familiar, Count Eugène Melchior de Vogüé. De Vogüé saw French civilization in decline and, like Rome in a similar situation, needing to turn to an inferior culture to revitalize itself through the employment of vigorous troops. The Africans were to be France's barbarian tribes, and Mangin one day their Germanicus.[15] Indeed, in his novel *Les Morts qui parlent*, de Vogüé pictures a colonial hero just returned from the western Sudan—a Marchand, or perhaps a Mangin—full of praise for the real and potential merits of Black African soldiers.[16] The hero speaks of redressing France's fortunes by means of an army of 100,000—even 200,000—men from the western Sudan, soldiers with "bayonets which do not reason, do not retreat, do not forgive."[17]

Ten years after de Vogüé wrote these lines, Mangin, by now a lieutenant colonel, had become the spokesman for an entire group of senior colonial officers, most of whom had served in command of *Tirailleurs*, who were arguing for an expanded *Tirailleurs Sénégalais* force.[18] Through appearances at various colonial and geographic societies, and in a series of articles in specialized journals, Mangin and his supporters lobbied hard for a larger Black Army.[19] They were not without opposition. Some came from French officers who had observed the mediocre performance of the *Tirailleurs Sénégalais* in North Africa;[20] Jean Jaurès led the political opposition of the socialists, fearing the domestic uses to which such a praetorian guard might be put.[21] In 1910, Mangin gathered together his various statements on the subject in *La Force Noire*, a book which received considerable attention at the time, and which has become symbolically linked to the modern birth of the *Tirailleurs Sénégalais*.[22] In his various writings, Mangin put forward two basic themes: first, that black Africa was an almost inexhaustible reservoir of men, and second, that by nature and history these men were ideally suited for military service.[23] Accustomed to thinking of manpower rather than technology as a critical measure of military strength, Mangin and his supporters hoped to counterpoise France's colonial populations against the more rapid population growth of Germany. How could France, with a stagnant or even declining birth rate, hope to win back the lost provinces in a future war against a Germany that already had a larger population and a higher birth rate? For Mangin West Africa provided the desperate answer. He estimated that the region alone could easily furnish 10,000 volunteers a year, and that the potential far exceeded that figure.[24]

As for the notion that Africans were "born soldiers," Mangin offered two explanations, one cultural and the other racial. On the one hand, Africans were superior to European workers and peasants because they were unspoiled by the difficult work of factory or field, and the healthier for it.[25] But more importantly, Africans were possessed of a nervous system sufficiently less developed than that of whites so as to make them far less sensitive to pain. Mangin stated: "The nervous system of the black man is much less developed than that of the white. All the surgeons have observed how impassive the black is under the knife."[26]

The reality, of course, was far different. It is easy enough to dismiss racist notions about West Africans' warlike traits and unique neurological systems. As for Mangin's demographic assumptions, French West Africa was in fact thinly populated, notwithstanding a few densely peopled regions such as the Mossi plateau, the Bambara belt south of the Niger, and the Futa Jalon mountains around Labé in Guinea. Early censuses were extremely crude guesses rather than systematic enumerations, but the published figures set the overall population at 10 million or less.[27] Even more significantly, French civilian administrators were already aware of a labor shortage. The peoples of West Africa, plagued by the political disorder accompanying the wars of French conquest and by the dislocations of the slave trade in the latter half of the nineteenth century, needed their labor to maintain subsistence food production in the household economy. Yet the growing appetite for labor of the French regime, especially with its large requisitions for the building of infrastructure, made for labor shortage, not surplus. Thus Governor-General William Ponty, while sharing with Mangin a belief in the military potential of West Africans to serve France, tempered the French military's demands for labor.[28] Ponty suggested to the minister of colonies that Mangin's minimum target of 10,000 soldiers a year should become the maximum annual levy if the economic development of the colony were not to be ruined. For 1909 Ponty proposed an intake of 8,000 men, four times the levy of any previous year's recruitment in FWA.[29]

By 1912, however, it had become obvious that the colony was having difficulty supplying even 5,000 volunteers. On Ponty's advice, the French government initiated a system of partial conscription by means of its Conscription Law of 7 February 1912, an important step on the road to universal conscription in FWA.[30] Ponty had been forced to admit that volunteers had simply not been coming forward in sufficient strength. The new law required conscripts, who could be anywhere from

Tirailleurs marching in Paris on Bastille Day, 1914

twenty to twenty-eight years of age, to serve four years, compared to five- or six-year engagements for volunteers. In case of war, reserves would be called up, an important clause in view of events that were to follow. Considerable discretionary power to administer this partial conscription system was left to the governor-general, in consultation with the CIC in FWA. Here again, this provision would prove important, as it was the vague 1912 law that prevailed for the huge wartime levies that were soon to follow.

Ponty's 1912 Conscription Law represented a significant setback for Mangin's original concept. He and his supporters had envisioned a professional army consisting entirely of volunteers. Additional men would come forward easily, Mangin assumed, because most West Africans were "martial peoples," and many soldiers of the recently conquered African states would welcome the opportunity to continue as professional soldiers.[31]

Despite the setback, Mangin persisted both in his ideas about a Black Army and in his enthusiastic leadership of the *Tirailleurs Sénégalais* in combat. *Tirailleurs* marched with him in 1912 during the conquest of Morocco, where he showed his preference for the bloody victory over the negotiated truce.[32] When the First World War brought Mangin promotion to general and senior command, thousands of African soldiers contributed to his mercurial military fortunes with their lives: at Verdun in 1916, on the disastrous campaign on the Chemin-des-Dames in 1917, and again in the final victorious offensive in the summer and fall of 1918. Nor was the fate of Africans limited only to Mangin's costly combat tactics of the all-out offen-

Mangin and the *Tirailleurs Sénégalais* in Morocco, 1913

sive. In 1917, he contributed significantly to a decision that brought thousands more to the front.[33] A desperate Clemenceau, remembering Mangin's advocacy of the Black Army, welcomed his opinion that FWA, far from being exhausted by its war effort, could still deliver hundreds of thousands more men.[34] A newspaper of the time commented: "It is worth asking whether the honorable General Charles Mangin was born to serve the Black Troops or whether the Black Troops were born to serve the honorable General Mangin."[35]

The African newspaperman who penned these lines in 1913 could hardly have imagined how ironic these words would be. After the Armistice of 1918, Mangin, now a war hero, persisted in his obsession to further the militarization of France's colonies. He argued in his war memoirs that the West African contribution of over 160,000 men was insufficient because colonial officials deliberately held men back.[36] He claimed rather preposterously that had the politicians and colonial administrators given him a free hand in recruiting, he would have raised a Black Army of a quarter million, even a half million troops.

Tragically, instead of being dismissed as a crank, Mangin was actually placed in a position in the years after 1918 where his extreme notions could affect policy. With his enthusiastic support, French military planners raised the size of standing

Tirailleurs Sénégalais forces to 55,000 men in 1920, with a projected force of 110,000 by 1925.[37] Part of the reason lay in the persistence of French demographic fears. The war had bled France of over two million men who not only would no longer join the labor market but would not play their part in population reproduction. As early as 1920, the CIC of FWA, General Gassouin, was warning of the "hollow" cohorts, a dearth of young Frenchmen for the universal military conscription in the years from 1934 to 1938.[38] The perceived need then was for an expanded colonial contribution in peacetime to replace missing French citizens.

In 1920 an interministerial commission was formed to study the demographic implications of continued and expanded conscription in West Africa. Mangin was made the chairman, representing the interests of the minister of defense. Officials of lower prestige from the Ministry of Colonies attempted to curb what they held to be excessive demands of the military. They represented the views of the colonial administration and of business interests who wished to keep a labor reserve in French West Africa. To resolve the potential impasse, the commission authorized what may have been the first "quasi-demographic" surveys of a segment of the French West African population.[39]

"Terror Out of Africa": Images of the *Tirailleurs Sénégalais* in Combat

When armed conflict broke out in Europe, most military planners anticipated a short war. If African troops had any role to play, it was strictly to relieve French soldiers as garrison troops in North Africa so that European units could cross over into Europe. In Africa itself, the *Tirailleurs Sénégalais*, along with their British allies, had a very easy time of it conquering German Togoland, where token resistance gave an entirely false picture of what was in store in German East Africa especially. As for the performance of the *Tirailleurs Sénégalais* on the western front, legend and fact are sufficiently intertwined as to warrant closer examination.

The performance of *Tirailleurs Sénégalais* units in Belgium in the opening stages of the war in 1914 is one high point of French military mythology. Appropriately enough, one episode involved Sergeant Baba Koulibaly, who served as General Mangin's personal bodyguard from 1903 until his death in 1922 while still in service. Koulibaly, a huge Bambara speaker from French Soudan, stood six feet four inches tall, and larger than life in French barracks talk as a result of his famous exploit early in the war. In the first days of August 1914 at Baronville, east of Givet, German patrols surprised General Mangin's motor escort in the village. Koulibaly and the three chauffeurs drove off the patrol of eleven Prussian dragoons. But, to hear Mangin describe events, this was but the beginning of Koulibaly's feats that day:

> Baba carried out a *sortie* entirely by himself and saw the second patrol of six German dragoons appear at the other end of the village. He charged by himself and fired on them, wounding one, and driving the others off. The noise attracted both French and German patrols; he linked up with French cavalry men and together they pounced on the Germans; Baba, unable to gallop after them, returned to the village square, where General de Vaulgrenant found him all alone, leaning on his rifle: 'what are you doing

here?' asked the General.—'I'm occupying the village', was his reply. This story and these remarks have become legendary in the entire cavalry corps.[40]

Koulibaly's individual exploits are matched by accounts of larger units. Thus, four battalions of the *Tirailleurs Sénégalais* were driven by car from France to the bend of the Yser in Belgium to form part of a hastily erected line against the German drive to the Channel. At the small town of Dixmude on the eastern side of the Yser, the Africans endured severe punishment. After waiting day after day for a German attack, and undergoing heavy shelling all the while, they were finally assaulted by the Germans. Only the fiercest pressure brought results: 40,000 Germans overran 10,000 defenders. Only thirty men and two officers from the First Senegalese Battalion survived the onslaught. The Third Battalion was reduced by one half, the Second had only one hundred left; the Fourth Battalion, made up only of Europeans, could muster only fifty men and one officer. Out of this debacle, the French could only reconstruct one battalion, the Twentieth Senegalese.[41] By holding out long enough however, the Africans did their part to hold the line and mute the German thrust of 1914.

Thus a gallant picture emerges. Both individually and as units, African soldiers are said to have covered themselves in glory. Nor is this a picture derived only from the epic regimental literatures recorded in the immediate aftermath of the war. The tradition is echoed in this recent, ringing account of the African defense at Dixmude: "Against the endless, disciplined, grey columns of the Germans, they fought with a desperate fury that the most eloquent panegyrist could hardly exaggerate."[42]

Underlying the myth of African praetorians defending France is a much less benign image, that of barbarian ruthlessness. In the hysteria of the First World War, African troops acquired the reputation of "headcutters who could, almost singlehandedly, scare *le boche* all the way back to Berlin."[43] The gory imagery was said to have become so popular among the French press and public that one authority found it necessary to caution an enthusiastic French journalist to describe African combat "without having recourse to massacres for which we reproach our enemies."[44]

On the other side of the propaganda war, the German press reduced Africans to the level of savagery. These accusations, of course, had another purpose in mind: to show the ruthlessness of enemies who were prepared to stoop so low as to employ cannibals, an accusation leveled frequently in the German press.[45] One account in June 1918 drew particular attention to the determined resistance of *Tirailleurs* around the city of Reims as being conducted by mindless blacks drunk on reserves of brandy and all brandishing the "*coupe-coupe*, the big combat knife."[46] Thus the image of a head-cutting barbarian who took no prisoners was cultivated by some elements on the French side, and given credence by equally doubtful sources on the German side.

The myth of African military achievement in the First World War can be contrasted with a series of countermyths, opinions which emphasized the negative aspects of African service. Opponents of the Black Army argued that African soldiers were undisciplined, unreliable, and essentially ineffective elements in France's war effort.

Such negative views were most common among France's British allies. Even before the war, British opinion had been mildly unfavorable to the Mangin plan to create a Black Army. H. G. Wells appeared to have struck a popular chord when he projected a situation by the year 2100 in which black police would be imported from Africa to quell civil disturbances in Europe.[47] Wells thus invoked a widespread British aversion to the employment of colonials in positions of authority over Europeans, a dangerous precedent for a European imperial power ruling over millions of African and Asian subjects.

This British aversion to France's army surfaced frequently during the war itself. General Haig, for example, was known to be notoriously contemptuous, and with little justification, of the entire French armed forces. Needless to say, he was not fond of French colonial troops. He took perverse pleasure from the news of any poor performance from his French and colonial allies. When he learned of the poor showing of the Thirty-seventh African Division at Verdun, he noted in his diary that Moroccan soldiers "used to do the same thing on my right on the Aisne."[48]

An interesting illustration of the countermyth derives from widely shared Canadian opinion of African soldiers. The Canadians had barely landed in France in February 1915 when they found themselves thrust into the lines at Ypres. There, they found on their left flank "French African" troops, who had similarly only recently arrived on the front. Together these unfortunate colonials of the two empires were unlucky enough to receive the first German gas attack of the war on 22 April. Thoroughly unprepared and unequipped, men who saved their lives by retreating can surely be understood. Yet a respected Canadian military historian, G. F. G. Stanley, echoes the Canadian consensus when he states that the Africans

> broke and fled in terror, thus opening the way for a German breakthrough. In spite of the burning effects of the gas, the Canadians closed the gap and saved the situation until the arrival of reinforcements.[49]

African soldiers do not receive kind treatment in Alistair Horne's brilliant and scholarly study of the conflagration at Verdun. He repeats the common accusations, without providing evidence or examples, that African troops had "a summary way with captives."[50] He also dwells at length on the breakdown in discipline of men in the Thirty-seventh African Division, who, upon being taken prisoner by the Germans, were said to have screamed at their officers and to have "spit at the captured officers of other French regiments."[51] To be fair to Horne, however, he does attempt to account for this behavior, as will be shown below.

Perhaps the ugliest incident involving African troops, found not only in British accounts but in French ones as well, was a setback that occurred within the general context of Allied failure on the Chemin-des-Dames (known in British accounts as the second battle on the Aisne). In this battle, African units of Mangin's attacking force were mauled at Heurtebise on 16 April 1917. According to one critic, the Second Colonial Corps especially was said to have panicked and fled at the beginning of the battle.[52] Another military historian, Leon Wolff, notes that Senegalese troops broke and ran, "even boarding hospital trains to speed their departure."[53] Lastly, the French politician Paul Painlevé, anxious to discredit Mangin, noted:

> ... And I could reply to him [Mangin] that the invasion of the French ambulance train at Courlandon by undisciplined *Tirailleurs Sénégalais*, right

under the eyes of his Command Headquarters, did more to demoralize public opinion than the statistics on losses which were never published.54

German contributions to the countermyth were largely consistent with British views, but for quite different motives. Before the war, the German press quailed at the prospect of thousands of Africans and other colonials being imported by France in order to line up along the German frontier. Advocates of a stronger and larger German army, however, used this threat of colonial hordes on their doorstep as food for their campaign. Maximilian Harden, for one, was a vociferous supporter of an enlarged German army to confront blacks.55

The most dramatic German response to African troops came, in fact, after the Armistice, when Africans and other colonials formed a significant component in Mangin's Tenth Army of Occupation in the Rhineland. Known in the German press campaign as *die schwarze Schande*, "the black shame,"56 the argument suggested that the French had deliberately selected colonial troops to occupy the Rhineland in order to humiliate further a prostrate Germany, and that these troops were guilty of numerous sexual assaults against German women. The echo these charges received in the British and American press was so significant that France was forced in the end to transfer black African soldiers to the Levant, despite repeated and documented refutations of the sensational charges against colonial troops.

Even in France itself a visible minority of opinion contributed to the countermyth of African military ineptitude. Despite the success of Mangin in establishing a particular image of the African soldier, he did have his opponents, men with equally strong views about the harm colonial troops had done to France. Generally speaking, this opposition came from among French officers who had served in North Africa, men who had opposed Mangin's attempt to elevate the black African soldier to a premier place among the French empire's troops. Occasionally, as in the writings of Paul Azan, the counterarguments ran that North African troops were better soldiers and should therefore be given the primary emphasis in the Colonial Army. At other times the argument concentrated on the observed inefficiency of black Africans while on campaign in Algeria or Morocco in the years preceding the outbreak of war in Europe. Thus General de Torcy noted that in the experience of officers in North Africa, *Tirailleurs* were found to be adequate under fire but hard to maneuver and mediocre shots.57 De Torcy further held that the *Tirailleurs Sénégalais* were of doubtful value in cold climates, however glorious had been their performance in the tropics.

The experiences of African troops in the first winter of the war lent credence to de Torcy's views. Africans suffered acutely from the effects of the cold as well as from the German offensive. Large numbers of frostbite cases, together with their past experience in North Africa, led the French command to adopt a practice that came to be known as *hivernage* (the French term for winter season in West Africa). It involved the transfer of black units from the front to southern France at the beginning of each winter, with military camps for Africans being established for the most part in the Midi, and centered around Fréjus. While there were no complaints from the men or from the junior officers thus transferred away from the front, the French High Command did find it inconvenient to have to pull thousands of soldiers back from the front each winter.58

After the war, when the debate resumed in France over whether to continue with a large *Tirailleurs Sénégalais* force, critics once again surfaced. Significant among these was Réginald Kann, who had served as an officer among four mixed regiments in the Dardenelles. Each regiment had one white battalion and two black ones. This caused no problems in the initial engagements of the Dardenelles campaign, but the next time out the Africans were said to have lost their taste for action and to have "contaminated" the white units.[59]

While it may not be possible to resolve these conflicting images by reconstructing combat events, one valuable approach is to attempt to view the events from below, as John Keegan has put it in his stimulating *The Face of Battle*.[60] Keegan stresses such issues as the state of soldiers' morale, how they behaved in battle, and how they viewed the enemy, prisoners included. With his approach in mind, it is worth re-examining aspects of the three battles in which Africans were said to have performed poorly, those of Verdun, Ypres, and Chemin-des-Dames.

The behavior, as prisoners, of men of the Thirty-seventh African Division at Verdun has been described but not yet explained. Prior to combat, they had been split into small units, deprived of their familiar officers and placed under new ones, and, most importantly, provided with far less artillery preparation than was assigned regular units of the French army.[61] They were expected to hold a line devoid of any prepared positions. Subjected to bitter cold during a night exposed in this fashion, their morale plummeted. When they observed thousands of aimlessly wandering wounded, shattered remnants of the initial German shelling and onslaught in the last week of February, they were extremely vulnerable. Finally, the German bombardment without cover was too much. An entire battalion wavered and broke at the sight of German infantry coming at them. One regiment, the Third Zouaves, seems to have entirely vanished, perhaps not entirely at German hands. A French captain ordered machine guns to open up on the backs of the fleeing men. It is no surprise, then, that the shattered remnants who were taken prisoner turned on their officers. Horne reasonably notes that the real failure was not that of the Africans but of the High Command. Whenever infantry was left exposed in unprepared positions to enemy artillery, with no artillery support of their own, bad results were inevitable all along the western front.

If the absence of cover helps explain African behavior at Verdun, the absence of planning tells us something about the seizure of ambulance trains on the Chemin-des-Dames. Generals Nivelle and Mangin, architects of the ill-fated offensive in the spring of 1917, were so confident of smashing success that they expected only 10,000 wounded in the offensive. The French Medical Services, grown cautious over the excessive optimism of the High Command, made provision for 15,000. As it turned out, Chemin-des-Dames produced 90,000 wounded.[62] Given the acute shortage of space, it would not be surprising if ambulance crews gave preference to French soldiers over colonial ones. While Africans may normally have accepted some discrimination as part of the colonial dispensation, their unwillingness to be left to die of their wounds on the battlefield is readily understandable.

The failure of Africans to hold their positions at Ypres while undergoing the first poison gas attack of the war has already been noted. It is not an easy matter to resolve except to note that retreat in some disorder would be the likely response of most men, given their lack of preparation to deal with such a frightening weapon. It is, rather, the Canadians' ability to hold the line that would seem to require explanation.

The ability of a group of men to sustain the gas attack on Ypres is largely a question of ideological preparation and collective will. Beyond this, the First World War's transformation into the worst killing ground in military history is at issue. Rank-and-file soldiers and generals, factory workers and factory owners, capitalists and socialists, all became caught up in the national causes which were held to be at risk. This mystification of the nation was extended to colonial worlds as well, and Canadians were a clear example. Later in the century, on the eve of the Second World War, members of the *Tirailleurs Sénégalais* would become similarly subject to a Jacobin nationalist ideology stressing the virtues of patriotism.[63] In the First World War, however, it may be argued that the *Tirailleurs Sénégalais* were either mercenaries or unwilling conscripts, men who did not yet share in the myth of the nation-in-arms. Like Renaissance mercenaries, Africans were still capable of making a calculated assessment of their prospects for survival on the battlefield.

The Canadian response was different. They had been taught to see themselves as enthusiastic defenders of an imperial ideal. This sentiment is reflected in the story of their achievement at Vimy Ridge, where Canadians performed brilliantly on the first day of the ill-fated offensive of 1917, the same campaign that saw the Africans humiliated on the Chemin-des-Dames. The glorification of this Canadian episode bears extraordinary comparison with those epic accounts of African successes in the French literature. As recently as the centenary of Canada's birth, Stanley reflected accurately the thinking of an entire generation of Canadians when he proudly noted that Vimy Ridge won the Canadians the honored reputation of being "first class Shock troops."[64] Stanley added that Canadians were thereafter deployed in this manner for the remainder of the war.[65]

Unfortunately, Canadian losses all too clearly verify this practice. More recent studies have begun to adopt a realistic view of the sacrificial nature of Canadian losses.[66] McWilliams and Steel effectively demonstrate the terrible suffering of the Forty-sixth Canadian Infantry Battalion (the South Saskatchewan) at Vimy, the Somme, Amiens, and Passchendaele. In the latter battle, losses were 403 killed and wounded out of 600. Overall, of 5,374 men serving in the battalion, 4,917 were either killed or wounded, most often in a futile engagement.

The myth of the African as bloodthirsty barbarian can be much more readily exposed. Propaganda lay behind both French efforts to strike fear in the enemy and German efforts to project Africans as savages so as to discredit the French for having used these troops. Both sides seem to have succeeded. German troops were said to inquire when moving into a new sector of the line of trenches, "Are there any Africans opposite?"[67]

One French account offers a graphic description of the terror felt by the German soldiers during an attack of a mixed company of African and European troops:

> But they raised their hands, they were terrorized. Their corpsmen waved the Red Cross flags, yelling 'protektion sanitat.' The prisoners [about 120] are so scared of the blacks [of the assault force] that they run to the rear lines of the French. They are so scared that they even tried to climb over the top [of the trenches] while keeping their hands raised at all times.[68]

The same fear is reflected in the memoirs of Bakary Diallo, one of the few firsthand African accounts we have of the "face of battle":

> A German who had mistaken our lines for his own, was captured with his load of coffee by a Senegalese sentry. When he saw himself surrounded by *tirailleurs*, he began to shake. Poor man, couldn't you have expected this possibility just as well as the gold of glory? The blacks that you took for savages captured you in war, but instead of cutting down your life, they made you a prisoner. May your fear not prevent you from proclaiming in your country, tomorrow, after the battle, the sentiment of justice which will rehabilitate the name of the human race of which we are all savages.[69]

While it is difficult to know how typical these two instances were, both accounts make it plain that African units did in fact take prisoners. It seems fair to conclude, following Keegan and S.L.A. Marshall, that African soldiers were no different from most men who have ever fought in battle. They argue that most men are equally afraid to die, but that men also fear being thought a coward by their comrades. Bakary Diallo offered an African voice on the subject of courage in battle, which would support these opinions. Just before his first taste of combat in Morocco he observed:

> Courage in battle has a high value for the *Tirailleurs Sénégalais*. If by some misfortune they were able to see how my insides were churning with fear, I would have lost their esteem forever. . . . Despite my trembling heart, I maintained an exterior of iron.[70]

On another occasion, noting that soldiers at the western front were fed a full meal before battle, he said: "Everything was prepared during the night, and one has to eat well to have the strength to die."[71] Bravery, then, is a complex human trait. It would seem to depend on such measurable variables as leadership, training, and motivation, not ethnicity and race.[72]

The Career of Major Abdel-Kader Mademba

Abdel-Kader Mademba, son of Mademba Sy and nephew of Captain Mahmadou Racine Sy, was the only African subject to reach the rank of major and hold a regular commission in the French Army before the Second World War. A captain during the First World War until his promotion to major in the early 1920s, his military and political career was sufficiently important to warrant closer examination.

On his father's side, Abdel-Kader was a grandson of M'Baye Sy, a noteworthy member of the Torodbe Fulbe clan of Senegambia. On his mother's side, his grandfather was Al-Hajj Umar Tal, the famous Tijani religious leader and opponent of French expansion. M'Baye Sy is said to have placed his entire family in the service of France in 1847, in preference to the Tijani empire of Al-Hajj Umar Tal.[73] M'Baye Sy's oldest son, Mademba Sy, born on 3 March 1852 at Saint-Louis, was to become one of the heirs to the Tijani empire in the Soudan, thanks to a French appointment as ruler of Sansanding, opposite Ségou in the Niger Valley. Other sons of M'Baye Sy served the French in other ways. One was Mahmadou Racine Sy, the first African soldier to reach the rank of captain in the *Tirailleurs Sénégalais*.[74]

Colors of the *Ier Régiment des Tirailleurs Sénégalais*, First World War

Another son, Ahmadi Coumba Sy, reached the highest level of interpreter in French service.

Mademba Sy sent his seven sons to Algiers in 1901 to have them educated in a lycée. The different disease environment and climate of North Africa took its toll, and three of the brothers did not survive to adulthood. Of the remaining four, Abdel-Kader and Cheikh became officers in the *Tirailleurs Sénégalais*, Racine an agricultural engineer, and Ben Daoud a teacher in the colonial service.[75] Racine served during the First World War as a Supply Officer in Bordeaux, and Cheikh was killed near Soissons in 1918.[76]

Abdel-Kader Mademba was probably the best known African military intermediary during the interwar period both in French and African circles. Each milieu,

Captain Abdel-Kader Mademba

for somewhat different reasons, thought highly of him. To the French, he was Captain (and later Major) Abdel-Kader Mademba, the protégé of General Mangin, a hero of the Dardenelles campaign, and an important aide to Blaise Diagne in his wildly successful recruitment drive in 1918. To Muslim Fulbe in FWA, his name was Abd al-Kader Mademba Sy, a man with a distinguished heritage. He was, in fact, a grandson of Al-Hajj Umar Tal, even if he clearly did not share his grandfather's hostility to these Christian intruders. When Colonel Archinard had captured Ségou in 1890, Mademba Sy had been a member of his entourage. Archinard rewarded the ambitious young man with a spoil of war, one of the young daughters of Umar, who was unable to flee the Tijani capital of Ségou in time. Mademba Sy married this Fulbe noblewoman and Abdel-Kader was born to her three years later.[77]

Despite his reputation, Abdel-Kader's military career was, in fact, short and unpleasant. In 1914, at the age of twenty-one, he volunteered for the *Tirailleurs Sénégalais*; he was wounded in the chest at the Battle of Koum-Kalé in the Dardenelles, his first campaign, in 1915. One of his lungs was removed, which, coupled with a congenital family weakness (his brothers had died of lung ailments and his father of pneumonia), left him in poor health the rest of his short life. Abdel-Kader's wound and family medical history would have made him a good candidate for discharge, but that did not happen. Instead he enrolled in an officers' training course at Cassis, from which he graduated as a second lieutenant in time to participate in the protracted Verdun campaign, where he was wounded once more. This seems to have ended his direct combat activity but not his utility to the French. Named lieutenant and awarded the Cross of the Legion of Honor, Mademba became a celebrity.

Some of his correspondence with General Mangin has survived, offering a glimpse of his commitment to France and his military career. He pined for promotion and action, and felt his health problem would ruin his career. He wrote to Mangin while in a sanatorium in 1917, one of the many times he was hospitalized, "The thought that my entire life is subject to suffering as a result of this terrible disease worries me constantly, and I believe that the very fact of being here causes me to lose valuable opportunities."[78]

In fact, he had just returned from FWA, where earlier in the year he had conducted an important fact-finding tour. After the widespread revolts against conscription and French rule in 1915 and 1916, the French Military wished to determine for itself whether the view of civilian officials like Van Vollenhoven that no further recruitment could take place was correct.[79]

Abdel-Kader Mademba's report lent support to his patron, Charles Mangin, in the effort to persuade Clemenceau to push ahead with a new conscription drive in 1918. Mademba's mission had been kept secret from the civilian administration and was undertaken while he was home in the Soudan ostensibly on sick leave. His brief eight-page report argued that recruitment was only one of the numerous causes of the 1915–1916 revolts. First and foremost, he laid blame on forced recruiting of labor for road building. Next, he cited exactions and abuses of the *cercle* police who went so far in some cases as to requisition millet and rice seed grains. Finally, the decision to allow seriously wounded soldiers to return to their communities without either monetary compensation or artificial limbs had left a bitter taste throughout FWA. Further military recruitment was still possible, he concluded, but it had to be conducted equitably in every village so that all could see that not only the unprivileged were required to serve.[80] In his private letter to Mangin, Mademba noted that private commercial houses were fierce adversaries of military recruitment and that their claim of labor shortage was a great lie. In his view the French Soudan alone could furnish 20,000 additional men, if the recruitment were properly organized.[81]

Not only did Blaise Diagne take seriously Mademba's recommendations during his successful recruitment tour of FWA in 1918. The African deputy asked Mademba to join the entourage, along with two other African officers, Henri Gomis and Amadou N'Diaye Duguay-Clédor. With Diagne himself carrying the title of Commissioner of the French Republic, equal in rank to that of Governor-General of FWA, and with the three Africans smartly dressed in officers' uniforms, the entourage must have caused a sensation. To say the least, European colonial officials were not accustomed to defer to a black man and his entourage of three young African officers. Notable African families who had hitherto provided military recruiters with their clients and slaves could not have failed to take notice. Recognizing that a new generation of African intermediaries was emerging through the military experience of the war, the African elite now decided to put forward their sons and nephews for military service in significant numbers for the first time.[82]

The efforts of the Diagne mission sapped Mademba's strength once again. In

October, 1918, too ill to be transferred back to the front, he remained in the Soudan to help in the next recruitment drive scheduled for the fall and winter of 1918–1919, a levy that was postponed when news of the armistice reached FWA.[83] After the war, Mademba was promoted to captain and served tours of duty with the *Tirailleurs Sénégalais* in Syria and Morocco.[84] He continued his correspondence with Mangin, and agreed with him that FWA was the "reservoir" of the French Army. He wrote a pamphlet entitled "*Recrutement de Tirailleurs et main d'oeuvre au Soudan Français*" in the early 1920s, in which he repeated his sentiments of 1917: civilian officials and the private companies lied about the labor shortage in order to keep their best employees and their influence.[85]

Mademba's views on the *Tirailleurs Sénégalais* and on military issues generally were indistinguishable from those of his mentor, Charles Mangin, after whom he named one of his sons. Like all soldiers, he found it humiliating that the Thirty-sixth *Bataillon* of the *Tirailleurs Sénégalais*, an outstanding combat unit, was used as a labor force in Lyons. He wrote to Mangin in December, 1917: "I fear that because of all this manual labor, they will forget their drill with rifles and bayonets."[86] In this same letter he deplored the "defeatist crisis" of 1917, a euphemism of military loyalists for the great mutiny of the French Army in 1917, and congratulated colonial workers from Asia and Africa who refused to join unpatriotic French workers who threatened to strike during that same crisis.[87] He embraced entirely the paternalist ideology of the French Colonial Army, arguing that all the African soldier needed was some recognition of his devotion to his unit. Citing the example of one *Tirailleur*, Scriba Coulibaly, whom he had met in the hospital at Menton, he deplored the failure of higher military authorities to treat such loyal soldiers fairly. This soldier, who had seen combat from the war's beginning on the Marne and again in the Dardenelles, and who had been paralyzed in a leg after being wounded at the Somme, feared most having to return empty-handed to his village, and wanted a ribbon to take back with him.[88]

Mademba became a major in the early 1920s, the highest rank attained by any African until the 1950s. With this rank came French citizenship, a regular officer's commission in the French army, and the higher salary and benefits that went with it. Unfortunately, he did not live long enough to enjoy this success. He died of respiratory disease on 21 September 1932, aged 38, at the military hospital in Briançon. The African Deputy to Paris and Mme. Blaise Diagne attended the official funeral there. After a year, according to the obituary notice, his body was to be exhumed and the remains transported to Sansanding to rest beside the grave of his father, Mademba Sy.[89]

The Conscription Debate

Until the First World War the *Tirailleurs Sénégalais* was essentially a mercenary army. It was composed of volunteer soldiers attracted to the profession by a variety of incentives, whose numbers were supplemented by partial conscription through the terms of the Conscription Law of 1912. The First World War changed the *Tirailleurs Sénégalais* forever. Not only did French officials themselves admit that wartime recruitment was conducted by coercive methods reminiscent of the repudiated era of the slave trade.[90] Even more dramatic was the decision of the Clemenceau government in 1919 to maintain and expand a conscript rather than a volunteer army by the introduction of universal peacetime conscription in FWA.[91]

The conscription issue was tied to France's perceived need to maintain a large standing army immediately after the war.[92] France's losses in manpower during the war had been enormous, and the pressure for rapid demobilization of the survivors was irresistible on several counts. On the one hand, people were weary of war and all it meant, including army service; on the other, there was pressure to get rank-and-file Frenchmen back to work in the French economy. Yet the need to maintain a large army was apparent, given the policy goals of French leadership. France would occupy considerable parts of German's industrial heartland to guarantee payment of war reparations. The acquisition of former German colonies and parts of Turkey's empire had considerably extended the limits of France's colonial empire, and this called for an increase in occupation troops. Thus, if French soldiers were to be demobilized, a substantial number of colonial troops could not be.

Demographic considerations also influenced the French decision to keep a large colonial army. In the late nineteenth century, fears had been expressed about the existing demographic imbalance of France in relation to Germany. The lower French birth rate promised to make this imbalance so serious as to give Germany an insurmountable military advantage. Mangin and his supporters had invoked precisely this argument in justifying the need to incorporate colonial subjects into military units. The effects of the First World War temporarily diminished France's concern about German military rivalry but did not destroy the demographic argument. Even if Germany was not a threat at that moment, France's general military posture as a great power still rested heavily on a large standing army.

While this may explain why a large section of the *Tirailleurs Sénégalais* was not demobilized after the war, it does not account for Clemenceau's decision to maintain conscription rather than return to the older professional tradition that the army itself favored. A series of factors combined to persuade Clemenceau. First, conscripts were cheaper than volunteer professionals, who had to be attracted with higher premiums to join up and stay in military service. Second, there was some doubt whether enough colonial subjects would sign up at any price given the cruel evidence of risk provided by the First World War.[93] Third, the lesson of the First World War conscription levies was clear. Despite the cries of alarm from self-interested private companies and the civilian colonial administration, FWA had furnished the manpower with only a modest degree of protest, and could logically be expected to continue to do so now that annual levies would be systematic, and of lower magnitude.

The Clemenceau government did not arrive at this conclusion without practical experience. In 1917 it had been forced to weigh the risk of economic stagnation or even armed revolt against its pressing need for more troops. When the government made inquiries about a large new levy in 1917, it received a chorus of protests from FWA, and especially from the Governor-General Joost van Vollenhoven.[94] He produced a long and passionate document on the subject, but his argument could be reduced to two basic points. First, extensive and brutal recruiting of men was ruining the economy of the colony, both in its traditional subsistence base and in its fledgling modern sector. Labor was in short supply and the people hopelessly demoralized by forced recruitment, shortages of food, and related difficulties. Not only were thousands of men being drained off by the war, but thousands more were fleeing, sometimes with their families and even entire communities, across the frontiers to British West Africa to escape the recruiters. This demoralization had even spread to the overseas French community in West Africa. Understaffed and overworked civilian administrators were depressed by the unpleasantness associated with forced recruitment; private companies were discouraged by the lack of labor for their enterprises and the unhealthy commercial climate for the future.[95] If nothing else, conscription was hurting FWA's economic contribution to the war effort.

Van Vollenhoven's second point was apocalyptic. He argued that France was in imminent danger of losing her colony through armed uprisings. Recruitment was held to have pushed the population to the brink of revolt, and over it, in substantial regions of Upper-Senegal-and-Niger and Dahomey.[96] Had these revolts spread to densely populated communities like those of the Mossi or Bambara, an understaffed and preoccupied France might not be able to hold on to the colony. Another large levy of men might very well produce this disaster. Van Vollenhoven recommended a halt to conscription in favor of a big economic effort; this was French West Africa's most effective way of helping in the war effort.

In 1917 France needed soldiers more than peanuts. Clemenceau's cabinet decided to take the risk and ordered the largest levy of the war, 50,000 men to be conscripted in early 1918.[97] Van Vollenhoven's promising colonial career was over. He resigned in disgust and marched off to the trenches, perhaps in search of martyrdom. If that was his wish, it was fulfilled; he was killed in action early in 1918.[98]

Meanwhile, a new actor in French West African politics was making his presence felt. He was Blaise Diagne, Senegal's first African deputy, a man with a considerable political interest in recruitment issues. When war had broken out in 1914, Diagne had been a member of the Chamber of Deputies for only six weeks. Recognizing in the French declaration of war an opportunity to secure beyond all doubt the right of *originaires*, that is, Senegalese born in the Four Communes[99] to French citizenship, he immediately telegraphed his assistant, Galandou Diouf, in Dakar:

> In the light of yesterday's events, I appeal to you to ask the administration that we be allowed to defend our own territory by enlisting for the duration of the war in the French corps stationed in Senegal. . . . [They should] recognize that all our people are worthy of the rights and duties of citizens, especially after my election. Notify the Governor-General and our compatriots of this [plan] while I tell the Ministry.[100]

Diagne was seeking two rather circular objectives. First, he wished to establish that *originaires*, because they were citizens, should serve in regular French army units rather than in the ranks of the *Tirailleurs Sénégalais*. Second, he hoped that their willingness to volunteer to defend France would secure for the *originaires* and their descendants the right to remain French citizens. Although he met with initial resistance from the colonial lobby in Marseilles, the deputy from Senegal managed to win permanent citizenship for his *originaire* constituents by the terms of the so-called Diagne Laws of 1915 and 1916.[101] Not only did this victory consolidate Diagne's popularity in the Four Communes and help him to win continuous reelection to the Chamber until his death in 1934, it also established him as a political broker to whom French authorities might turn for assistance on recruitment or indeed other political issues that could be handled better by intermediaries than by the French themselves. After the Second World War, Diagne's successors as African deputies continued to include African military issues as part of their political mandate.[102]

In his determination to secure more African troops, Clemenceau now turned to Blaise Diagne. In 1918, the Senegalese deputy was named Commissioner General of Recruitment, a prestigious rank equivalent to Governor-General, and was sent out to West Africa to bring in the military levy.[103] To the astonishment of the civilian administration that had supported van Vollenhoven, Diagne had startling success. He brought in a recruitment of over 60,000 men with virtually no armed resistance. A complex series of reasons has been given to explain Diagne's triumph: the strength of his personality and his prestige as an African with the top colonial rank of Governor-General; the presence in his contingent of three or four African officers with rank of lieutenant or captain; the much better advance preparation carried out by colonial officials; and especially the unwritten promise given to many traditional African authorities that this would be the last big recruitment, the last drive that would bring French victory and peace.[104]

While all of these factors were important elements in the Diagne mission's success, the rewards for Africans *outside* the Four Communes of Senegal were far less significant. This may be demonstrated by a brief examination of military recruitment in the colony of Dahomey, where an educated African elite had emerged by the turn of the century. This elite had responded enthusiastically to the outbreak of European hostilities in August 1914. In that same month, the *Journal officiel du Dahomey* published the following petition addressed to the governor from the youth of Porto-Novo:

> Monsieur le Gouverneur,
> France is at war with Germany. Our place as her adopted CHILDREN is at her side, under her FLAG. The glorious call which has sounded is a call to filial devotion for us as well. We respond to it by putting ourselves at the disposition of the Mother Country. We offer to France all the vital force of our ardent youth, even its sacrifice. Would that SHE welcomes us in the ranks of her soldiers.[105]

As romantic images gave way to increased taxes and prices, forced requisitions and recruitment, opinions changed. By 1915-1916, resistance to military recruitment was held to be the main cause of two major revolts in the north of Dahomey, and a year later, it led to the dismissal of Governor Noufflard.[106] An educated elite of Dahomeans, emulating Diagne's tactic in Senegal, sought to enlist in the regular French army, but this was completely rejected, except for the few who had otherwise qualified for French citizenship already. One notable figure in the Dahomean national movement, Jean Adjovi, the son of a wealthy landowner from Ouidah, took a different approach. The head of a large clan after his father's death, he volunteered for the *Tirailleurs Sénégalais* in 1915 and encouraged other members of his extended family to emulate him.[107] In 1918 he was promoted to lieutenant and returned to Dahomey to provide valuable assistance to Blaise Diagne in persuading notable Dahomean families to offer up their sons for military service.[108] Adjovi was demobilized in 1919 and a year later became a French citizen. Nevertheless, Adjovi's tactics proved unsuccessful in securing a broader extension of citizenship to Dahomeans or other French West African subjects.

Diagne's success in 1918 contributed significantly to Clemenceau's decision to maintain conscription after the war. If there had been no significant revolts in 1918 when French military personnel were in short supply, it was less likely (and in any case less dangerous) there would be revolts in peacetime. Whatever the logic, events proved the French right. The interwar period was not marked by serious unrest or uprising attributable to annual peacetime conscription.[109] Perhaps the same logic suggested to unhappy Africans that the risk of armed revolt was too great. If the uprising of 1915 and 1916 had been brutally crushed and did not spread, what chance was there of success in peacetime? And who would lead such revolts? The traditional elite was compromised by its participation in the conscription system, and the *évolué* community was under the tight control of Blaise Diagne, one of conscription's strongest supporters.

The successful recruitment of 1918 seriously discredited the civilian administration in the eyes of French decision makers in Paris. Van Vollenhoven and his subordinates were dismissed as administrative Cassandras. From the perspective of the French empire as a whole, the government in Paris came to regard FWA as an exporter of soldiers, much to the chagrin of the West African private sector.[110] For employers, peacetime conscription meant competition from the army for the labor of the most able young men in African society.

The First World War exacted a heavy toll on West African soldiers. Over five years of combat, FWA furnished 170,891 men, including the massive Diagne levy of 63,000 in 1918.[111] Using 1921 census data to measure recruitment against total population, Marc Michel has found that while all regions with the exception of Mauritania and Niger contributed significantly to this recruitment, Senegal bore the heaviest burden.[112] Casualties for black Africans ran at approximately 185 per thousand, or 30,000 killed in action. These very substantial rates can be compared with the overall mortality rates of the French army of around 260 per thousand, or 1.3 million dead among the 5 million French citizens and subjects in uniform.[113]

The war also transformed the *Tirailleurs Sénégalais* into a mass army recruited from every social level of West African society. To be sure, the social origins of most soldiers were rooted in slavery. As late as 1918, one colonial official, Kersaint-Gilly, estimated that the army was still 75 percent of slave origin, and he may very well have been correct.[114] On the other hand, new factors were at work to leaven the social mix. Blaise Diagne's recruitment drive of 1918 had persuaded African notables to allow their sons to volunteer, or else risk a future in which their former slaves would outrank them. In addition, wartime conscription brought recruits into the army from acephalous populations where there had been fewer slaves.

Yet of all these changes, the one that would have the most profound consequences was the decision to maintain military conscription in peacetime. The next two chapters examine the consequences of this fateful choice.

4

Military Conscription in Theory and Practice

> *In short, military service will no longer be an exceptional contribution requested of our subjects, but an annual and universal obligation. You should immediately devote all your efforts as well as those of your commandants de cercle towards preparing your subjects to accept this blood tax.*
>
> Governor-General Angoulvant, 1918

French military conscription in West Africa was indeed a tax in blood and sweat. Hundreds of thousands of young Africans were conscripted to serve overseas in European and colonial wars during the first half of this century. Still thousands more were drafted into what was called the "second portion," a reserve of potential soldiers who were also obliged, in some cases, to serve a three-year term in dreaded labor brigades. This chapter examines the imposition of the Conscription Law of 1919 in theory and in practice upon the population of FWA.

Demographic Imagination and Numerical Reality

As Chapter 3 has indicated, civilian and military officials held sharply contrasting opinions as to the demographic potential of FWA to supply military labor. On the one hand, Governor-General Van Vollenhoven portrayed FWA as a heavily burdened population hardly able to feed itself, let alone export foodstuffs *and* military labor in addition. On the other side stood General Charles Mangin, who persisted in his belief well after the First World War that FWA could easily sustain an expanded *Tirailleurs Sénégalais*.

Unfortunately for French West Africans, Mangin, the war hero, was powerfully placed to influence policy. With his enthusiastic support, French military planners raised the size of the standing strength of the *Tirailleurs Sénégalais* to 55,000 men in 1920, with a projected force of 110,000 by 1925.[1] Part of the reason lay in the persistence of French demographic fears. The war had bled France of over two million men who would not only no longer join the labor market but who would not play their part in population reproduction. As early as 1920, Gassouin, Commander-in-Chief for FWA, was warning of the "hollow" cohorts, a dearth of young Frenchmen for the universal military conscription in the years from 1934 to 1938.[2] The perceived need then was for an expanded colonial contribution in peacetime to replace missing French citizens.

In 1920 an interministerial commission was formed to study the demographic implications of continued and expanded conscription in West Africa. Mangin was made the chairman, representing the interests of the Minister of Defense. Officials of lesser stature from the Ministry of Colonies attempted to curb what they held to be excessive demands of the military. They represented the views of the colonial administration and of business interests who wished to keep a labor reserve in FWA. To resolve the potential impasse the commission authorized what may have been the first "quasi-demographic" surveys of a segment of the French West African population.

The interministerial commission ultimately produced a set of preliminary findings in January 1922[3] and a more definitive report in July 1923 based upon a survey and partial census.[4] The reports reveal the underlying demographic assumptions of French planners. The commission identified five variables it felt necessary to measure or estimate.

The first variable was called *"the overall population."* The commission began with earlier crude censuses that were little more than estimates: 10,758,000 people as of 1906, increasing to 11,878,000 by 1916, finally 12,494,000 by 1922. No formal census was taken and the commission did not evaluate the quality of the data.[5] It might be reasonably assumed that the 1906 figure was a crude guess, the 1916 a slightly more refined one, and the 1922 estimate a somewhat more accurate one.

The second variable was *"the recruitable population."* Various assumptions reduced the size of the population subject to military recruitment. Nomadic peoples such as the Moors, Tuareg, and pastoral Fulbe were declared unsuitable for formal military duty in the *Tirailleurs Sénégalais* and their numbers deducted from the total. In practice, as military conscription came to be applied in FWA, the nomadic peoples of the Sahel and Sahara were exempted from serving in the *Tirailleurs Sénégalais*. In addition, areas of recent dissidence such as the lower Casamance in Senegal were exempted from the draft, while peoples inhabiting the lower Côte d'Ivoire were declared to be "too feeble and unwarlike" to provide anything but a tiny number of soldiers. These largely subjective assumptions about which elements of the population were useful militarily reduced the net population as of 1922 to 10,638,000, almost two million less than the estimated overall population.[6]

The third variable for the commission was *"the male population."* The preliminary report of January 1922 recognized the absence of data to estimate sex ratios in the West African population. The commission report attempted an extrapolation, nevertheless, and produced the very low ratio of 72.4 males per 100 females based upon fragmentary ratios from the war period (an estimate from Guinea for 1917 varying from 71.2 to 83.2, and one for Upper-Senegal-and-Niger set at 69.2).

Demographic Imagination and Numerical Reality 49

Although those figures were speculative, the migration of males as part of the growing industrial army of reserve labor in West Africa was already a reality by the 1920s, as thousands of young males and somewhat fewer females left FWA to work on the plantations of the British West African colonies.[7] Given the already high rate of migration, it is entirely possible that demographic surveys would have revealed actual sex ratios as low as these or even lower, if what was being measured was the number of males present in a community at the time of the survey.[8] But the commissioners were entirely silent on this issue and instead offered the figures on sex ratios as absolutes.

The commission's final report in 1923 offered different figures, this time based upon a survey of 5,000 people in each colony, selected as being "as typical as possible" (see Table 4.1). With sex ratios ranging from a low in Senegal of 84.1 to a high in Niger of 100.0, the commission opted to use the mean of 92.3. But the decision was not actually based upon the partial data collected in West Africa. Having noted that the sex ratio of males in France was 92.7, the commissioners felt satisfied that their new estimate was reasonable.

The fourth variable was "*males in their twentieth year.*" The preliminary report of 1922 took the entirely perfunctory step of accepting French demographic findings in their estimate of males in the military cohort. In France, however, males in their twenty-first year, rather than in their twentieth year, did military service. Not bothering to make this correction, the commissioners used the French male cohort coefficient for the twenty-first year of .015 (that is, 1.5 percent of the total French population), as the norm for West Africans in their twentieth year.

The final report did try to correct this absurdity by incorporating results from the 1922–1923 West African survey. For the first time the commissioners admitted to the dilemma they faced in analyzing the uneven data. In Soudan and Upper Volta, for example, officials simply averaged the coefficient as 10 percent of all males between the ages of 19 and 28, while in Niger local commandants took 2.2 percent of the total male population. Others made errors of arithmetic or logic. In one *cercle*, for example, of 213 males between 19 and 28, it was claimed that no less than 154

TABLE 4.1 Male-to-female sex ratios based on the survey of 1922–1923 by colony (and by region and ethnic group for Upper Volta)

Colony	Ratio	Upper Volta by Region and Ethnic Group	Ratio
Senegal	84.1	Ouagadougou, Mossi	100.0
Soudan	91.6	Kaya, Mossi	91.2
Guinea	93.1	Bobo-Dioulasso, Bobo	96.1
Niger	100.0	Dédougou, Bobo	100.0
Dahomey	92.3		
Côte d'Ivoire	91.2		
Mauritania*	98.8		

*Black Africans only.

Source: Military Cabinet of the Governor-General of FWA, "Etudes sur le recrutement en AOF. Résultats des sondages effectués dans les colonies du Groupe en 1922," Dakar, 5 July 1923, ANS, 4D72 81.

were 19 years old. Despairing of accurate coefficients based upon such shoddy work, the commissioners fell back on European data once again. This time they at least bothered to find coefficients for males in their twentieth year.

Noting the figure of .0163 for France and .0164 for Germany, the commissioners opted for the French figure for West Africa. Was the German number included to show the logic of a comparative figure or the legitimacy of choosing the French one? It is not possible to say.

The fifth and final variable of the commission was *"males physically fit for military service."* The war years were held to be unreliable statistically because of the haphazard recruitment methods.[9] Instead the final report used percentages of men declared unfit for military service as determined by the annual military recruitment of 1921. These ran at the exceptionally high rates of 722 per thousand for FWA as a whole.[10] Guinea, held to be the model colony for reliable military statistics, had a rejection rate of 800 per thousand, with some regions such as Futa Jalon running as high as 936 per thousand. The Guinea rate of 800 per thousand was used for the colony as a whole.

When the computation based upon all the variables was made, it produced the following estimated annual military cohort: 10,638,000 population x .48 male coefficient x .0163 males in their twentieth year x .20 physical fitness ratio – 16, 646 as the maximum annual cohort. The final report of the commission concluded, therefore, that a range of from 12,000 to 14,000 soldiers a year was a reasonable burden to apply to the West African population.[11]

However suspect the methods and the data, civilian officials regarded the results as something of a victory over the French military. Colonial officials and business interests had agreed that Mangin and Gassouin's projected army of 80,000 or more would be a disaster. Instead, by setting what they regarded as a more moderate annual levy of from 12,000 to 14,000 men, the interministerial commission were projecting a standing *Tirailleurs Sénégalais* force of roughly 55,000 men, exactly the figure civilian Governors Ponty and, after him, Clozet, had suggested. The Minister of Colonies, Albert Sarraut, in his grandiose development project for the French empire, indicated his satisfaction with the commission's findings, noting that black Africans would make up roughly half of a projected Native Army of 100,000 men he deemed necessary to "relieve the military burden on the metropole."[12] To an already heavily taxed African population, however, an annual levy of 12,000 men or more was heavy enough.

This essentially political decision of the interministerial commission confirmed an underlying reality of French policy. Colonial governments before the Second World War did not have demographic policies any more than they had labor policies.[13] The pseudoscientific element of the demographic survey conducted by the interministerial commission was a shoddy exercise that in the end produced a result consistent with what the conflicting competitors for West African labor needed. The sham nature of the demography was self-evident. Not only were the methods slipshod, but no effort was made to determine the size, for example, of a cohort of young females of child-bearing age, or even to suggest that fertility was a key variable in any attempt to estimate the current and especially the future demographic potential of West African society. The commissioners were never able to estimate accurately two major characteristics of the population, the age/sex structure or, even more fundamentally, the overall size. It is ironic that this early demographic exercise should ignore women so completely in its attempt to estimate the population of FWA. By measuring only young males, the most mobile group and the one that most immediately interested them, they were dooming their efforts to failure.

Mobile Draft Boards in Operation

The Conscription Law of 1919 set conditions for the annual draft in FWA for the entire interwar period.[14] The system was in practice similar to the lottery technique used under Napoleon III in France. In each *cercle*, a list of eligible males was drawn up, initially with the help of the African canton chiefs, but after 1926 by the *commandant de cercle* using actual census lists, which were systematically maintained in most *cercles* for males only.[15] So long as canton chiefs held some power to establish these lists, they could not resist the numerous opportunities to play favorites in their own societies. Yet in so acting, chiefs often alienated their constituents to such an extent that their ability to serve their French masters was called into question.[16] Thus conscription became yet another example of the dilemmas of African chiefs co-opted into the French colonial system.

Another feature of the modified French conscription system unique to West Africa was the mobile draft board.[17] Each colony had at least one and sometimes three such boards. The chairman was a French military officer, usually with the rank of captain and usually posted at the regimental headquarters of the colony. He was seconded by his civilian counterpart, the *commandant* of the particular *cercle* in which the board was meeting. Below these men were two clerks, one from the military and one from the civilian administration, who were in charge of the voluminous paperwork associated with each year's class of conscripts. Finally, each board was assigned a physician to conduct the medical examinations. In theory, the physicians were military doctors from the *Service de Santé* of the army, but the board often hired civilian doctors, and frequently even male nurses, since fully qualified physicians, whether military or civilian, were never numerous in FWA.[18] No Africans served on the boards in any decisive sense, although one of the clerks might have occasionally been an African.

The draft boards reflected competing interests.[19] The civilian administrators wished to keep the ablest young men in the *cercle*, to serve as clerks if they were literate, or as a vital source of the best labor available. The army wanted these very same people. Although an army man chaired the boards, the civilian administrator had a practical advantage over him. He knew his own *cercle* better than an officer who came in only for the week or so it took to conduct the board hearings. More importantly, the civilian official could give out student deferments throughout the year. In this sense, only a part of the draft board was truly mobile. The army officer, the physician, and one clerk would have to visit perhaps ten or fifteen *cercles* in the colony from January to March when recruitment took place.[20] Only the civilian membership of any particular board had continuity and local knowledge.

Quotas were established according to fixed procedures. In Paris each autumn the minister of defense indicated the number of soldiers required by the *Tirailleurs*

Mobile draft board in French West Africa, interwar period

Sénégalais. This figure was then submitted to cabinet for approval, with the minister of colonies able to reduce this figure if his ministry could make a strong case.[21] The approved quota was then immutable and was sent out to the Governor-General in Dakar to be filled. The Governor-General had the power to distribute the quota among the eight territories of FWA. He followed a fixed formula developed by trial and error and with some attention paid to population densities during the early 1920s. When the Lieutenant-Governor of a particular colony received his fixed quota from Dakar, he in turn could distribute it among the *cercles*. Similarly at the *cercle* level the *commandant de cercle* could split the *cercle* quota by subdivisions or districts.

Figure 4.1 provides numerical expressions of how a levy of 10,000 men recruited for the *Tirailleurs Sénégalais* would have been distributed among each *cercle* during the interwar period. By and large, these numerical expressions correspond well with population estimates, suggesting that French authorities did apply conscription quotas equitably among the sedentary populations of FWA.[22] Figure 4.2 compares regional quotas across FWA and indicates that most regions remained close to the federation average, that is, between 90 and 110 men per 10,000 recruited. The regions above this average were Senegal valley, Casamance, and central

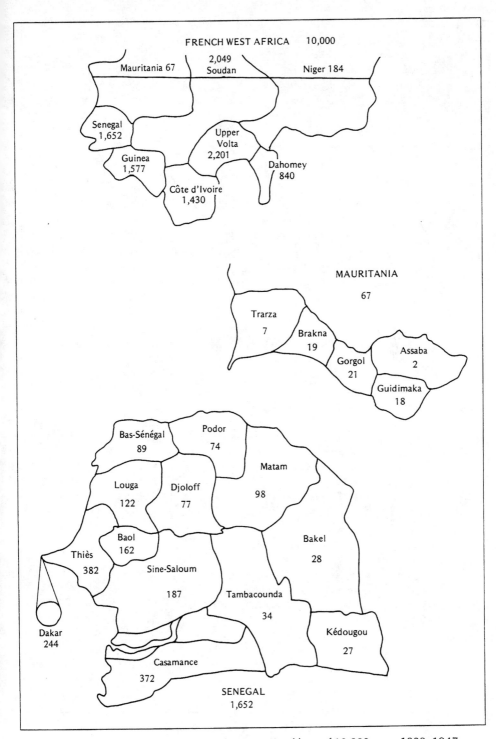

FIGURE 4.1. Quota averages to produce an annual levy of 10,000 men, 1920–1947
Source: Compiled from various Annual Recruitment Reports in ANS, Fonds Modernes, séries 2G, 4D, and 5D.

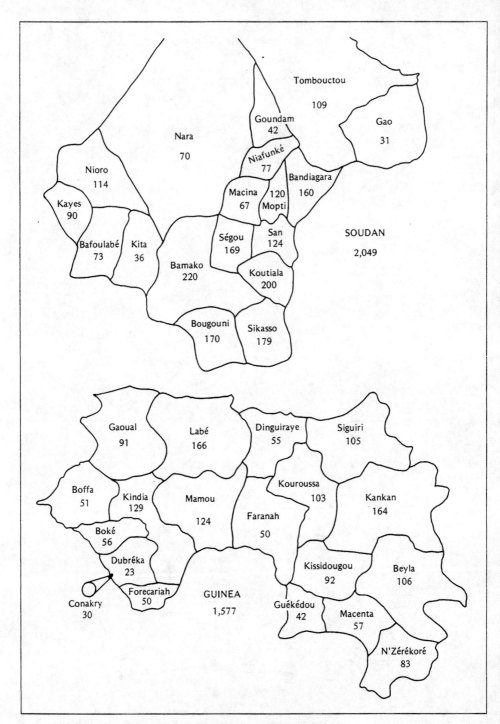

FIGURE 4.1. (continued)

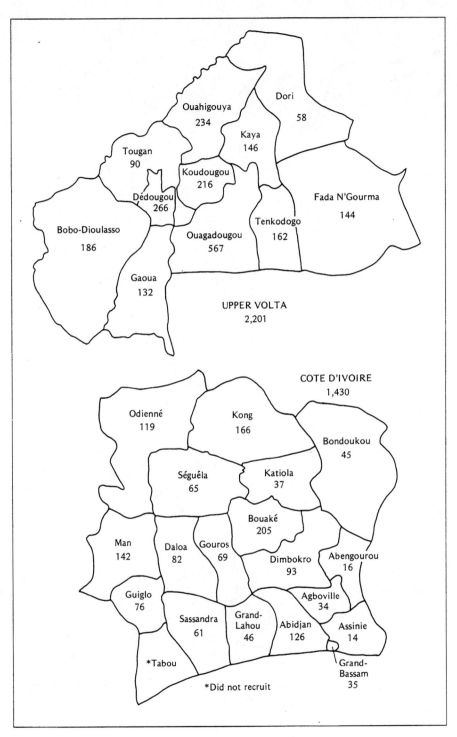

FIGURE 4.1. (continued)

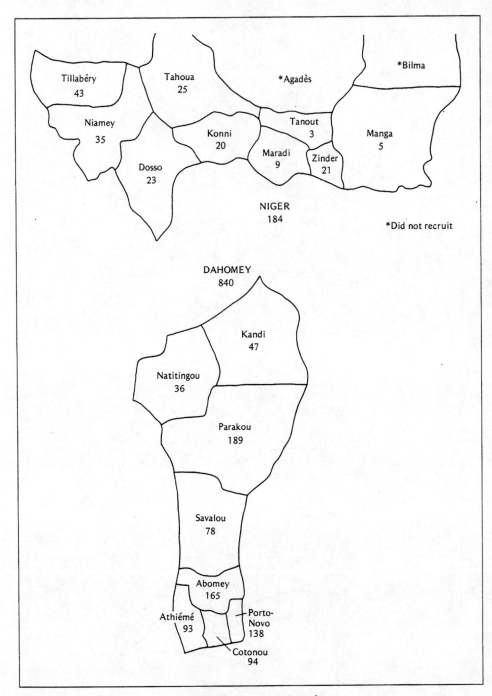

FIGURE 4.1. (continued)

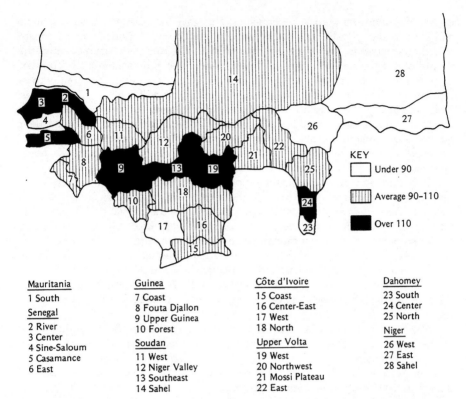

Mauritania	Guinea	Côte d'Ivoire	Dahomey
1 South	7 Coast	15 Coast	23 South
	8 Fouta Djallon	16 Center-East	24 Center
Senegal	9 Upper Guinea	17 West	25 North
2 River	10 Forest	18 North	
3 Center			Niger
4 Sine-Saloum	Soudan	Upper Volta	26 West
5 Casamance	11 West	19 West	27 East
6 East	12 Niger Valley	20 Northwest	28 Sahel
	13 Southeast	21 Mossi Plateau	
	14 Sahel	22 East	

FIGURE 4.2. Quota averages to produce an annual levy of 10,000 men, 1920–1947, by regions

Source: Compiled from various Annual Recruitment Reports in ANS, Fonds Modernes, séries 2G, 4D, and 5D.

Senegal; upper Guinea; southeast Soudan; western Upper Volta; and central Dahomey. Considerably below the average were all of Mauritania and Niger; southern Dahomey; Sine-Saloum in Senegal; and western Côte d'Ivoire. The only anomaly in these results was Sine-Saloum. As one of the densely populated regions of Senegal, it might have been expected to have produced a larger share of soldiers in the *Tirailleurs Sénégalais* relative to its population.[23]

Both in relative and absolute terms, men from Senegal were underrepresented in the *Tirailleurs Sénégalais*. Because French officials used total population as a general principle for recruitment, Soudan and Upper Volta had the largest contingents in the *Tirailleurs Sénégalais*.[24] In addition, Senegalese youths had several practical opportunities to avoid military service as a result of their colony's special status in FWA. A minority of Senegalese, those born or living in the Four Communes, were recognized as French citizens after 1917 and served in units of the regular French army, not in the *Tirailleurs Sénégalais*.[25] Thirdly, Senegalese subjects lived in proximity to the center of patronage in the federation, the city of Dakar and its neighborhood, the Cap Vert peninsula. This gave them better access to schools and jobs in the private and public sectors, and better opportunities to secure military deferments and even exemptions.[26] Lastly, and most significantly, the timing of the annual levy favored lower Senegalese representation. Each year, recruiters began their work in the colonies of the interior during the months from January to March. Only after these results were complete did conscription begin in Senegal, usually in April and May. This allowed colonial authorities to adjust the Senegalese quota downward should colonies of the interior have exceeded their quotas, a circumstance that did arise on occasion. The converse was never true. Colonies of the interior always at least matched their quotas, so it was never necessary to make up deficiencies in Senegal.[27]

Important as it was, establishment of the quota was only the first step in the yearly recruitment process. By the time the *commandant* received his quota, perhaps as late as December or even January, preparations for the annual draft board were well under way.[28] All during the year the *commandant*'s office had to keep working on the preparation of annual census lists of young men who would be nineteen in the next year and who therefore formed the total potential of the *cercle*. It was the responsibility of each individual, and implicitly of his village or canton chief, to see to it that he made it to the board at his own expense. For many this involved a short distance to travel, but in remote or dispersed *cercles*, it might take five or ten days' march to reach the board.[29]

The conscription process at the local level began with a medical examination of all the men present. Once the French introduced the census list technique to replace the interim method of presentation by the chiefs, some men were always absent without permission. These were termed *bons absents*, or absentees.[30] Table 4.2 illustrates the conscription results for a typical cohort, that of the year 1926, for each colony.[31] While a total of 12,637 men were recruited for the *Tirailleurs Sénégalais* in 1926, the actual quota as determined in Paris had been 12,500. Table 4.2 puts the total recruited at 12,637, but it was not uncommon for the quota to be exceeded slightly for the federation as a whole. Figure 4.3 indicates percentages in the various categories. Not only does it reveal high rates of absenteeism, but the percentage of men unfit or exempt was remarkably high. According to French officials, the great majority of these men were excused from military service for medical reasons.[32] Thus, to take figures from Bougouni (Table 4.2), while 2,514 men started out on the recruitment lists in 1926, only 325 men, or 12.9 per cent, remained from which to fill the quota.

Such astonishingly low fitness numbers invite speculation. Did civilian authorities insist upon totally unrealistic medical standards in order to keep the numbers of soldiers down? We cannot know, but whatever the reason, French colonialism in West Africa displayed the same inability to measure or to extract significant human resources for the army that it had for the labor market. One report from FWA in the 1930s, when only 2 percent of the population was working for wages, expressed the fear that labor recruitment had already reached its limits.[33]

Once the men fit for service had been determined, the common unwritten procedure called for the chairman of the draft board to address this group to call for volunteers.[34] In the early 1920s higher premiums were offered depending upon whether a man volunteered for three, four, or six years' service.[35] This factor helps

TABLE 4.2 Conscription figures for 1926 by colony (including cercle of Bougouni, Soudan)

	Men on Lists	Men Examined	Unfit or Exempt 1926	Absentees	2nd Portion	1st Portion Drafted	Volunteers	Total Recruited
Cercle of Bougouni	2,514	2,377	2,052	137	179	145	1	146
Soudan	38,043	32,280	24,433	5,763	5,047	2,740	60	2,800
Guinea	42,921	36,314	31,349	6,607	2,865	1,221	879	2,100
Côte d'Ivoire	24,692	23,081	16,417	1,611	4,964	1,644	56	1,700
Upper Volta	32,815	25,401	13,025	7,414	8,776	2,916	684	3,600
Dahomey	23,660	12,877	8,065	10,783	3,830	751	231	982
Niger	5,077	2,323	1,584	2,754	441	134	164	298
Senegal	20,204	15,626	12,281	4,578	2,288	1,044	13	1,057
Mauritania	1,043	874	594	169	180	89	11	100
FWA	188,455	148,776	107,748	39,679	28,391	10,539	2,098	12,637

Source: Lieutenant Governor of Soudan, Report on recruitment for Soudan in 1926, Bamako, 7 April 1926, ANS, 4D69 81; the CIC, General Peyregne, Report on recruitment for FWA in 1926, Dakar, 9 September 1926, ANS, 4D70 81.

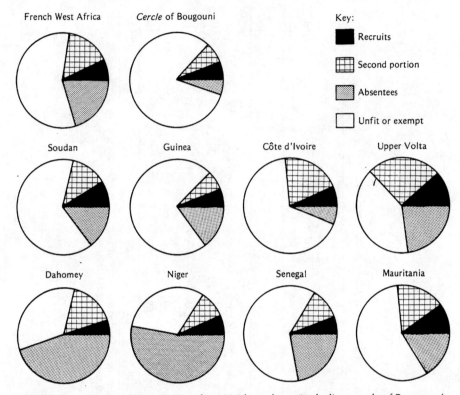

FIGURE 4.3. Conscription averages for 1926 by colony (including *cercle* of Bougouni, Soudan)

Source: Lieutenant Governor of Soudan, Report on Recruitment for Soudan in 1926, Bamako, 7 April 1926, ANS, 4D69 81; General Peyregne, Report on Recruitment for FWA in 1926, Dakar, 9 September 1926, ANS, 4D70 81.

account for a falloff in the number of volunteers as this practice was gradually phased out. There was also some deception in the way the system was applied. In the early years of conscription, some young men, perhaps confused by the complexity of the system, may have come forward at this stage because they were unaware that passing the medical did not necessarily guarantee they would serve in the army. By 1926, however, Africans all over FWA knew the system and would not have volunteered by mistake. In Bougouni, for example, only one solitary volunteer came forward in 1926.

After the volunteers were subtracted from the quota a lottery was held to determine the balance. Those who drew a "bad number" went into the army as conscripts; they were called the first portion. The remainder whose numbers had not been drawn became part of the second portion. The second portion was in theory an inactive reserve, the first to be called up in case of partial or total mobilization. No such mobilization occurred until 1938, when the threat of war in

Europe grew greater.[36] Instead, for several years men in the second portion went back to their villages and were considered to have fulfilled their military obligations. In the late 1920s, however, especially in Soudan and parts of Upper Volta, men of the second portion were conscripted into labor brigades to serve for two or three years in works declared by the French authorities to be of a public nature.[37] In practice this meant work on the Port of Dakar, on the Dakar-Niger railway, and especially on construction of dams associated with the *Office du Niger* irrigation scheme at Markala in the Soudan.[38] Placed under control of private companies and working under deplorable conditions generally, Africans found service in the second portion in Soudan so arduous and death rates so substantial that many willingly began to seek entry into the first portion, guaranteeing them military service as opposed to labor duty.[39] This is how Dramane Sarambé, who served in the military brigades building the installations at Markala, Soudan in the 1930s, recalled his ordeal:

> To put it bluntly, it was labor at the cheapest possible price. We were paid 30 francs CFA a month. We were poorly fed but we received work clothes as if we were really soldiers. . . . Any Malian or Voltaique family still has a fresh memory of the building of the bridge [at Markala] where men worked in the rain and under the whip without respite. The lazy and the revolutionaries were thrown into the river to intimidate the other workers . . . .[40]

The net effect of this labor mobilization of the second portion in Soudan, but not anywhere else in FWA, was, as Table 4.3 shows, to double the "blood tax" on Africans in Soudan for eighteen long years in the middle of this century.

While Africans learned the basics of the conscription system as a matter of knowing their interests, the process was sufficiently alien to leave many perplexed. Yessongui Sekongo, an *Ivoirien* ex-soldier from the Korhogo region, explained his recruitment in an interview years later by placing it in his own cultural context:

> I was a cultivator and I also went to pull the wood, twice, before my military service. Three of us went from here and I was the only one recruited. I don't know why they took me. I did know that God wanted me to do my military service. The white had said that people with sores on them would not go, but then the white chose me even though I had sores. Then I was a handsome boy—I was pleased with myself.
>
> Also, I had to work with the *marabout* so that I would be left alone. I went to other villages where they had them. He wrote something in Arabic on a piece of paper so that they wouldn't take me. I carried it with me, but they took me anyway. It was like that before, it was usual to go to the *marabout* and then they wouldn't take you. I wasn't lucky. It was because it was a white who made the selection. It would have worked if it had been an African . . . .The *gris-gris* from the *marabout* were for things that couldn't be done right at the moment. That's why we went to the *marabout* for the army. But the *marabout* couldn't help you escape from a fact—something happening right now at the moment in the field—right away.[41]

Sekongo's account is valuable for its references to the practice of seeking insurance against the draft from a local Muslim religious specialist, to the narcissism no doubt common among many young men who were pleased in spite of themselves as being selected for military service, and for his awareness that in the end, the process was beyond any individual's control.

TABLE 4.3 Recruitment of second-portion workers and first-portion soldiers in French Soudan, 1928–1946

Year	First-Portion Soldiers	Second-Portion Workers*	First and Second-Portion Totals	Second-Portion Workers on Thiès-Niger Rail†
1928	3,100	1,500	4,600	500
1929	3,200	1,500	4,700	500
1930	2,400	2,000	4,400	800
1931	2,800	1,265	4,065	265
1932	2,285	3,400	5,685	1,400
1933	1,850	700	2,550	—
1934	3,100	3,100	6,200	2,600
1935	2,200	600	2,800	
1936	2,290	3,000	5,290	
1937	3,580	1,000	4,580	
1938	2,116	3,000	5,116	
1939	1,583	1,580	3,163	
1940	6,950	—	6,950	
1941	2,858	4,700	7,558	
1942	3,000	5,500	8,550	
1943	3,000	8,000	11,000	
1944	2,825	3,604	6,429	
1945	2,111	2,998	5,109	
1946	1,300	4,156	5,456	
Total	52,548	51,653	104,201	
Annual average‡	2,766	2,719	5,485	

*In most years these workers were assigned almost entirely to the *Office du Niger* project.
†The workers assigned to railway construction work.
‡The total population of Soudan in 1935 was stated to be 3,421,429.
Source: Annual reports on recruitment, Dakar, ANS, 2G, 4D, and 5D.

Like obligatory draft systems elsewhere, the system in FWA was biased against rural, less privileged groups. The typical recruit was a physically fit peasant youth of low status, and without traditional skills such as those of artisan, blacksmith, or trader.[42] Two interest groups worked to protect the more highly skilled or more highly ranked. The African family and the village as a traditional economic unit was one, and the modern sector, represented by the civilian administrator and private company, another. The CIC in 1928, complained:

> Quite clearly, French speakers are down this year owing to the complicity of the commercial houses and certain administrations which are not keen to have the military recruitment of natives deprive them of their best employees.[43]

Against this alliance was the outnumbered army officer, hampered by his lack of familiarity with local society and the system.

Some Africans had reasons for wanting others to enter military service. The same groups that normally helped keep one young man out of the army might very well wish to see another man replace him. In short, sending a rebellious youth off to do military service was an obvious method of social control.[44] Similarly, vendetta or personal grudges could also play a part. Among a noble family for instance, an older brother might wish to eliminate a younger sibling from contention for political power.[45] Indeed, well-born families often wanted at least one of their sons to serve in the army.[46] A good service record and the acquired ability to speak some French might very well help a returning veteran get appointed to a chiefship over another brother who had not done military service.[47] In many regions sending a son into the army served as an insurance policy for the traditional elite. The army, for its part, also favored the incorporation of sons of notables and earmarked them during the draft board operations as potential noncommissioned officers.[48]

Peacetime conscription between the wars drew on a majority from the bottom and a minority from the top of African society. Those who were consistently underrepresented were the new middle sector, the upwardly mobile, French-educated Africans. In time the army became reconciled to losing such men, and rationalized their loss by arguing that these so-called *évolués* might be tainted by nationalist ideology and thus a bad influence upon the rest of the troops.[49] Later this imbalance had serious consequences for the kind of training given to Africans in the Colonial Army. Soldiers of the *Tirailleurs Sénégalais* began and ended as infantrymen. Their lack of French language skills made it difficult for the army in the short three years of service to do much teaching of trades that might have later civilian application or benefit.[50]

While conscription in general was unpopular, it may have been less so in some regions of FWA than in others. In remote areas, especially during the depression years of the 1930s, military service might have been the most attractive, or perhaps the only, employment available.[51] It would seem, also, that in time certain regions developed a tradition of military service. For example, densely populated ethnic communities like those of the Mossi and Bambara dominated the army through sheer weight of numbers. Because the French military in fact placed a high value on the "warlike" aspects of these two cultures, many Bambara and Mossi soldiers rose rapidly to the rank of noncommissioned officers.[52] In time this tended to make conscription and military service somewhat less alien an experience for young men from these societies. They were entering an institution where, even though they spoke no French, they were confident of finding significant numbers of their fellow language speakers in the ranks. On the other hand, this same tendency explains why service was all the more unpopular for Africans from far less populated linguistic communities, particularly those from the forest zones of West Africa.[53] For such men the army was truly an alien experience, dominated by French officers and African strangers. Thus, it did not take long before recruitment patterns became self-reinforcing, as did French assumptions become self-fulfilling, as to which communities were the "most warlike," or made the "best" soldiers.

Although French perceptions about ethnic "aptitudes" were important, they often masked political factors. In Niger, for example, recruitment among the Hausa-speaking peoples in the eastern parts of the colony had to be approached delicately to avoid driving youths south across the border to northern Nigeria, where millions of other Hausa-speaking British subjects lived. One colonial governor of Niger went so far as to comment, when noting the draft had been frequently cancelled in Hausa *cercles*, that it was no great loss to the *Tirailleurs Sénégalais* because this ethnic group made poor soldiers.54 In Nigeria, however, British military recruiters drew their military recruits for the West Africa Frontier Force from the so-called approved tribes living north of the line drawn from Okuta in Borgu to Yola in the Nupe region. This meant, of course, that Hausa-speakers figured very prominently in the West Africa Frontier Force.55 Indeed, the term *Hausa* was often applied to soldiers in British West Africa in much the same manner as *Sénégalais* was used in FWA.

Like ethnicity, race was another political variable within the *Tirailleurs Sénégalais*. Most soldiers were recruited from the black African agricultural population living in sub-Saharan Africa. But French West Africa also included such Caucasoid pastoral peoples of the desert and sahel zones as the Moors, Tuaregs, and Berbers. These populations were exempted from conscription into the *Tirailleurs Sénégalais* on the grounds that they served in specialized desert units such as *goums* and *spahis*; that is, camel and cavalry corps. In reality, the French were concerned with two factors: the costs of conscription in far-flung desert constituencies, and the racial tensions that would result within the *Tirailleurs Sénégalais* should desert peoples find themselves under the command of a black African sergeant or native officer. The desert peoples held strong racial ideologies stemming from centuries of slavery and the slave trade. In the colonial period, many pastoralists continued to keep black slaves and clients, and would not accept being commanded by people they regarded as racially inferior.56

African Officers in the Interwar Era

Recruitment of African noncommissioned and native officers was ineffective in the interwar years. This section first examines French policy towards African officers in the *Tirailleurs Sénégalais* and then goes on to profile four Africans who did serve in this capacity: Captain Charles N'Tchoréré, originally from Gabon; and three Senegalese officers, Lieutenant Pierre Diémé from the Casamance, Lieutenant Moustapha Diouf of Dakar, and General Ahmadu Fall of Saint-Louis.

For better and for worse, Charles Mangin clearly established himself at the beginning of the century as France's leading advocate of the *Tirailleurs Sénégalais*. While he continued to push for an expanded black African army in the years immediately following the First World War, it must be acknowledged that unlike most French officers, Mangin also wished to see African officers play a significant role in this army. As head of an interministerial commission empowered to survey the entire question of the Black Army, Mangin was able to place the creation of a corps of native officers on the commission's agenda.57 Noting that in the past Africans had been named officers for great merit under fire or as a reward at the end of their service, Mangin now saw them being called upon to help lead African troops as a genuine cadre. It had become a moral duty, he would persuade the commission to argue in its *projet de loi*, to provide for native officers because of the great war effort of the colonies. Nor were black Africans to be held back from this advancement on the racist ground they were congenitally incapable of exercising military leadership:

> The West African black, even given his coarse intelligence, is susceptible to learning (there are many examples); above all, he possesses all the qualities of character—bravado, courage, authority, love of command, devotion and

loyalty—which transform him easily into a remarkable war leader, knowing how to command and how to obey.[58]

Mangin envisioned four routes to the creation of native officers. The first was to be identical to that followed by the cream of French youth, and would be open to those Africans who could meet the high standards of admission through competition at one of the *grandes écoles* which produced select French officers: Saint-Cyr, Ecole Polytechnique, Lyons, or Bordeaux. A second and more common route for Frenchmen to acquire an officer's commission would also be open to Africans along with other colonials: by competition after attending one of the special schools for noncommissioned officers and officer cadets at Saint-Maixent, Saumur, Fontainebleu, Versailles, or Vincennes. Either of these two routes would give the native officer French status and citizenship, and the choice of promotion either through a so-called lateral native cadre of ranks from second lieutenant to lieutenant to captain or, in rarer cases, up through the French cadre of ranks to the highest reaches an individual's talent could carry him. Lateral cadres would not receive automatic promotion through seniority.

Of the two remaining routes, one was the time-honored method of promotion of battle-hardened veterans from the ranks. The other avenue was to be created exclusively for native officers in the lateral cadre. It involved the creation of special schools, to be established in France near to personnel and equipment needed for instruction, and open to competition among natives of the entire French colonial empire. Youths wishing to apply to these schools would need local authorization; thus colonial authorities would be able to screen out those from families of "doubtful loyalty." Lateral cadre officers would have their own statutes stipulating pay scales for duty and pensions below the rates offered to metropolitan cadres. A French officer of similar rank would, moreover, always have authority over a lateral cadre officer.[59]

Mangin's opinions about African officers' training encountered significant opposition. The 1921 report had been preceded by a questionnaire sent out by the Minister of Colonies to the Governor-General of FWA asking for reactions.[60] In 1922, Mangin's interministerial commission again discussed the matter, and again differences of opinion were expressed. One of the few sets of minutes surviving from the commission, those for the meeting of 9 June 1922, revealed a fundamental disagreement among commission members.[61] Colonel Auroux, for example, stated that the gulf between native and regular officers would be best kept large because highly trained and intelligent native officers might represent a danger to France. He did not want colonials to have access to the best military schools in the mother country.[62] Similarly, Generals Mordrelle and Puypeyroux preferred to keep colonials out of all military schools except the *grandes écoles*, arguing that the few colonials who would pass this highest test of assimilation would be reliable.[63]

The opposition to Mangin extended beyond the commission. The CIC for FWA, General Gadel, also saw matters in racial terms. Asians and even Malagasys were more intelligent than Africans and made better native officer material. With the exception of a few races and social classes who were "born to command," the mass of black Africans were best suited to infantry rank-and-file:

> Nevertheless, overall (and especially in the interior of FWA), the black man must be considered to be of a low intelligence and a man who forgets matters very quickly.

As a child he shows intellectual aptitudes that are full of promise, but these skills disappear rapidly at the age of adolescence.[64]

Mangin lost his case, though whether the few Africans who qualified for entry to the *grandes écoles* would have chosen a military career is a moot point. After 1921, French emphasis was entirely on the "special," and inferior, routes of officers' training in the lateral cadres. In November of that year, on the recommendation of Mangin's interministerial commission, the *École Spéciale des Sous-Officiers Indigènes* at Fréjus first came into being.[65] This institution represented a compromise. Mangin had originally hoped to send Africans to the standard four-year programs at Saint-Maixent or Saumur which were designed for talented French noncommissioned officers. These programs, however, granted regular French officers' commissions to its graduates, and French authorities rejected this possibility for the majority of Africans.[66] The special school for Africans at Fréjus used entrance examinations to select roughly 25 to 30 youths from all over the French empire each year for a two-year program, after which they were given the rank of sergeant major, with promotion to second lieutenant normally acquired only after an unspecified amount of practical experience in the Colonial Army.[67] In 1931, it was reported that only 5 of the cohort at Fréjus were from black Africa.[68] The Fréjus center was closed down in 1939 because of the war.[69]

Another set of schools with some links to the military were the special primary schools created in the 1920s called the *Écoles des Enfants de Troupe*, or EETs. Linear descendants of Faidherbe's *Écoles des Otages*, the EETs were designed both to assure a steady supply of subalterns for the *Tirailleurs Sénégalais* and as a means of rewarding active and retired career soldiers with special opportunities for their sons. Three of these military-style academies were established, at Saint-Louis in Senegal, at Bingerville in Côte d'Ivoire, and at Kati in Soudan.[70] Each school had between 70 and 120 pupils, with priority given to sons of veterans of the First World War.

Educational standards at the EETs were not high. Pupils were not expected to advance beyond the basic primary school curriculum represented by the granting of the *certificat d'études primaires*, or CEP (a grade school diploma), at the advanced age of eighteen.[71] At age nineteen, even if the CEP remained uncompleted, the cadets were inducted directly into the *Tirailleurs Sénégalais* with the rank of corporal after six months of training, and with the promise of a career as a noncommissioned officer or even as an African officer, if promotion proved to be warranted. Parents were required to sign a statement of awareness of a clause which enabled the government of FWA to recoup half the cost of education should a boy quit school or refuse the commitment to enlist for a five-year term upon graduation.[72] The EETs, of course, were designed to produce African sergeants; only a handful of these men were ever subsequently promoted to native officer.

Except for a small minority of Africans, these represented the only routes to officers' rank. For those Africans who were French citizens, or for the rare African subject with high educational qualifications, access to officers' training usually reserved only for Europeans was permitted. For example, in 1930-1931, Amadou Fall, an *originaire* of Saint-Louis, and therefore a French citizen, was permitted to attend one of the French noncommissioned officers' schools, the *École Militaire de l'Infanterie et des Chars de Combat* at Saint-Maixent, along with fifteen other French candidates.[73] He graduated as a second lieutenant and went on to a distinguished military career.[74]

The minority of Africans who became officers found themselves the victims of a form of discrimination less obvious than that experienced by rank-and-file *Tirailleurs* but nonetheless very real to them. General Guignard reported that French corps commanders were uncomfortable having African officers in their units because army statutes were vague about the status of such native officers.[75] Two of the three African officers in his division in 1924 each had two or three wives and five or six children back in FWA. The third had met and married a German woman during the time the *Tirailleurs Sénégalais* was stationed in the Rhineland. This officer received a higher family allowance than did his two colleagues despite their having "remained faithful to their race," as Guignard put it.[76] The General added further that African officers had unrealistic expectations of advancement they could never expect to achieve. Graduates of the short native officers' course at Fréjus, they nevertheless saw themselves the equals to French officers of Saint-Cyr, and were convinced that they were being held back by their color, despite the obvious shortcomings of their rudimentary training.

There were other differences, real rather than imagined. Unlike French junior officers, no African officer was ever given sole command of a company.[77] Like commoners before 1789, Africans could not hold regular commissions. French policy allowed native officers first-class travel, and this meant extending the same privilege to their families. But after an incident involving the family of a Senegalese officer sent with him to Morocco, the army thereafter obliged African officers to leave their families behind when posted abroad.[78] Last but by no means least, Army salaries were unattractive. In 1929, according to the CIC of FWA, General Benoit, store clerks and hotel waiters made more than sergeants, and a native officer earned little more than an administrative orderly.[79]

Given all of these encumbrances, it is hardly surprising that the army found it difficult to recruit African cadres of noncommissioned and of native officers. In the year 1928, the French command received only seven nominations of Africans for sergeant major, and could count only 40 Africans at that rank in the entire *Tirailleurs Sénégalais* force. African officers were slightly over half that number.[80] General Benoit argued that the *Tirailleurs Sénégalais* remained attractive as a career only to men from the most backward parts of FWA, men who, while brave and loyal, were extremely limited professionally. To attract better-educated men, he persuaded the military command in Paris to reduce the term of service of African citizens and those subjects who were graduates of *grandes écoles* to one year in the hope that at least some might be willing to stay on and make the military a career.[81]

It would seem that General Benoit's initiative had little effect. While statistical evidence is not available, the profiles of four Africans who did serve as officers in the interwar period would suggest a pattern of careers consistent with the limits of French policy. The first of these men was Captain Charles N'Tchoréré. His military career began like many an African youth's from the remote corners of France's African empire, even if it ended tragically and atypically. One of the rare *Tirailleurs* originally from French Equatorial Africa, N'Tchoréré was born in Libreville on 15 November 1896.[82] There he attended Ecole Montfort, a boys' primary school run by Roman Catholic missionaries,[83] but otherwise nothing else is known about his family background or early years. In 1916, at age twenty, he enlisted in the *Tirailleurs*

Sénégalais and fought in the First World War, in which he distinguished himself sufficiently to be promoted to sergeant.[84] In the 1920s he was sent to the *École Spéciale des Sous-Officiers Indigènes* at Fréjus and eventually became a native officer in 1927. By 1939, when war broke out, he had retired with the rank of lieutenant and had been named as commandant of the EET at Saint-Louis. At age forty-five he volunteered for combat duty and was promoted to captain. By May of 1940, he had been placed in command of the seventh company of the 53rd RICMS, a distinguished *Tirailleurs Sénégalais* regiment assigned to defend the lower Somme in the region of Airaines. On 7 June 1940, after three days of valiant resistance, his company and regiment were overrun and he was taken prisoner. A young German officer manhandled N'Tchoréré, ordered him to join the other enlisted men, and shot him in cold blood when he insisted on his right to be treated as a captured French officer.[85]

The career of Pierre Diémé was typical of those of other young soldiers of modest origins who found in the *Tirailleurs Sénégalais* a means of upward mobility. Born to Diola-speaking agriculturalists in the Casamance region of Senegal on 1 November 1917, Diémé was in fact conscripted in his twentieth year and sent to Casablanca. Promoted to sergeant by the time his three-year term had expired by 1940, he decided to stay in service, saw action in Europe in 1944 and in Indochina and Algeria afterwards. He was promoted to second lieutenant in 1954 and to lieutenant in 1956. At independence he chose transfer to the Senegalese National Army with the rank of captain, and in 1964 at the age of forty-seven he took compulsory retirement with a French army lieutenant's pension. Awarded the *Croix de Guerre* and named a *Chevalier de la Légion d'Honneur*, Diémé was by his own admission poorly educated when he entered the *Tirailleurs Sénégalais* (he had not completed the elementary school certificate at the Roman Catholic school he attended in Ziguinchor) but an accomplished and respected person when he left military service. The veterans of Casamance elected him president of their regional association in 1971 and clearly regarded him as one of the notables of their region.[86]

The third and fourth profiles represent officers coming from more privileged strata of French West African society, namely the *originaire* citizens from the Four Communes of Senegal. Moustapha Diouf was the son of Galandou Diouf, Blaise Diagne's successor as Senegalese deputy in the National Assembly. He may not have had any intention of a military career until the impending crisis in Europe in the late 1930s obliged him to consider military service as a patriotic duty. In a manner reminiscent of those young patriots of 1914, he volunteered for military duty in 1939, on the eve of the Second World War, and eventually became a lieutenant in the Free French Army of General de Gaulle. He is said to have been the first black African parachutist in the Army of Liberation in 1944.[87] When the war ended however, he did not renew his commission, as his name does not appear on officers' lists after 1945. Presumably he returned to his civilian position in the colonial civil service.

One of the few Africans in this period to have set an officer's career for himself at an early age was Ahmadu Fall. He was born in Saint-Louis, Senegal, to an important African family of merchants, on 17 June 1906. His military career spanned three decades and would culminate with his appointment as CIC and general of the Senegalese National Army in December, 1960.[88] According to family history, his merchant father had fought briefly with the French at Médine, and his grandfather had been one of the interpreters to the Archinard mission.[89] Growing up in Saint-Louis, in what he described as a military town, he played at war with other boys and dreamed of a military career. (Five or six cousins also served in the army.) After attending Lycée Faidherbe there, he volunteered for five years' service to launch his military career in November of 1925. Next came a tour of duty in FEA, during which time he became a sergeant. In 1930, he was sent to the officers' training school for regular French noncommissioned officers at Saint-Maixent, from which he graduated two years later as a second lieutenant, the only African among his cohort of fifteen men.[90] Fall's next postings were to Mauritania and then to Upper Volta; from there he was sent to the European front at the outbreak of war in 1939, in time to be twice wounded at the Somme and the Oise, cited in the order of the day, and awarded *Palme et Croix de Guerre*.[91] Repatriated to Senegal during the Vichy interlude, Fall was back in uniform in 1943, served in Italy in June 1944, and was with a Free French unit of engineers landing at Saint-Tropez in August of that same year. Like many other Africans, he spent the winter behind the front lines in the Midi guarding bases and German POWs.[92]

Ahmadu Fall remained in France until 1946, and his rise was very rapid after the war. In 1949 he was named a major and was appointed as military attaché to the FEA cabinet of High Commissioner (as governor-generals came to be called in the French Union) Cornut-Gentille. After a short posting to Morocco in 1950, he served again as military attaché after 1951, this time in FWA, for the cabinets of high commissioners Cornut-Gentille, Cusin, and Messmer. At independence, he accepted a transfer to the short-lived union of Senegal and Soudan in the Federation of Mali but reverted to the Senegalese army as a full colonel when the federation failed in 1960.[93]

The only one of the four officers profiled to have acquired a French commission, General Fall was a philosophical conservative who had grown comfortable with colonial ideology.[94] He shared with many other French officers trained in the interwar era an admiration for the strong paternalism in the *Tirailleurs Sénégalais* that other, more progressive individuals found excessive.[95] Fall believed that the *Tirailleurs Sénégalais*, even as a conscript army, had benefited Africans by transforming raw soldiers into disciplined and useful members of their society. Although as a senior officer in the 1950s he was obliged to support efforts to make the *Tirailleurs Sénégalais* more attractive to bright young African civilians, he shared with many other French officers of his day the view that the new African generation was tainted with African nationalism and generally much less dependable than the previous one had been.

French policy makers in 1919 had insisted on viewing FWA through their own ideological smoke screen. Their assumptions about the French demographic situation, for example, conditioned their judgment of how FWA could best serve the mother country. When they imposed a peacetime "blood tax" on West African youths, they reached a decision that was to have profound implications for virtually the entire population of FWA. Despite the exceptional performances of outstanding officers like N'Tchoréré, Fall, Diouf, and Diémé, most Africans who served in the interwar period found the army an unattractive institution. The next chapter examines these consequences of conscription, a process which can be described as military migration.

5

Consequences of Conscription: Military Migration

The former slave population has also accepted the principle of conscription with less difficulty than have the aristocratic element. The former slaves correctly regard military service as an excellent means of levelling social classes. They have everything to gain in the future in seeing their masters serve cheek by jowl in the same ranks with them. It can be said without exaggeration that certain population groups did not really achieve their complete autonomy until the day when their children, definitively liberated by military service, returned from duty to their homes, thereby instructing all as to the equality of every one before the white man's law.
Preliminary report of the Inter-Ministerial Commission on Recruitment in FWA, Dakar, 1922

As the last chapter has demonstrated, the "blood tax" became a concern for every French West African, whether or not it was paid. This chapter now examines the far-reaching implications of the recruitment system by focusing on three patterns of what may conveniently be called military migration.

Flight from the recruiter constituted the first and largest migration. The general average of absentees for FWA between 1923 and 1946 was slightly below 20 percent of the men listed on the census lists (see Figures 5.1 and 5.2 below). The annual cohort of men called to serve in the *Tirailleurs Sénégalais* made up the second migration. While some of these men might fulfill their three-year military obligation in FWA, a majority would be shipped overseas to France itself, or to garrisons anywhere in the vast French empire. Lastly, the third migration involved the return of demobilized soldiers. In their movement from barracks to civilian life, some men

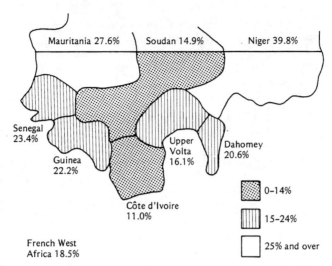

FIGURE 5.1. Average percentages of absentees per 100 men on draft lists, 1923–1946
Source: Compiled from various annual reports on recruitment in ANS, 2G, 4D, and 5D.

chose to settle in the administrative or regional centers of FWA rather than to return to their ancestral villages.

Migration as Resistance to Conscription

Migration as a form of resistance to conscription became a significant factor during the First World War.[1] The large levies applied without warning, the rumors that Europe had become a killing ground and French officers butchers of men,[2] as well as the brutal methods of the recruiter, all combined to encourage flight as the basic means of avoiding military service. Migration as resistance to the military recruiter mainly involved youths in their twentieth year, but refugees could also include older and younger brothers and other kin who feared being taken as hostages or substitutes.

Proximity to colonial frontiers was an important permissive factor. Among the Agni of eastern Côte d'Ivoire, for example, entire villages picked up and moved across the border as a result of recruitment demands. On the other hand, many members of this community, including its leadership, returned home after the war when they were guaranteed special treatment and lower quotas in the annual draft by the lieutenant governor of Côte d'Ivoire.[3]

Figures 5.1 and 5.2 provide absentee rates over the period from 1923 to 1946. Figure 5.1 shows that the general average of absentees for FWA was 18.5 percent of all men in the age cohort, that is, roughly 185 men per thousand.

Not all of these absentees can be said to be draft resisters exclusively, let alone emigrants. Some might indeed have been hidden inside the community, though this entailed great risk. As the civilian administration was quick to point out in

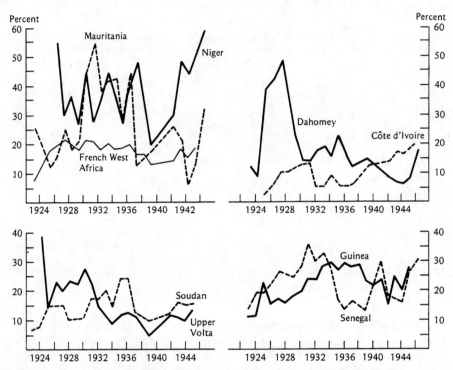

FIGURE 5.2. Percentage of absentees per 100 men on draft lists, 1923–1946
Source: Compiled from various annual reports on recruitment in ANS, 2G, 4D, and 5D.

apology for these high rates, some absentees were seasonal migrants, known in French-speaking West Africa as *navétanes*, in the peanut production of Senegal, for example, who happened to be away from their homes during the annual military levy.[4] Others may have been Dyula long-distance traders or Fulbe herdsmen temporarily earning their livelihood elsewhere. The Lieutenant Governor of Guinea went so far as to claim that the population of his colony was so mobile that fully 10 percent of its people could be away from their homes at any given time.[5]

While there was a grain of truth to some of these qualifications, they basically represent efforts of administrators to explain away damaging evidence of their misrule. The *navétane* migration, for example, took place during the planting, cultivating, and harvesting months from May to October.[6] Conscription, on the other hand, was deliberately timed to occur in the *hivernage* months from January to April, before *navétane* workers from Soudan, Guinea, or Upper Volta had set out on their annual migration. (It is true, however, that migrant workers from the savanna seeking employment in the cocoa plantations of the Gold Coast or Côte d'Ivoire were absent during the *hivernage* months because cocoa, a forest product, had a cycle different from that of the peanut, a product of the savanna.)

Military recruitment was not the only factor prompting men to leave their homes either temporarily or permanently. Migrants from savanna to forest and coast were attracted by the prospects of better wages as well as pushed by the French tax collectors and recruiters. But, given that the army called up physically able young men in their twentieth year, who for the same reason were sought for agricultural and plantation labor elsewhere, it seems clear that avoidance of conscription constituted an important reason for a young man to absent himself in his years of highest vulnerability. While the absentee living in the Gold Coast may have had mixed motives for having left FWA, he certainly had a compelling reason not to return.

Some men who were absent from their districts may not have fled across colonial frontiers. A certain number temporarily settled in other parts of FWA where they were not known by the administration and where they could be protected from the recruiter.[7] Others, perhaps less well connected, ran a risk in showing up in a new district, especially during the recruitment season. Substitution was a common practice in many regions.[8] Local notables either hired a substitute or used force to coerce unprotected young men to present themselves before the draft boards. In this way, the district or village could not only make good its quota but also reduce their overall number of absentees.[9]

Such illegal practices were not limited to backward and remote areas. In Dakar, right under the noses of the highest-ranking officials of the colonial state, there occurred an incident involving a young man from Gambia. This British subject, who had been rounded up in Dakar as a vagrant, was assumed by authorities to have been an absentee, and was dragooned into military service under the region of Dakar's quota for that year.[10] Only because the young man happened to be the son of a prominent trader from Bathurst who was known to British authorities was he able to secure his release. He somehow got word to his family, who notified the British consul in Dakar. The consul, in turn, requested through diplomatic channels that a British subject who was being held against his will and forced into the French colonial army be released.

Absentees risked sanctions and penalties to themselves and others by their actions. Theoretically, they and their families, their villages or their chiefs, could all be held to be in violation of the Conscription Law of 1919, which imposed prison sentences for obstructing the conscription process. This might have been a powerful constraint against flight, but the penalties were rarely imposed. In the early postwar years, few questions were asked as long as FWA as a whole was able to meet its quotas without difficulty. The private sector and even public officials often looked the other way rather than turn a valuable employee over to the army, justifying their actions on the grounds that the quotas were being met.[11] But by the end of the 1920s, under pressure from the military, the colonial administration began to watch for absentees and to attempt to capture them for military service.[12] Young men could be seized and declared absentees in the absence of proof they had been before a draft board. Such men would be given a medical examination and, if declared fit, were immediately inducted into the *Tirailleurs Sénégalais* and taken off the top of the quota for the following year.[13] These *bons absents recupérés*, as they were called, numbered approximately 2,000 men per annum by the 1930s. It was hoped that this method would deter evaders on the grounds that if men reported as required, the lottery draft system might not require them to serve. Not surprisingly, African youths continued to see flight as a risk worth taking. The numbers of absentees remained consistently high throughout the interwar period.

Even with allowances for high population mobility, substitution, and administrative laxity expressed in benign penalties, the high overall rates of absenteeism suggest a basic dissatisfaction with military service. A closer examination of Figures 5.1 and 5.2 would seem warranted. While most regions conformed closely to the very high FWA average, further broad generalizations are difficult to make. There does not seem to have been a simple dichotomy between regions, such as high rates in savanna versus low rates in forest zones, for example. Thus, Soudan, which falls within the savanna zone, and Côte d'Ivoire, largely in the tropical forest area, both had relatively low rates of absenteeism, lower than the savanna colonies of Senegal and Upper Volta, as well as the substantially forested colony of Guinea.

The regions with the highest rates of absenteeism were clearly Niger and Mauritania. Both were frontier backwaters where the colonial state's power was relatively weak. Both were thinly populated and therefore marginal to the overall process of recruitment. It is not surprising that absentee rates would be high in regions where it was more trouble than it was worth to enforce the Conscription Law. More noteworthy are the above-average rates that appear within colonies which formed the core of the conscription system. The following *cercles* had particularly high rates: Bondoukou and Guiglo in Côte d'Ivoire; Bakel and Podor in Senegal; Natitingou and Parakou in Dahomey; Gaoual and Labé in Guinea; Kayes and Satadougou in Soudan; and Ouagadougou in Upper Volta.

On the other hand, absentee rates were considerably below average in Côte d'Ivoire generally, as well as in selected *cercles* in other colonies. While no single-factor explanation can account for all of these regions, it is entirely possible in many of these cases that military service had become an accepted vocation. The *Tirailleurs Sénégalais* may very well have become the preferred, if not the only, alternative for earning wages with which to pay annual taxes. The following *cercles* had particularly low rates of absenteeism: Kong and Assinie in Côte d'Ivoire; Tougan and Batié in Upper Volta; San and Koutiala in Soudan; and Beyla and N'Zérékoré in Guinea.

Fluctuations in the absentee rates over time are indicated in Figure 5.2. In the case of Dahomey, there occurred a dramatic rise in the rate of absenteeism in the late 1920s, a period during which this colony was experiencing the difficult aftermath of political repression of its leading intellectuals.[14] Also, more concretely related to conscription, colonial authorities brought down changes in legislation in 1926 allowing for the possible mobilization of the hitherto inactive second portion of the annual cohort into labor brigades, dressed in uniform and under military discipline, to be handed over to private contractors.[15] The unusually high absentee rates were the responses of Dahomeans to the very well-founded rumor that the second portion was about to be activated. It is significant that the protest by Dahomeans proved effective. Under this pressure, the French prudently decided to abandon plans to activate the second portion, and in the early 1930s Dahomean absentee rates returned to a stable pattern. Conversely, labor brigades were indeed activated in Soudan (the only colony where this occurred) in order to provide labor for the *Office du Niger* construction works.[16] This highly unpopular form of forced labor, in turn, produced two effects on military recruitment. First, the rates of volunteers increased; volunteering guaranteed an able-bodied Soudanese youth that he would at least serve in the orderly conditions of first portion, in regular units of the *Tirailleurs Sénégalais*. Second, the rates of absenteeism in the Soudan increased sharply in the 1930s.

Some *cercles* and colonies might be said to be bellwethers of conscription and military service. Dahomey is a good example of such a colony, and Casamance, in Senegal, one such region. Because it bordered on both Gambia and Portuguese Guinea, Casamance continually presented potential conscripts with the opportunity for escape across the frontier. When absentee rates in Casamance were low, this suggested that conditions of pay and military service in general were acceptable, and terms and risk of overseas service tolerable. High rates of absenteeism conversely would indicate the reverse: low pay in relation to economic opportunities at home, combined with high risks of service abroad.

Finally, some absentee rates remain anomalous, as, for example, in the case of high rates for Cotonou and low rates for Porto-Novo, two almost contiguous cities in Dahomey. Cotonou's absentee rates in fact corresponded closely to those of other cities such as Conakry, Guinea. The proximity of Nigeria and the ease with which that frontier could be crossed would explain opportunities for high absentee rates. What is surprising, then, is that Porto-Novo's rate should have been so low. It was a town predominantly peopled by Yoruba-speakers, a majority of whom were Muslims. Powerful "hajji" traders, adept in the past as well as after independence at smuggling, were no doubt equally adept at concealing young men and at distorting population statistics generally.[17] It may very well be that the local skills in manipulating the conscription system were greater in Porto-Novo than in Cotonou.

Soldiers' Migration

Another form of migration, also involving significant numbers of men, consisted of the actual process of incorporation and distribution of troops. In stages, this migration took place first from village to the *cercle* administrative center, then to garrison towns distributed throughout FWA, and lastly, to destinations as far afield as Madagascar, France, and Indochina.

The first step, from village to *cercle* center, may have been a simple matter of half a day's travel for some. Given the enormous areas of some *cercles*, however, this displacement could represent as much as eight days' travel, with the recruit expected to meet his own food and lodging expenses en route.[18] Some recruits often turned up at draft boards having traveled across *cercle* and perhaps even colonial borders. In Niger, for example, it was common practice for recruiters to look the other way and sign up recruits as *"Nigériens"* when in fact the men might be Sara peasants from Chad, or from Cameroun, both parts of French Equatorial Africa.[19] Indeed, Cameroun was a League of Nations mandate and conscription was forbidden there. It can thus be seen why French authorities might look the other way if *Camerounais* of military age decided to travel to Niger to substitute themselves for locals in return for payment of a fee from the family. Indeed, statistics which indicate high rates of volunteers in regions which had a reputation for disliking the draft—places like Abengourou and Assinie, for example, in eastern Côte d'Ivoire—may also have been districts where substitution of outsiders, perhaps from the north of the colony and from Upper Volta, may have been a factor.

One of the peculiar mechanisms that the French imposed helped increase this tendency. Both because the French authorities came to view the premiums paid out for volunteers as a great expense and because of pressures resulting from the liberal ideology of a universal male conscription system,[20] the French gradually imposed a limit on the number of volunteers that could be permitted from any one recruitment district. This ceiling came to be one-third of the contingent; as a result, in some areas volunteers were actually turned away, whereas elsewhere a significant number of reluctant soldiers were being conscripted.[21] It did not take long for enterprising French officials as well as African middlemen to see the possibilities in this. Soon, French officials began to look the other way as a clandestine traffic in substitutes developed, especially in Niger. French reports indicate that they were well aware that many young men in the Niger contingent were not locals but from Chad, Congo or even Nigeria. Premiums of up to 1500 francs, ten times the military's bonus for volunteers, were said to have been paid.[22]

Sometimes these frauds went too far to be tolerated. One such case involved a soldier who had served three years but had then been refused in his bid to re-enlist on the grounds that he was a discipline problem. Apparently, this same man appeared in a district of Niger, and for a certain fee, passed himself off as a local volunteer. Only when an African noncommissioned officer happened to recognize this man at an induction center was the fraud discovered.[23]

Once the contingent was recruited at the *cercle* level, the men were dispatched to their induction centers. Again, this might involve a short journey to the capital city or, often, to the garrison town located close by the capital but apart from it, such as Kindia near Conakry, Thiaroye outside Dakar, or Bingerville close to Abidjan. But local recruitment always went well beyond the simple replacement needs of the local army units, and a significant number of recruits were directed many miles away to regiments on the coast.

The distribution of the *Tirailleurs Sénégalais* within West Africa was a complex matter. For one thing, it raised questions about the desirability of soldiers of the same ethnic origin as the civilian population serving as their occupiers. However tempting it may have been to move soldiers around deliberately, such planning became too costly and complicated, given the administrative capabilities of a limited number of French officials. Another reason for avoiding such movements was that attrition rates were high when new recruits, especially, were marched long distances.[24] Figure 5.3 shows the composition of a typical contingent of new recruits in 1926, mainly destined for overseas duty, who were attached to the First African *Tirailleur* Regiment, whose home base was in Saint-Louis, Senegal. While Senegalese recruits constituted 38 percent of the contingent, the largest minority, 44 percent, came from Upper Volta while smaller percentages were contributed by the colonies of Guinea and Soudan.

The induction march itself was a matter of some debate among French policy makers. The distances could be enormous, yet no motorized transport, whether trucks or the railroad, was employed until the 1930s, when military complaints at the wastage led to the voting of credits for transport.[25] Until that time, the men were expected to walk. This introduction to infantry life was too much for many, however realistic it may have appeared on paper. Poorly clothed to bear the cold nights spent in tents during *l'hivernage*, inadequately fed at stopovers by local *commandants de*

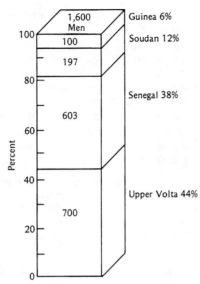

FIGURE 5.3. Colonies of origin for the 1,600 men of the First *Tirailleurs Sénégalais* Regiment (Saint-Louis, Senegal), in percentages, 1926
Source: *Historique du Ier Régiment des Tirailleurs Sénégalais* (Saint-Louis: n.d.).

cercle who were obliged to absorb the costs of feeding these military transients, and under close guard to assure that they did not desert, many young conscripts must have found these literally forced marches a horrible introduction to military life. A significant number never lived to see garrison duty or combat abroad. The marches resulted in high attrition rates, largely through pulmonary infections. Apart from the deaths, the sick who landed in Thiaroye army hospital, for example, had then to appeal to a parsimonious colonial administration to help pay the costs of travel home upon discharge.[26]

By far the most dramatic soldiers' migration was their movement abroad for garrison and sometimes combat duties in the French empire. As indicated in Chapters 2 and 3, overseas migration of African soldiers had begun under France's First Empire, and constituted an important part of Mangin's rationale for proposing the creation of *La Force Noire* in 1910. The numbers and proportions of African soldiers serving abroad varied through time but was always considerably larger than those remaining in West Africa. Setting aside the vast numbers of the First World War, in 1922 some 15,000 men constituted the Army of Occupation for FWA, while over twice as many men served overseas. By 1929 the differential had been reduced to 20,000 in FWA versus 28,000 abroad.[27]

For most of the colonial period, military migration overseas involved only adult males. In the years prior to the First World War, however, when the first large contingents of black Africans were being dispatched to Morocco to take part in the French conquest there, the soldiers were permitted to take their families with them. The dispatch of some 8,000 men to Algeria and Morocco between 1908 and 1912 meant that at least this same number of dependents were also displaced from West Africa. During and especially after the First World War, this practice was gradually halted on the grounds of cost.[28] Thereafter, soldiers were not permitted to take spouses abroad, and had to remain celibate, frequent prostitutes, or more rarely, either marry or cohabit with women of the local overseas population. Only after the Second World War did the army again provide housing and allowances for wives and families in major cantonment areas.[29]

To provide some female companionship for African soldiers stationed in France, army authorities relied on the institution of the *marraine*, or godmother. More like a pen pal or surrogate sister than a mother, *marraines* were much appreciated by lonely soldiers. The *Ivoirien* ex-serviceman Gmbale Soro stated:

> I had a girlfriend in France. She went to the office and took my name and number. At the end of each month she would send me a little package—with a shirt and some biscuits. After the war I saw her. I arranged it myself. I asked where the village was and got some leave and went and saw her. I showed her my name and number and then she knew it was me. She was very pleased to see me. Yes, I sent her letters. I still have her photograph.[30]

Such relationships could sometimes lead to misunderstandings. Military censors intercepted the following letter from a French woman in Melun to an African lieutenant living in Niamey, whom she had met during the Second World War:

> I am pained to have to reply that my affection for you is like that of a sister or god-mother, . . . if I wished to meet you in Paris it was only because you felt yourself understood and comforted by a young French woman who loves her colonial empire . . . "[31]

Tirailleurs who were stationed in France did not have wide latitude to leave their camps. According to the Guinean First World War veteran Kandé Kamara:

> People—special police—were given orders to look for soldiers, particularly black soldiers. If they saw you walking around town—and obviously there was no reason for you to be in town because no soldier would be sent there by himself—you would be arrested.[32]

A different problem occurred in North Africa. There lonely *Tirailleurs Sénégalais* occasionally found themselves involved in quarrels with the local population over women.[33] Perhaps to mitigate such disputes, the French officials in North Africa took some interest in the sexual practices of their troops stationed there. One army physician who was assigned to *Tirailleur* units in the 1940s recalled having regularly inspected the women of a house of prostitution to prevent the spread of venereal disease.[34]

Sexual relations between African soldiers and French women provoked a racist discourse. As recently as 1980, long after the *Tirailleurs Sénégalais* units had been disbanded, a popular collection of old photographs of Marseilles was published which invoked racist stereotypes. Beneath a period photograph of a *Tirailleur Sénégalais* standing arm in arm with a Marseilles prostitute was this caption: "White women exercise an undeniable attraction for the *Tirailleur Sénégalais*."[35]

Despite discouragement from army authorities and the racial bias of European society, liaisons between African soldiers and white women not only took place, they sometimes led to common-law and even civil marriages. Many such arrangements developed after the armistice of 1940. In the immediate years following the Second World War, all African soldiers in France were ordered demobilized back to FWA whether or not they were leaving dependents, wives, and jobs behind. To compensate financially for women and children who were left behind, officials decided to allocate 100,000 CFA francs out of the FWA budget.[36] When the separated wives enlisted the help of their mayors and deputies to protest to military and colonial authorities that they wished to be reunited with their African husbands, officials first checked on the marital status of the ex-servicemen in Africa, and on the marketable skills, solvency, and general character of the man. A final inquiry sought to determine the willingness of the ex-soldier to acknowledge the woman and their relationship. If the man was unwilling or of dubious character, the matter was dropped. In a few cases, notably involving Dahomean Christians with trades such as mechanics or chauffeurs, officials recommended that the African husband return to France rather than have a white wife join him in FWA. The Governor-General of FWA was explicit:

> I am not in favor of allowing these European women to settle in villages in the bush. Their life in common with the autochthonous inhabitants cannot but have deplorable effects. What is more, these women are certain to be deceived.
> As far as retired career soldiers who have contracted marriages and who are still living in France are concerned, I would much prefer to see them remain in the Metropole, with the local *département* seeing to their placement.[37]

A series of cases which can be partially reconstructed from censor's files illustrate the extent of army paternalism.[38] For example, the governor of Dahomey advised against a young Dahomean of Cotonou, aged thirty, who lived comfortably with his brother, a jeweller with claimed assets of 125,000 francs, marrying a fifty-two-year-old French woman from Dijon on the grounds that the disparity of age only compounded their differences in culture. Similarly, the governor of Guinea refused the request of an ex-soldier in his colony to divorce the African wife he had married in Ségou in 1938 in order to become engaged to a French woman. Not only could the governor find no grounds for the divorce, he was keeping the ex-serviceman to his first marriage "for his own good." Even when, by the standards of the day, the administration may have had grounds for such advice, their power to intervene was another glaring illustration of how the hierarchical authoritarianism of colonialism was indistinguishable to Africans from the militarism they had endured during their service.

One case that actually resulted in the return of an ex-soldier to France was resolved through the intervention of the mayor of a small town in Meurthe-et-Moselle *département*. An African sergeant had married a young woman, legitimized her two children, and was proving to be a reliable employee at the local foundry. Suddenly the army ordered him shipped back to his home in Bamako, where he was earning only 35 francs a day. The wife and her father were both in poor health and unable to work, and the entire family forced onto town welfare, at which point the mayor intervened with the Ministry of Colonies. When the Governor of Soudan cabled to indicate that the husband was willing to return to France but was without funds for his passage, the mayor was obliged to turn to local sources to make good the fare.

Only in one recorded instance did the French administration set aside their prejudices and allow a French woman to join her husband in Africa. The case involved the marriage of the daughter of a poor French laborer to a Dakar soldier who was described as having a primary school education and who "had shed most of his native customs and superstitions. He remained nevertheless attached to the Muslim religion." The son of a prosperous merchant, this ex-soldier earned 8,000 francs a month and was willing to have his wife join him in Senegal. While this sum was not sufficient for a European household, according to the governor, it would allow the young woman to live as well in Dakar as she could in France, and so he granted her permission to emigrate to the colony. In fact, however, the presence of the rare North African, Syrian, Vietnamese, or French woman who had married an African soldier and, against all pressures, had actually returned with her husband to live in one of the cities of FWA, was as rare as it was remarked upon at the time.

It was by no means the case that Africans were always willing to be reunited with their former lovers. When approached by authorities on the initiative of a French woman from Fréjus, one ex-soldier from Senegal would have nothing to do with her, describing her as a "pensioner in a whorehouse in Draguignan." In another case, a corporal from Saint-Louis told a French inquiry that his father, a strict Muslim, disapproved of his marriage to a Christian and had ordered him to refuse a reunion. This last case served as a reminder that life in FWA for so-called mixed marriages was liable to encounter both the overt racism from Europeans and a more concealed racial prejudice from Africans.[39]

Veterans' Return Migration[40]

The third and last form of military migration involved the return of ex-soldiers from service abroad. The reintegration of three-year conscripts and career servicemen into French West African society is of considerable historic and demographic significance. The difficulty lies in describing this process in anything other than speculative terms, or in measuring this movement of people quantitatively.[41] If they ever were produced, no official statistics enumerating returning soldiers have survived. Perhaps the French military felt that the rates of attrition in the *Tirailleurs Sénégalais* would have provided arguments for their opponents. In any event, historians are obliged to estimate the rates of return as best they can. The French army did produce general mortality rates for the French Colonial Army in the interwar period (Table 5.1).

These figures reveal a mortality rate for the *Tirailleurs Sénégalais* running three times that of Europeans. Highest *Tirailleur* death rates occurred in Morocco in the 1920s and in France in the 1930s;[42] and exceptionally high death rates were encountered in 1925, when severe epidemics struck African soldiers in both Morocco and the Levant. Unfortunately, these rates are not correlated with accurate figures on the total size of African units in any of these theaters of operation.

No French calculations have as yet been found estimating the rate of return of African soldiers to their homes. Elsewhere, I have attempted this calculation based

TABLE 5.1 General mortality per 1,000 men in the French Colonial Army, 1920–1938

	1920	21	22	23	24	25	26	27	28	29	30	31	32	33	34	35	36	37	38
Colonial Army																			
West Africans										16	12	13	13	12	12	10	11	11	9
French										5	4	4	5	4	4	4	3	4	3
Metropolitan Army																			
North Africans										7	5	5	5	5	4	4	4	4	3
French										5	3	4	4	3	3	3	2	2	2
Colonial Army in France																			
West Africans										16	13	14	13	14	16	13	16	13	12
French										6	4	5	6	4	4	4	3	4	3
Colonial Army in the Levant																			
West Africans	9	29	31	28	23	72	38	15	15	14	16	18	13	9	11	6	4	4	7
French	15	7	9	12	13	56	43	26	9	2	7	6	6	2	3	3	3	7	6
Colonial Army in Algeria and Tunisia																			
West Africans										14	8	8	7	8	8	7	6	9	9
French										3	2	4	4	5	4	4	4	4	5
Colonial Army in Morocco																			
West Africans	28	55	51	36	29	71	25	21	21	16	15	14	21	14	11	8	9	7	7
French	11	15	12	13	12	57	21	17	8	5	7	4	5	5	7	5	6	3	4

Source: Compiled from data in Ministère de la Guerre, *Statistique médicale de l'Armée metropolitaine et de l'Armée coloniale,* Vols. 50 to 68 (Paris: Imprimerie Nationale, 1919–1938).

upon known rates of induction into the *Tirailleurs Sénégalais*, the size of the standing army, and estimates of attrition during induction marches and in various combat and garrison zones of the French empire.[43] What remains entirely uncertain are the rates of distribution of soldiers throughout these zones, the rate of re-enlistment of soldiers, and the numbers who, French regulations to the contrary, may have slipped through the net during the demobilization process that was to take them back to their villages. Within these considerable limitations, my calculations suggest that for every 10,000 men absorbed annually into the army in the period from 1920 to 1938, 440 would die during the course of their three years of military service, another 1,100 would re-enlist at the end of their first tours of duty, leaving 8,400 to return to West Africa.[44] Of this group most might return to their ancestral villages, but an indeterminate number might settle in larger provincial towns and cities. This pattern clearly held true for career soldiers and veterans who preferred the Western amenities of larger centers. The capital of Dakar, for example, was a powerful magnet. By the 1950s it counted four times the number of army veterans among its population as it had furnished soldiers in the preceding decade.[45]

Another way of looking at these figures is to estimate the loss of labor that military migration represented to the domestic or household mode of production in FWA. Again using fragmentary data, I have estimated elsewhere that at least one in every three conscripts never returned to his rural home.[46] By coincidence, since he had no data to justify his prognostication at the time, Maurice Delafosse, an experienced observer of French West African colonial administration, had predicted a loss of at least one-third to the village economy when he spoke out against systematic peacetime conscription in 1922.[47]

The loss of 4,000 men on average per year (one-third the contingent) was not the only drain of labor away from the village economy. Wage labor migration was a prominent feature of FWA between the wars, and the general flow was from the densely populated savanna lands of the interior south or west to the forest and coast. By the 1930s, a small but growing pattern of wage labor migration to urban centers was also in evidence.[48] For those who remained behind in the villages, this absent labor had to be made good by the fewer hands of older and younger males, including the elderly and the less fit, and especially by women.

Not all of this estimated loss to the domestic economy was the result of mortality, of course. An untold number of returning veterans did choose to abandon their rural homes in favor of town life in FWA, illustrating once again how closely military migration paralleled labor migration in FWA. Unfortunately, it is not possible, in this respect either, to rely on firm statistics. What can be attempted is a comparison of average rates of recruitment for the *Tirailleurs Sénégalais* with the numbers of veterans registered in the various colonies as of 1952. If the percentage of veterans is higher than the percentage of soldiers recruited, this would mean either a differing rate of attrition overseas or a differential rate of return to home colonies. If a uniform rate of attrition is assumed, then Figure 5.4 offers a rough indication of patterns of urbanization in FWA by 1952. One pattern shows a dramatic rate of urbanization away from rural Senegal and towards Dakar. Another illustrates that Soudan registered a net gain of veterans while Guinea suffered a net loss. It is tempting but risky to suggest a migration of veterans from Guinea to Soudan based on these figures. For most of the other territories, the data suggest a

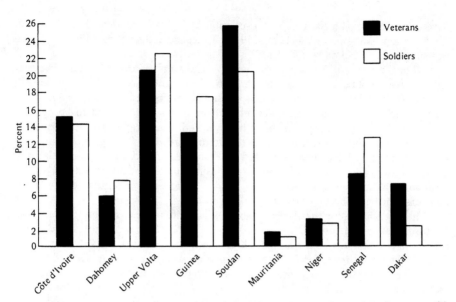

FIGURE 5.4. Distribution of veterans by residence, and of soldiers by colony of origin, in percentages, 1952

Source: Report of Lieutenant Colonel Vatinelle, Dakar, 21 December 1952, ANS, 4D178 144, for veterans; various recruitment reports between 1925 and 1948 for soldiers, in ANS, 2G, 4D and 5D.

relatively close correlation between soldiers and veterans. This would not rule out the possibility that urbanization could have been regional, with veterans gravitating from country to city within the same colonial territories. Such a pattern did occur in Upper Volta, where veterans became a basic element in the new population of such provincial towns as Tougan, Ouahigouya, Kaya, and Boromo, as well as larger centers like Bobo-Dioulasso and Ouagadougou.[49]

Conscription and Morality

In France ever since the Revolution compulsory military service has been closely linked to the democratization of society.[50] Not every nineteenth-century government lived up to this ideal by any means, but especially after the Third Republic reform of 1 January 1873 abolished the practice of substitution and reduced the term of service from seven years to five, conscription came to be seen as a legitimate duty owed by citizens. No doubt a second reform in 1889 reducing service from five years to three further encouraged this tendency.[51] In exchange for the benefits of citizenship in a democratic state, male citizens were expected to perform their military obligations. It became easy enough in the colonies, however, for French authorities to slip into the injustice of applying the obligation without the accompanying democratic privileges. For Africans, conscription was but another example of how their situation compared to France of the *ancien régime* rather than to France of the

Revolution. No egalitarianism was ever intended or extended with the African Conscription Laws of either 1912 or 1919. The purpose of the entire military exercise was simple: to obtain more soldiers.[52]

When French authorities attempted to justify their actions by claiming that Africans were not being asked to do what Frenchmen could avoid, they were clearly dissimulating.[53] West African subjects of France performed compulsory military service under dramatically different terms from those that applied to French citizens. As of 1919, the French citizen in his twentieth year faced twenty-four months of compulsory military service. In 1926 this term was reduced to eighteen months.[54] French conscripts were usually posted in France itself and were given relatively generous leave and food allowances. Many youths took the opportunity to learn a profitable civilian skill, or perhaps to pursue a career as a regular officer if so motivated and qualified.

Military service for the African subject was quite another matter. Service was for three years and often involved combat in colonial wars or garrison duty in remote and uncomfortable outposts of empire. The pay was considerably lower than for French citizens, the food of lower quality, and the perquisites fewer. Only a minority re-enlisted and became career soldiers; of these fewer still moved up to a career as a native officer, and even in these cases the maximum grade they could attain was captain. As for skills acquired in military service, most were geared to the tasks of infantrymen and did not allow for the development of applied technical knowledge that could have benefited ex-soldiers seeking a promising civilian career. Moreover, the African recruit faced indignities of a sort that French citizens would not have tolerated. Medical exams prior to induction took place in public with the men completely nude. Conscripts could sometimes be required to march hundreds or even thousands of miles on foot to reach their induction centers. Standard uniforms for the *Tirailleurs Sénégalais* were not suitably designed either for colder climates or intensely hot tropical ones.[55]

The burden of conscription reached well beyond soldiers. African families were separated or divided, thousands of men were lost forever, and opportunities for injustice and divisiveness abounded at the local level of authority. Even if many families were spared the loss of a son—because he drew a lucky number, because the family had influence, because he dared to be absent from his draft board, or because he successfully made it across the frontier to British West Africa—anxiety over the annual drafts touched millions of families in FWA during the colonial period.

The militarization of FWA through conscription had other effects that are, as yet, poorly understood. Returning soldiers were agents in the spread of communicable diseases. Because all migrants contributed to this spread of disease, it is not possible to separate soldiers from others among the sexually active migrant workers. Yet, because they were the group that traveled abroad in greatest numbers, their impact on general demographic variables of mortality and fertility must have been considerable. French medical reports noted that infectious pulmonary diseases, such as tuberculosis and pneumonia, represented a far greater threat than venereal disease to the health of serving soldiers, and presumably of returning ones.[56] In those areas of FWA where rates of tuberculosis increased dramatically, it is possible that ex-soldiers returning in large numbers may have been the unwilling agents.[57] Finally, an indirect impact of military migration on West African fertility can be suggested. The removal of large numbers of the physically most able males for periods of three years and more from their village communities must have affected such variables of family life as the age of marriage and the rate of polygyny, while the value of bride-price was often altered by the returning soldier, especially if he had a large and steady military pension. That these variables were dramatically altered by the overall patterns of labor migration has been ably demonstrated by the work of Piché and Gregory.[58] What should be noted here is that soldiers, and those who fled to avoid becoming soldiers, were also migrants on a large and systematic scale. Along with migrant workers, soldiers helped change the course of West African demographic history in the interwar years.[59]

It would, of course, be an error to see Africans as docile victims of this unjust system of military conscription. When geography and other circumstances permitted, Africans successfully resisted military service in numerous effective ways. Flight from the recruiter was the most important method when alternative forms of disposing of their labor proved more advantageous. Other techniques were substitutions and even self-mutilation to thwart the draft boards. So extensive and systematic was avoidance of military service through absenteeism and flight that it became a significant motive in the patterns of regional migration that emerged in the interwar era.

The conscription system was unpopular even among those obliged to enforce it. Civilian administrators and the private firms deplored the waste of resources it so patently revealed. How could they think otherwise of a system that processed 188,000 men each year in order to produce 12,500 recruits?[60] Waste was evident in other respects as well. Expensive and understaffed civilian administrators found themselves spending weeks of each year preparing for the annual draft rather than getting on with more pressing matters. Even higher-paid medical personnel spent months separating the fit from the unfit, rather than diagnosing or even beginning to treat their medical problems. Private firms noted anxiously the flight of youths across colonial borders to escape the recruiter.[61]

Even the French army was unhappy. As the conscription system developed, it moved further away from the original premise of Mangin and other founders of the Black Army. These men had envisioned a crack army of highly motivated volunteers. Constant objections and bickering from civilian officials had led to a series of small "adjustments" to conscription that satisfied no one. For example, the emerging practice in the 1920s of ensuring that absentees would be the first to be inducted raised serious questions about the military motivations of up to 10 percent of later annual contingents.[62] Later in the same decade French officials took the further step of limiting the number of volunteers to no more than one-third of any one contingent, in order, it was argued, that the burden of service be shared across FWA as equally as possible.[63] While the actual number of volunteers in the federation as a whole did not reach this figure, in some regions of Upper Volta and Soudan especially, draft boards had reluctantly to turn away volunteers in order to comply with this decree.[64] Actual reasons for such a step are difficult to establish, but the consequences are at least clear. As the system evolved, there was less place in the army for those who may have wished to volunteer for whatever reason and more for those who found military service so distasteful that they would try to avoid it by being absent from their draft board calls.

Anxiety about conscription was also more than a nuisance factor to civilian officials and to firms in the private sector. It is both puzzling and tragic that the French government did not consider fundamental reform of the conscription system by the late 1920s. Given that some African military service was inevitable in view of overall French priorities, it would have been in the interest both of the French rulers and the African subjects to have returned to a volunteer system once the memory of the First World War had receded. By offering decent salaries and concentrating their recruiting in regions that had developed a tradition of military service, the French would have served their own interests better and lifted a heavy burden from the backs of the Africans in the bargain.

6

"Morts pour la France": The Tirailleurs Sénégalais and the Second World War

> I have here in front of me your last letter; how proud I am to find in it the phrase, "Whatever happens papa, I will always be ready to defend our dear country France."
> Letter from Captain Charles N'Tchoréré to his son, Corporal Jean-Baptiste N'Tchoréré, no date

This chapter focuses on a surprisingly obscure episode in the Second World War—the experience of African combatants in France.[1] Not only were significant numbers of *Tirailleurs* manning the front lines during the terrible days of May and June 1940, many of the survivors of these units were interned in the *Fronts-Stalags* (German labor camps), located largely in northeastern France, for periods up to four years. Still other Africans landed in Provence with the Free French forces as part of the Allied invasion of southern France in the summer of 1944. The repatriation to FWA of these men stretched over a period from November 1944 well into 1946, often under the most trying of circumstances.

Two major reasons for neglect of this important topic come to mind. First, the military archives for the campaigns of 1939–1940 became available to the public only in the early 1980s.[2] Secondly, until recently many Frenchmen have preferred to avoid the painful story of France's defeat and later reentry into the war as a minor player. Fortunately, French attitudes to this traumatic period are changing. Roger Bruge, a popular military historian who has done much to demystify the events of May and June 1940 has also drawn attention to the efforts of colonial troops to stem the German tide.[3]

While comparisons with the better documented events of the First World War must therefore be tentative, it seems reasonable to suggest that the Second World War's impact on Africans was at least as great, and probably even greater insofar as political consequences are concerned. In the First World War some 175,000 Africans served France in Europe over five years of war, with approximately 17 percent, or 30,000 men, killed in action.[4] These were, of course, terrible losses, especially when the moral question of conscription without consultation and often by force is considered. Nevertheless, Africans participated and died at rates that were proportionate to overall figures which give a mortality rate of 26 percent, or roughly 1.3 million dead among the 5 million French citizens and subjects in uniform.

During the Second World War, on the other hand, FWA's military contribution was disproportionate to that of metropolitan France. Accidents of war and French policy combined to make the African contribution larger this time around. There were seven African divisions and three other colonial divisions out of the eighty French divisions defending the borders of France in 1939, a much higher ratio of colonials to French than for the First World War.[5] There are complications in such a rough comparison, but in simple ratios, Africans in 1940 constituted almost 9 percent of the French army in France, whereas for the entire First World War they may have approximated only 3 percent.[6]

The proportions changed more dramatically after the armistice of June 1940. The neutralization of the regular French army as well as its partial dismemberment effectively removed most French soldiers from combat. The Free French forces of de Gaulle took a good deal of time to build up their numerical strength. In the period from June 1940 until the liberation of France in the summer of 1944, black Africans constituted the main elements of the rank and file in the Free French Army, with North Africans also an important element. As late as September 1944 the 20,000 black Africans in France as part of the French army participating in the Allied landings represented 20 percent of de Lattre's total force.

While it may be impossible to assemble definitive figures on African combatants for the entire course of the Second World War, some rough estimates of African recruitment directly into military units is possible. Between the outbreak of hostilities in September 1939 and the fall of France in June 1940, some 100,000 men were mobilized in FWA alone; three-quarters of these were already in France or *en route* to the European theater.[7] At the armistice in June, anywhere from 24,000 to 48,000 Africans were declared missing, of whom perhaps 15,000 to 16,000 became prisoners of war (POWs).[8] Meanwhile, under Vichy rule in FWA, to protect the colony against potential British invasion, the French increased the peacetime size of their standing army of Africans from 50,000 to 100,000.[9] Again, under the Free French after November 1942, recruitment for the war effort was stepped up, as 100,000 Africans left for the front between 1943 and 1945.[10] At a conservative estimate, then, the French recruited in excess of 200,000 black Africans during the Second World War.[11]

Total mortality rates are even more difficult to estimate. If 10,000 men were killed in May and June 1940, and half the POWs did not survive their incarceration,[12] this would place total deaths at 17,500 men. To this figure would have to be added deaths in the campaigns of the Free French in North Africa through 1943, in the Italian campaigns of 1943–1944, and in the landings in southern France in 1944, which could give a figure in excess of 20,000 and perhaps as high as 25,0000 overall. In other words, 12 percent of all those recruited could have ended up *morts pour la France*.

Ideology and Mobilization

The contrast with the First World War extends to many other domains of military life, not the least of which is ideology. In the era of the First World War, French authorities made very little effort to transmit an ideology to the black African soldier. In his memoir, Bakary Diallo described the announcement of France's declaration of war in 1914 to his *Tirailleur* unit in Morocco in these terms: "The sergeant-major read a report. It was somewhat long and we only understood two things: 'Germany had declared war on France. . . . And France was calling on all her children [to defend her].' "[13] The idea that Africans were in some sense citizen-soldiers was certainly not widely held, either by Frenchmen or by West Africans outside the Four Communes of Senegal or coastal Dahomey until much later.[14] In the absence of a Jacobin ideology stressing the virtues of patriotism, African soldiers either relied on traditional perceptions rooted in West Africa's precolonial social relations or else established close personal ties of a patron-client type with their African and French officers.[15] In the case of such paternalist senior officers as Gouraud or, especially, Mangin, these ties were maintained at least symbolically all the way up to the top field command level of the French army.[16]

By 1939, however, the French African army had become institutionalized. The cumbersome peacetime conscription system ushered in by means of the Conscription Law of 1919 had made military service a universal male obligation. Expanded services in education, communications, and medicine had brought considerable social change to FWA. When, as the threat of war mounted in the late 1930s, it came time to mobilize far more Africans than the normal cohort, the French authorities could no longer safely ignore the ideological preparation of their subjects.

Nowhere was this more evident than in a remarkable publication entitled *La Gazette du Tirailleur* that appeared fortnightly from 1 January to 1 June 1940.[17] The paper ran war news from an Allied perspective. These ranged from anti-Hitler jokes to stories asserting that morale was very low among the civilian population of Germany. But the most consistent vehicle for communicating war aims to semiliterate and nonliterate African soldiers was a comic strip entitled *Mamadou s'en va-t'en guerre*.[18]

As the saga of Mamadou unfolded each fortnight, the anonymous authors addressed a series of major issues: French war aims, Hitler's aggression, Nazi designs on African colonies, and the significant part Africans were expected to play in aid of the motherland. The comic strip followed Mamadou's adventures from his West African village to France and the northern front. Mamadou's training was highlighted by good food, attractive uniforms, and dreams of glorious combat followed by a decoration for killing Germans (episode no. 2, 1 February 1940). Mamadou and his friend Ibrahima, being good Muslims, drink lemonade, not alcohol, in the cafes of Dakar and flirt with pretty African girls (no. 3). Their girlfriends bid them a tearful farewell and the African public gives them a heroes' send-off from the port of Dakar (no. 4). The crossing is uneventful save for glimpses the two friends catch of enemy submarines observing their convoy (no. 5). But French reconnaissance aircraft protect the fleet and they reach Casablanca and then Gibraltar safely; in the Straits they express their admiration for the naval might of England, France's staunch ally. Finally their ship reaches Marseilles, where the

The first episode of "*Mamadou s'en va-t'en guerre*," 1939

African soldiers are warmly greeted by a French population showering them with gifts of cigarettes and flowers (no. 6). The next three episodes (nos. 7–9) depict life in the barracks of southern France, with much emphasis on the bonds of friendship Mamadou and Ibrahima forge with a French family. On a tourist excursion by boat to the Château d'If, Mamadou plunges into the cold winter water of the Mediterranean to rescue a young girl, Mireille, who has accidentally fallen overboard. The two African soldiers are befriended by Mireille's grateful family and invited home for meals and human warmth. They meet her brother Marius, a corporal who turns out, by coincidence, to be attached to their company at their new camp "somewhere on the Côte d'Azur." There is time for more tourism and the obligatory visit to a perfume factory at a town whose name is censored, but which is probably Grasse. By now the three friends are anxious to go to the front and deal with the hated enemy once and for all. The last caption in the last issue published, on 1 June 1940, depicts the three friends taking a train "somewhere to the northeast."[19]

Hyperbole aside, *Mamadou s'en va-t'en guerre* is essentially consistent with the realities experienced by African soldiers as they were mobilized for the Second World War. To be sure, Mamadou and Ibrahima were model soldiers, but the Dakar command would hardly have chosen, for example, heavy-drinking roustabouts. As for recruitment, while it was an exaggeration to say that Mamadou's voluntary enlistment was typical, the French Colonial Army had, by the late 1920s, "entered into the African way of life," as one French officer put it.[20] For the majority, if compulsory military service was painful and unpleasant, they were obliged now after almost twenty years to regard it as an inevitable burden of colonialism. With the growing crisis in Europe, and the declaration of war on Germany, many Africans, especially those who had retired as career soldiers, rushed to reactivate their commissions, if they were officers, or, more commonly, to re-enlist as rank-and-file soldiers and noncommissioned officers. One such example was Captain Charles N'Tchoréré, born in Gabon, who had retired from the French colonial army after twenty years' service, including the last years of the First World War, and who had been named commander of a military academy for African cadets in Saint-Louis, Senegal.[21] N'Tchoréré volunteered in time to be transported to France and to be killed in June 1940.

The French public received the African troops well. Learning from the experiences of the First World War, when centers for African soldiers were established in the south of France in particular, French officials systematically organized welcome committees to help African soldiers adjust to their tours of duty in France.[22] From 1928 on, African and other colonial troops came to be posted regularly in numerous French garrisons scattered throughout the south of France primarily. Visits to tourist attractions were common for African soldiers, and some soldiers were delighted to use free railway passes to travel throughout Provence.[23]

The most important change from First World War practices was the decision of the French army to "amalgamate," or integrate, African regiments with metropolitan French ones.[24] As the *Mamadou* comic strip suggests, roughly half the African regiments were broken up, down to company level, and merged with French ones to form the *Régiments d'Infanterie Coloniale Mixtes Sénégalais* (RICMS). French regimental diaries indicate that this dramatic social experiment came off rather well. From the reports of intelligence officers who censored the mail of young French soldiers to their families, we note complaints about the arrival of Africans in their units, but these are directed more towards the army than the Africans themselves, and are quite free of racism:

> In my company we ate well at first but since the Senegalese arrived they only give us rice and lentils; and for the meat, you need a scythe to cut it. . . .
> Sunday the Blacks gave a concert with native dancing; their instruments were empty jerry cans. Talk about your tam-tams and yet I would have liked you to have seen them dance, it's really worth seeing.[25]

If African soldiers were carefully prepared ideologically for their important task of helping defend France, and genuinely made to feel welcome in carrying out the job, nevertheless one glaring omission in their preparation stands out. While there is much lampooning of the enemy in the pages of *La Gazette du Tirailleur*, nothing at all is said of German strengths. This striking absence is entirely consistent with the smug attitude of the French general staff in 1939–1940, safe behind an illusory Maginot Line of defense, and falsely encouraged by the long waiting of the Phony War into believing that its defensive strategy would succeed in holding back the Germans with a minimum of French losses. In short, the African troops were as ill prepared as their French commanders for the mechanized Panzer divisions about to launch their stunning attacks of May 1940.

Combat

Two colonial divisions, each with large proportions of Africans in their midst, the First and the Sixth, stood on the Aisne and the Argonne in May 1940, unlucky to be hit by the full force of the Panzer invasion.[26] Two other colonial divisions, the Fourth and the Fifth, bore the brunt of the German attacks on the Somme from 23 May on. In the main, Africans fought tenaciously and retreated in orderly fashion before overwhelmingly superior forces. Of course, valor was not sufficient by itself to defeat an opponent displaying technological and especially tactical superiority, and there was nothing objectively that African troops could have done to have altered the course of events in this phase of the Second World War. Nevertheless, it is noteworthy that many Black Africans came to believe that they fought vigorously to defend France while many French units succumbed to demoralization and defeatism.[27]

Veterans' accounts collected by Nancy Lawler in Côte d'Ivoire offer a variety of opinions. Laqui Kondé from the region of Korhogo had completed his three-year term from 1928 to 1931 and was called up from the reserves in 1939. In May of 1940, the eve of German invasion found him serving in France with the Eighth RTS. While he was unsure about the locale of the combat or its duration, his account stressed poor French leadership:

> It was miserable. When we went to sleep and woke up again we thanked God. The war started. A light went on, every one got up. It was very confused. People were running. We were trying to find a place to shoot that would be safe. Bombs were dropping. Every one was for himself. There was no one in charge then. We were at the edge of the Bandol Sea, in France. We were on that course for one year. We were always on the move. Sometimes we were stronger, sometimes they were and we ran . . . .
>
> Yes, we were with the French all the time. We were stronger than the whites. That bullet that hit my tooth would have killed a white. When the shooting came, the whites ran. They knew the area and we did not, so we stayed. Our officers? They were behind us. I didn't know or think about anything except life or death. We were between the two. Yes, the officers said: Stand. We will win. But it was the Germans who were stronger so we forgot what they said.[28]

A common view of many *ivoirien* veterans was that General Gamelin, commander in chief of French forces, was primarily responsible for the defeat. Here is Peleguitamnadio Yeo, a veteran from Korhogo region with his version of how Gamelin betrayed his trust:

> Now, we were stronger—we pushed them back. They retreated and we followed them. Then General Gamelin said we couldn't continue capturing Germans because politics had entered the war.
>
> General Gamelin had made the French soldiers gather all together in one place. He moved us from all different locations around the battlefield. He was a German who was doing his military service in France. We didn't know that before this happened, but afterwards a captain of the *Gendarmerie* discovered his secret and told us. They sent Gamelin to Paris and de Gaulle took his place. Gamelin went to France and de Gaulle took over as our leader and ten days later we were taken.[29]

Edmond N'Guetta, an *ivoirien* ex-POW, offered a more considered account of combat and capture:

> In April 1940, I was in the 24th company, 24th BTS. . . . It was the big German offensive. We were sent to the front. It was awful. So many were killed, wounded. We retreated all the time. We couldn't stop. We couldn't rest. By June we were in the Chartres region. There were refugees along the road everywhere. We marched faster than they did. Every one—a mixture of all kinds of people—was on the road.
>
> In the region one day we were completely encircled by the Germans. We tried to make a stand—there were planes, bombardments all day. Tanks, cannons. Many were wounded. We had no more cars, no more trucks, nothing. We couldn't advance or retreat. We waited all day. Towards the evening of June 6th a great cavalry attack was made against us. Tanks and planes. They opened fire. We couldn't do anything. Our unit was separated. We were all separated. We tried to resist but we were submerged all at once. One could not resist, could not see—they were everywhere.
>
> We could see dead French soldiers everywhere. The Germans had cut the wires. There was no place to hide—just wheat fields. We tried. The Germans said: Come. We said: No. The general headquarters was supposed to be in the town. In reality the generals had all left. We were jammed together. Machine guns firing on us. So many killed as they tried to escape—to hide themselves. I hid in a kiosk all night waiting for the French troops to come. I stayed there all day and all night. The next day we wanted to leave—there were three French soldiers with me. We had to leave the wounded. We had to tell them we couldn't carry them.
>
> The town was deserted. We went to a deserted farm, took a bit of breadfish. Then we hid in the cellar. Two days we hid in that cellar. One of the servants returned. She told us Paris had fallen. There was no more hope. We decided to surrender. It was the only thing to do. We still had our rifles. We hid them and our ammunition in the hayloft. We told no one, not even the servant. We left on the main road. Every one was on it. Some Germans passed us and asked if we wanted to surrender. The soldiers said to keep going towards Chartres. We left for Chartres with a German guard. A truck came by with other prisoners—other *Tirailleurs*.
>
> Things were all right at Chartres. After a while the Germans decided we weren't so bad because we never tried to escape. With black skin it would have been difficult to do.[30]

The subjective debate over whether African soldiers outperformed their French counterparts in 1940 is part of a mythological colonial discourse. Whatever utility it may have had for French imperialism, it obscures several important dimensions of the Battle of France. Recent revisionist work by Roger Bruge has helped correct the notion that metropolitan French units performed badly in retreat.[31] If panic and disorder may have characterized some aspects of the French withdrawal towards the south, Bruge offers compelling evidence of bitter French resistance in eastern France towards the Swiss border. The same historian also adds evidence which may help explain the tenacity of African resistance. Oberst Nehring, Guderian's Chief of Staff, signed the following note regarding the treatment to be reserved for certain categories of prisoners of war. One paragraph concerns Africans as "colonial soldiers":

> It has been established that French colonial soldiers have mutilated in bestial fashion our German wounded. Towards these native soldiers, all kindness would be an error. It is rigorously forbidden to send these prisoners towards the rear without a guard. They are to be treated with the greatest rigor.[32]

Such harsh orders were bound to be "rigorously" interpreted down the line. A French officer, describing events on the Somme on 5 June 1940, gave graphic testimony of one ugly incident:

> At 18 hours 15 we are prisoners. The Germans, young ones wearing the black uniforms, are very excited. They shoot in the air, on the ground, steady bursts of machine guns. The sight of *Tirailleurs Sénégalais* exasperates them still more. I sense a massacre. Second Lieutenant Sauze beside me, receives a machine gun burst in the stomach and drops down, asking to be finished off. A few *Tirailleurs* are killed as well. We are searched. Pistols, wallets, pens, the lot are taken. It's over. We march under the ironic eye, and through the insults of the Boches. Our captivity begins.[33]

Cruel treatment of African troops had a long history, and French officers must bear at least a small share of the responsibility. Guderian's chief of staff, Nehring, had accused French colonial soldiers of having mutilated German wounded, though no specific incident was cited. Perhaps this was an invocation of German fears that had their origin in the First World War, when some French officers unscrupulously manipulated the image of ferocious African soldiers, all brandishing the *coupe-coupe*, the large combat knife.[34] The most dramatic German response to African troops occurred in the years following the armistice of 1918, when Africans and other colonials formed a significant component of Mangin's Tenth Army of Occupation in the Rhineland. The German press campaign, known as *die schwarze schande*, or "black shame," claimed that the French had deliberately selected colonial troops to occupy the Rhineland so as to humiliate further a prostrate Germany, and that these troops were guilty of numerous sexual assaults against German women.[35] The echo these charges received in the British and American press was so loud that France was forced in the end to transfer its colonial soldiers to the Levant, despite repeated and documented refutations of the sensational charges against them. These events, and the use generally of colonial troops in European wars, added more fuel to the race prejudice so explicit in the mind of Adolf Hitler, and arguably of many other German figures.[36]

A *Tirailleur* helpless during the Wehrmacht offensive, May–June 1940

It is difficult to untangle the web of cause and effect. African soldiers came to resist so fanatically because they had been told that the Germans would be extremely harsh with them if they should be taken prisoner. German soldiers went to extreme lengths against Africans partly, perhaps, for ideological reasons, but also because the tenacious resistance of Africans cost the Germans heavy casualties. A German press account of the Pomeranian infantry's clash with Africans in 1940 makes both these points probable:

> The French fought tenaciously; the blacks especially used every resource to the bitter end, defended every house. To break them, we had to use flame throwers, and, to overcome the last Senegalese, we had to kill them one by one.[37]

After studying casualty lists, Roger Bruge drew a terse conclusion: "The facts, however, are stark; mixed regiments in the line in May and June 1940 systematically took higher casualties than did all white regiments."[38]

An African view has been provided by the *ivoirien* veteran and former POW Lielourou Tuo. According to him, the brutality of his captors sprang from a desire to settle old scores from the First World War and was directed equally at European and African prisoners:

> We left on foot. It took us three days to get there. The Germans aren't men. They gave us nothing to eat on the road. Many were dying and wounded. The Germans were very bad. You needed a machete to cut their bread. It was very hard and very dry.

> There were many camps. We were still mixed with the whites. We slept together. They killed us in the camp. We were all mistreated. All of us—black and white. We thought we would all die. If it wasn't for the Americans we would have all died. In 1914–1918, Germans were captured by the French. Now it was the same thing. It was a debt that had to be repaid.
>
> We spent five years there. There were a lot of deaths. The Germans killed many prisoners. Because in 14–18, the French killed many Germans. They lined them up and shot them. It was the Germans who told us about that. They said they were going to settle the debt. So in camp, when morning came, they would pick a portion of the men—we were still in bed—and put them in a trench and shoot them. Others had their ears chopped off and their eyes taken out. This happened at the beginning. Of course they killed the French too. Why just us?
>
> Later, an old German who had fought in 14–18 stopped it. He said that no more men would be killed because they weren't the same men of 14–18. They came here region by region—not the same men. So they stopped killing us. It was terrible. They had forced the Africans, those with [tribal] scars to scrub their scars with soap and not to stop until they came off. We suffered so much. It was the old one who stopped it. They had forced the Africans to scrub their scars until they came out. So many suffered. It was the old German who intervened. Now, they made us work, chopping down fir trees. That was what we did until the Americans circled the camp and freed us.[39]

Prison and Transit Camps

As some of the above accounts suggest, many Africans did survive to become prisoners of war, though it is not possible to know how many were summarily executed by their captors or how many succumbed to the hardships of captivity.[40] It is possible that Africans were treated more harshly than French prisoners; they certainly were treated differently. The following passage, from David Irving's biography of Rommel, shows that at the highest levels of the Nazi state, Africans' rights and very lives as POWs were taken very lightly:

> Rommel spent part of August re-enacting for the movie cameras the Spook Division's crossing of the Somme. He had a great time playing movie director, and he schooled his troops in acting techniques. A battalion of French black troops was hauled out of the prison camps to stage the surrender of a village. Again, this time for the cameras, Rommel's tanks charged, guns blazing. He told the blacks to come out toward the tanks with their hands up and looking scared; but the men overacted, rolled the whites of their eyes and screamed with terror. Rommel cut the cameras, and patiently explained through interpreters that actors had to show their emotions more subtly than that. The battle scenes were finally filmed on such an epic and reckless scale that several more lives were lost, though through no fault of Rommel's. "No expense has been spared to show it as it really was," he wrote on the last day of shooting.[41]

While individual circumstances may have varied considerably, after a year or so in Germany most African soldiers were transferred to so-called *Front-Stalags*, work camps located just inside occupied France. Here they were forced to work at tasks that contributed to the German war effort, particularly mining and arms manufacturing in the industrial northeast. In return they were paid minimal wages. Some men were even allowed contact with French civilian families, and occasionally received news from home in this and other ways.[42] Others even went so far as to serve clandestinely in the Resistance when the opportunity presented itself. A few soldiers apparently were pampered by the Germans in the hopes they would prove useful collaborators. But for most, the incarceration, the forced labor, and the loneliness were sufficient to mark many men for life.

Prison life took its toll on African soldiers unfamiliar with the cold winters of northeastern France. Inadequate diet, clothing, and housing further weakened the men's resistance, and many succumbed to the great scourge of black African soldiers in Europe, pulmonary disease. To guard against the spread of infection, the Germans did allow French medical teams to inspect the camps and to transfer sick prisoners to Paris or even to Vichy jurisdiction in the south. In this fashion an untold number of Africans were able to end their incarceration before Liberation in 1944.[43] The most famous of African POWs, Léopold Senghor, seems to have escaped the Germans in precisely this way. By 1942 he had been released to Paris on medical grounds, and had resumed his profession of schoolteacher in the Paris suburb of Joinville.[44]

One of our rare sources for the experiences of Africans as POWs is the remarkable book by Hélène de Gobineau entitled *Noblesse d'Afrique*.[45] An anthropologist with some earlier familiarity with FWA, Madame de Gobineau was a member of a French commission of inspection for the work camps in the northeast. During her many visits she established a rapport with African POWs, and many of them contacted her when in Paris, after they had succeeded in one fashion or another in escaping prison. She describes the various subterfuges used by African soldiers, often with the support of French physicians, to deceive the Germans and save as many soldiers as possible. Devices ranged from feigned madness to fits of coughing, fasting, and the like, but always at considerable risk.

Despite the odd tale of success, it is the severity of life in and out of the camps that dominates Madame de Gobineau's account. German reprisal when subterfuge was discovered was predictably harsh, as in the execution of the African POW caught training French peasants in the handling of firearms. Many of those transferred out of the camp were fatally ill with tuberculosis or pneumonia and died soon afterwards in hospitals in Paris or in the Vichy sector. Worse, the French state did not take immediate responsibility for the care and ultimate repatriation of these soldiers. An estimated 5,000 black Africans, mainly ex-soldiers, drifted from one odd job to another in Paris, and generally lived by their wits in the tense atmosphere of an occupied and demoralized city.[46]

Then came Liberation. While their French brothers in arms returned home to tumultuous welcomes, African soldiers were less fortunate. Some, befriended by French families, shared vicariously in the joys of Liberation. Others formed parts of newly formed French military units and resumed the fight against the Germans. Perhaps because they were reluctant to waste the opportunity, French officials even went so far as to put some of the African ex-POWs to work in military labor units. Within a short period after their liberation, however, all African servicemen were rounded up and grouped in six centers in central and southern France to await eventual embarkation for home.[47]

Unlike French soldiers, who were quickly issued back pay and were discharged, the Africans languished in camps because shipping space proved difficult to obtain. Ex-prisoners are known to present special psychological problems for reintegration into military or civilian society, and tend to manifest a hostility even towards their liberators.[48] But these African soldiers, numbering in the thousands, had quite rational grounds for resenting the shabby and discriminatory treatment they received in the fall of 1944.

For the French authorities, the African ex-POWs represented part of a growing logistical problem. The ex-POWs numbered from perhaps 5,000 to 10,000 men. Soon they were joined by another 20,000 black African soldiers who had been very abruptly withdrawn from de Lattre's First French Army in September and October 1944, as part of the so-called "whitening" (the French term was *blanchissement*) of the Free French forces.[49]

The order to replace black African with French troops to "whiten" the French forces came from de Gaulle himself.[50] It was a decision reached on political grounds. Chadians and French West Africans had formed the rank and file for Leclerc's raids against the Italians in North Africa in 1940 and 1941. They were part of the Free French Army which was included in the Allied invasion of Italy in 1943.[51] To the Sixth RTS, commanded by Colonel Raoul Salan, of later Algerian notoriety, went the privilege of liberating Elba in the late spring of 1944. In the landings in Provence in August 1944, black Africans had seen considerable combat, especially in the taking of Toulon. Just as these 20,000 battled-hardened men were preparing to share in de Lattre's anticipated triumphs in France, de Gaulle ordered their replacement with young Frenchmen of the class of 1943, and by elements of various

Tirailleurs Sénégalais reach Tripoli with Leclerc, January 1943.

Partisan groups. For de Gaulle the reasons were clear. It was essential that young Frenchmen be given a taste of victory, a share in the Allied success in ridding France of its shame and humiliation. Equally important, the Partisans were often Communist-led, and de Gaulle hoped that the military discipline of regular army service would serve the dual purpose of controlling these groups and separating them from their political leadership.[52]

By the late summer and fall of 1944, de Gaulle's logistical problems were enormous. He was entirely dependent upon American largess to clothe, feed, and equip his army since the Germans in retreat were completing the destruction of the French economy. President Roosevelt, not an admirer of de Gaulle, and having to reconcile seemingly insatiable demands for war matériel from an American army fighting in two world theaters, as well as from the Soviets in the East, had set the upper limits of the Free French forces at 250,000 men.[53] De Gaulle could only integrate new French forces into his army as replacements for existing units. Invoking the old argument that African troops performed poorly in the winter campaigns served to raise an additional but not a primary consideration. Thus, literally without warning, serving black African soldiers were relieved of their front-line positions, of their arms, and even of the uniforms they were wearing, and were sent back to the south to spend the winter waiting for ships to take them back to Africa.[54]

During the First World War the *hivernage* system, which relieved Africans of trench warfare in favor of the relative comfort of winter barracks in Provence, had seemed a blessing.[55] In the fall of 1944 it was quite otherwise. These battle-hardened men had already been through years of hardship and combat. Now, when the spoils of victory were about to be theirs, they were suddenly being told that their services were no longer required. One rather special African soldier, Philippe Yacé, remembers refusing to accept this decision:

> I was with the 1st French Army. When we arrived, we headed for Besançon. In the winter, a general order was received saying that all men of color should be evacuated to the south where it was relatively warm. If the war didn't end during the winter, they could be brought back in the spring.
>
> I was among those who refused to leave the front, so I now signed up as a volunteer. Remember, we were among the Africans who were citizens and that order was for the *tirailleurs*. I was not a *tirailleur* but in the regular French Army. But when the officers saw the words—men of color—they tried to send us back too. When we said we wouldn't go, he asked us if we agreed to volunteer and we signed up. You understand, it was a question of our status, not love of war. We were French, not subjects. There were true volunteers only among the citizens.[56]

Tirailleurs experienced the winter of their greatest discontent in 1944–1945. They suffered, as did the civilian population of southern France, a shortage of food and clothing, but in addition also often lacked essential shelter, and were still subject to harsh military discipline. When some African soldiers, for example, sought to replace what had been taken from them by purchasing blankets from black American GI's, they were reprimanded and the clothing confiscated. Some African soldiers took to breaking into abandoned villas of the wealthy and using expensive furniture for firewood.[57]

Thiaroye and Other Soldiers' Protests

It is hardly surprising that the gathering together of so many justifiably discontented men should have led to serious incidents. Given the chaos of wartime France, sources to document these disturbances are scanty. Local newspapers had either suspended publication that winter or were subject to severe censorship. Prefectures were just beginning to resume administrative control over the civilian population. Military records for this period are still not open to access, but we are fortunate that two student administrators of the Ecole Nationale d'Administration conducted research into the 1944–1945 disturbances and left unpublished reports at ENFOM as part of their graduating assignments in 1946.[58]

All told, there were some fifteen recorded incidents, mostly occurring in the south of France, but one in Versailles, one in Monshire Camp at Huyton, near Liverpool, and one at the Thiaroye barracks on the outskirts of Dakar, Senegal. The soldiers involved in each instance were either ex-POWs or men recently discharged from de Lattre's army. Noncommissioned officers were critical to the leadership of each disturbance. Underlying issues of the uprisings ranged from aspects of daily life to ideological issues. Most common were complaints about poor food, clothing, and housing; failure to deliver back pay; bans on the sale of alcohol; and disputes over access to women. But the spark that set off the trouble was most often either a physical attack on African soldiers by French military personnel or else a racial slight. In most cases the disturbances were quickly ended without significant injury and no long-term sanctions. But two of the uprisings resulted in serious injury and loss of life. At Morlaix, seven African soldiers were wounded by the fire of French military forces, while at Thiaroye, by far the most serious protest, some thirty-five African soldiers lost their lives, and an equal number were seriously wounded.

The troubles which occurred at Monshire Camp near Liverpool were typical, even if the locale was not. French authorities were so hard-pressed for resources to accommodate colonial soldiers awaiting repatriation that in desperation they turned to their British allies for help. As an experiment, the British agreed to house temporarily some four hundred soldiers, with their officers, until shipping space could be found. The men arrived in late October 1944 and were already in an anxious and undisciplined state, according to the military's account of the affair. The soldiers wondered why ships could not be found to take them to West Africa in the same manner that vessels had been found to take them to England. At Monshire, the French African NCOs were furious at the refusal of the British to allow them to eat at the British Sergeants' Mess. Soon they took matters into their own hands and physically forced their way in. The French officers begged the British colonel commanding the camp to allow the integration of the mess, and he grudgingly agreed. Barred from taking furloughs in the town, the African soldiers went anyway. British officers were further displeased by the soldiers' neglect of their uniforms and their refusal to carry out daily drill. Much to their surprise, the African soldiers abruptly changed their behavior, cleaned themselves up, and took part solemnly in the Armistice Day marches of 11 November 1944. Immediately afterwards, however, they returned to their disobedient ways. Before the end of the month the British had arranged shipping to send the Africans packing, first to Casablanca and finally home to West Africa.[59]

Thiaroye and Other Soldiers' Protests

On 1 December 1944, at the barracks of Thiaroye, near Dakar, there occurred the most serious of the clashes between alienated African soldiers and the French state.[60] The uprising involved some 1,280 African ex-POWs in the first contingent to be repatriated from Europe in 1944. The affair drew the official label of mutiny because the men were still partially armed, uniformed, and under military discipline. They not only refused to obey their officers but for a brief time actually held hostage the CIC of French Forces in FWA.

The immediate cause of the soldiers' protest was the failure of the French authorities to provide them with back pay and demobilization premiums. The soldiers' claims were well founded because on 31 October 1944 the Minister of Colonies had written to Dakar authorizing a series of benefits for ex-POWs. While the provisions fell short of the treatment usually accorded ex-POWs, they were in any case simply not implemented.[61] The soldiers also were agitated by the refusal of the French authorities to allow them to keep the metropolitan francs they held on their persons.[62] These sometimes considerable sums no doubt added to the tensions between the men and the paternalist officials of FWA. Many officials saw no urgency to compensate soldiers already possessing substantial sums, while others assumed that the money had been gained unlawfully, despite evidence indicating that the men had received wages from their German captors. To the ex-POWs the authoritarian manner in which they were being treated was a bitter reminder that they were returning home to an unchanged colonial system, unappreciative of the great sacrifices they and their fallen comrades had made.

The uprising and its aftermath were shocking. The official report written soon after the bloody event states that thirty-five Africans were killed, an equal number seriously wounded, and hundreds more or less seriously injured.[63] Some thirty-four ex-POWs were arrested and tried on charges falling just short of mutiny. All were convicted and sentenced to terms ranging from one to ten years in prison. Before a general amnesty was issued on behalf of the Thiaroye victims in June of 1947, some five prisoners had died in jail. On the colonial side, no lives were lost; one African policeman was wounded and three French officers suffered lacerations.

There is no denying that Thiaroye sent shock waves throughout FWA. One *ivoirien* veteran, Dossigutta Silue, who was an orderly in the Dakar Hospital at the time of the disaster, recalled the event vividly forty years later:

> They fired on the mutineers. There were many dead that day. We were angry. We thought we might not get back to our villages. We might be killed too. The army fired upon the army—those were the orders. Our regiment wasn't far from them. They brought the bodies to us.[64]

Officials paraded the Thiaroye prisoners through the streets of Dakar after the uprising had been crushed, and the incident was on everyone's mind for months afterward. Friends wrote about it to African soldiers still in France and awaiting demobilization.[65] Others wrote to French families who had befriended African soldiers in hopes of enlisting French public support for a pardon for the men being held in prison.[66] For the vast majority of Africans the reaction was one of shock and indignation over the brutal repression of long-suffering soldiers whose only crime was to claim money that was rightfully theirs.

Thiaroye also served to bring out contradictions inherent in the African colonial situation. Thus, Seydou Nourou Tall and other conservative members of the Muslim hierarchy in Senegal supported the French repression on the grounds that men who had disobeyed and struck their officers could not escape punishment without producing severe troubles in the countryside.[67] Committed to collaboration with colonial authorities, the Muslim leadership was not prepared to make a political break with the French over Thiaroye, important an event though it might be. Seydou Nourou Tall, indeed, had been called in once before on the question of helping reintegrate African servicemen quietly into colonial society. In 1941, under the aegis of Vichy, he had produced a pamphlet in Arabic urging soldiers to respect their parents, their traditional chiefs, and French authority.[68] In the wake of Thiaroye, his immediate reaction was to criticize the French not for having acted repressively but for having failed to call in traditional Islamic leaders earlier to help calm down the men and prevent the uprising in the first place.

The conservative Muslim leadership would change its attitude toward veterans only gradually. The real political initiative was about to pass to the new politicians who would achieve prominence in the postwar period. Two examples of this new leadership were the Senegalese politicians Lamine Guèye and Léopold Sédar Senghor, both of whom became directly involved in the aftermath of Thiaroye.[69]

The victims of Thiaroye, determined to assert their innocence and their rights, turned for help to the incipient veterans' organizations that were beginning to spring up. Most prominent among the leaders of these groups was Papa Seck Douta, later to be head of the Federation of Veterans' Associations for all of French West Africa, who was in early 1945 president of the Senegalese veterans.[70] Indignant and shocked, Seck Douta saw in Thiaroye the undermining of all his efforts to promote the cause of veterans within the context of French assimilationist politics. A citizen himself, having been born in Dakar, Seck Douta was associated in French eyes with the new drive for political equality among African veterans who saw themselves as having been largely responsible for the Gaullist victory. Acting on instructions from the prisoners awaiting trial, Seck Douta contacted *Maître* Lamine Guèye, the best known African lawyer of the day, to lead the defense of the Thiaroye men. Guèye, about to launch his victorious campaign for mayor of Dakar, and with political ambitions to succeed earlier figures like Blaise Diagne and Galandou Diouf as the political spokesman of the Senegalese citizens, was pleased at the political opportunity inherent in an eloquent defense of the victims of Thiaroye.[71]

French officials put forward a variety of theories to explain the uprisings at Thiaroye and in southern France, in what had hitherto been an obedient and disciplined black African army. Some opinions tended to shift responsibility to the shoulders of others. Thus officials in West Africa attributed the poor discipline to mistakes made in France, especially the predominance of French reserve officers inexperienced in the command of African units.[72] When the harshness of the wartime experience was mentioned, it was argued without much supporting evidence that German propaganda had been at work among the prisoners during their captivity, although precisely what form this had taken was never made clear. Only a few officials were astute enough to realize that a fundamental change in attitude towards France and its colonial ideology had occurred as a result of their experiences:

> Separated for many years from their homes, placed in exceptionally difficult circumstances, these natives have acquired habits and ways of think-

ing, indeed an entire mentality that tends to make them a very special element. At the moment of advance of the Allied armies in France, the native prisoners found themselves suddenly liberated. A good number of them, in keeping with the preparatory work of the Secret Army, were then incorporated into formations of the French Forces of the Interior or in the Partisans' units and they participated in military actions against the enemy. Others were taken in by honorable French families. Still others, and this is to be deplored, lived by more or less illegal means and were sheltered by "godmothers" of doubtful morality. A veritable campaign of reorganization and recuperation had to be undertaken in order to put an end to this confused situation which did not always produce a favorable effect on the morale of these natives.[73]

What the French views, however perceptive, failed to recognize was the degree to which the African ex-POWs had acquired a heightened consciousness of themselves as Africans united by their shared experience in suffering. The notion that they had served France above and beyond the call of duty was deeply embedded in these men. Many shared the recurring African soldiers' sentiment that they had been better defenders of France's sovereignty during the debacle of 1940 than had metropolitan Frenchmen. While they had been enthusiastically received by the French public during mobilization in 1939 and 1940, circumstances were very different indeed in the aftermath of Liberation. Added to the understandable psychological pressure experienced by ex-POWs, who are said to feel that outsiders can never appreciate the suffering they endured, was the concrete evidence that Africans had been treated shabbily in France and in West Africa. It is significant that most of the uprisings in 1944-1945 grew out of this indignation at unequal and inferior treatment in comparison to fellow soldiers who were Europeans.

Whatever their interpretation of events, French officials shared a deep concern about the potentially damaging consequences of the malaise among veterans. Governor-General Cournarie instructed all local governors in FWA to subject returning servicemen and especially ex-POWs to particularly close surveillance, and to include in their political reports an indication of the African popular reaction to the events of Thiaroye.[74] Cournarie did not advocate concessions such as increased pensions for returning veterans. Only two years later, under the very different peacetime political circumstances of the Fourth Republic, did France begin to approach the veterans' issue in a new light, and then mainly as a result of political pressure coming from African political parties.[75] But a sense that a new political climate was approaching emerges from Cournarie's sharp warning that the use of force "could not be permitted to be repeated, under any pretext whatsoever."[76] Thus, the soldiers' uprisings, especially the one at Thiaroye, because its shock waves traveled throughout West Africa and knowledge of it could not be kept from the African people, served effectively to delegitimize naked force as a political instrument.

Thiaroye was not the last of the disturbances touching African ex-POWs. Two incidents in August 1945 on the Côte d'Azur, one at Saint-Raphaël and the other at Antibes, follow the familiar pattern.[77] In the Saint-Raphaël affair two civilians were killed and several wounded before the roughly three hundred African soldiers were subdued. The revolts were said to have been caused by delays in repatriation, by failure to include Africans in the first anniversary celebrations of the Allied landings in Provence (on the contrary, their leaves were canceled and the number of military patrols stepped up), and even by the rumor that instead of being sent home, they would be dispatched as part of a French force of 10,000 men bound to reestablish French control over Indochina. At Antibes, the minor revolt was sparked by the murder of an African sergeant who had discovered two metropolitan French soldiers attempting to burgle the Africans' barracks. Whatever the immediate cause, however, wretched men had been waiting over a year to return home, they had witnessed the ending of the war in Europe (the ending of the war in the Pacific might even have been known to them), and still they had not been sent home.

On one level, the impact of African soldiers on France was of considerable importance. De Gaulle's grudging promise at Brazzaville in 1944 to work out a new deal for Africans and for colonial peoples generally was partly a response to the military efforts of France's colonial subjects, partly an admission of the failures of earlier colonial policies, and partly a recognition of the new international climate ushered in by the Atlantic Charter. But for de Gaulle and the Gaullists, the Africans' military assistance was an embarrassment as well. True, the colonial subjects were also children of France, but having them active in the liberation of France was perhaps going too far. By and large, de Gaulle adroitly concealed the importance of the African military contribution from the Allies and from the French people, just as he was to exaggerate the importance of a Resistance inside France. His "whitening" of the de Lattre army in Provence, his care to make sure that the French forces in England preparing for their part in D-Day operations were all white, are consistent with this view. Still, when pressure mounted from African politicians on behalf of African veterans after the war, Gaullists supported their claims, and voted with the parties of the Left in favor of increased benefits for African veterans.

The Gaullists had reasons to be grateful. It is difficult to imagine what course events in France might have taken in 1944 and 1945 without the modest yet concrete military achievements of the Free French forces of Leclerc, de Lattre, and Koenig. Without the rank-and-file black African soldier, their victories would have been impossible.

Yet it was clearly in French-speaking West Africa that the tragic years of war burned most deeply. For all returning soldiers the war, and especially the hardships of 1944 and 1945, helped crystallize their group consciousness. The soldiers' protests provided a tremendous impetus for the growth of a whole series of veterans' associations which emerged after the war in each colony, centrally coordinated through Dakar, and with significant links to veterans' groups in France. The veterans' slogan, "equal sacrifices = equal rights," not only referred to pensions and benefits but could also be applied to the struggle for justice in a French community still wedded to the ideology of assimilation. The struggle of veterans is the focus of Chapters 8 and 9. Before that, Chapter 7 continues the story of the *Tirailleurs Sénégalais* in the changed circumstances of postwar French colonialism.

7

The New Army, 1945–1960

> Either we build a solid French Army or we make ready armies of liberation.
>
> Major Chailley, "L'Africanisation des cadres de l'armée"

As elsewhere in the world, the Second World War and its aftermath transformed all aspects of military as well as civilian life in FWA. Major reforms included the emergence of African electoral politics; the abolition of the *indigénât*, or forced labor; the multiplication of educational opportunities; and the growth of the economy and the wage sector through outside investment and increased tax levels.[1]

In the external arena, the end of British rule in India was being openly discussed, while in French Indochina nationalists under Ho Chi Minh were denying a resumption of French colonialism by force of arms. Closer to FWA, the immediate postwar years were marked by violent nationalist stirrings and brutal colonial counteroffensives in Algeria and in Madagascar.[2] To these external events could be added the tensions felt by African soldiers and veterans, for whom the bloodshed at the military camp of Thiaroye and its aftermath remained a widespread and ugly memory.[3]

In light of these new postwar realities, the French colonial military was obliged to change both its rhetoric and its policies. Despite some resistance from mainly older officers who preferred a paternalist colonial army where natives knew their place, official military rhetoric accepted the assimilationist view that the ultimate goal was not the disbanding of the colonial armies but rather their full integration within a larger French army.

This chapter examines the resulting transformation of the *Tirailleurs Sénégalais* from a large, unskilled troop of conscripts into a leaner, largely professional army in which volunteers came to outnumber conscripts, and where an increasing number of Africans would serve as junior and sometimes even as middle ranking officers.

French Policy

From a later vantage point, it is easy to assume that French planners immediately perceived the necessity of transforming the *Tirailleurs Sénégalais* after 1945. In fact, the change was neither smooth nor obvious then. The main purpose of the

Tirailleurs remained the same: to occupy FWA and to be ready to go abroad to defend the French empire, or union, as it would now be called. One important change in grand strategy, however—France's membership in NATO—did reduce the likelihood of African soldiers being required to help in the defense of Europe. This, in turn, reduced the need for a large force of *Tirailleurs Sénégalais* and enabled French military recruiters to become more selective. As Figure 1.1 in Chapter 1 indicates, the standing *Tirailleurs Sénégalais* strength fell from around 48,000 men before the war to roughly 38,000 after. Annual levies similarly fell from prewar averages of 12,000 to half and even a third that number in the 1950s. While it is true that the increasingly specialized and technical nature of modern warfare now required fewer, better-educated soldiers, in fact the high costs of training and the shortage of officers, especially noncommissioned officers, in the French army of the Fourth Republic were also responsible for smaller intakes.[4]

New though it may have been in some respects, the *Tirailleurs Sénégalais* army was not to be spared "dirty work." Ugly incidents involving African troops as strikebreakers in Nice in 1947-1948 aroused in the French Left the old fear that colonials would be used as Janissary style mercenaries. Another idea which did not die was the exploitation of the Colonial army's potential as a labor force. Under Vichy rule in late 1940, some 2,500 *Tirailleurs* had been designated to build a new highway, *Route intérnationale 2*, linking Senegal to Guinea and Côte d'Ivoire via Tambacounda, thus eliminating the longer train journey to Bamako followed by bus or truck to Kankan. Shortages of gasoline caused the government in Dakar to stop construction, but in May 1945 the commander in chief ordered the *Tirailleurs Sénégalais* out to resume work until the road was completed in 1946.[5] One enthusiastic French official, unaware perhaps of the impending abolition of forced labor, proposed a new name for the annual draft's second portion, the "pioneers' battalions," which would be commanded by Engineering Corps officers and maintained by the civilian budget of FWA.[6] Fortunately for the potential military laborers, nothing came of this idea once the National Assembly passed the "Houphouet-Boigny" law abolishing forced labor in the French colonies.[7]

One obstacle to change in the *Tirailleurs Sénégalais* was the attitude of the French military. As the Fourth Republic was experiencing a difficult birth, the French Army itself was the subject of much debate. Charles de Gaulle had gone public with his views on how major reforms should be undertaken to modernize the French army.[8] But when it came to the question of colonial troops, de Gaulle adopted a conservative approach. While acknowledging that they had served well in the past, he argued that political crises in Asia and in the Muslim world made it unsafe for France to rely exclusively on them. French troops, and he used the explicit phrase "troops entirely French," should also acquire experience in the overseas colonies. His tone, in the new climate of African political expression, was offensive to many Africans, and brought a scathing attack in the radical West African press.[9]

Others in the French army had difficulty accepting the new era of Franco-African relations as well. An incident in the early fall of 1945 in Zinder, Niger, is illustrative. With the Algerian uprising at Sétif that summer very much on the minds of Europeans in West Africa, two European noncoms started a rumor about an impending African uprising inspired by the Algerians.[10] The rebels would attack the White Fathers, the hotel in Zinder, and all Europeans they could find. Panic so gripped the Europeans that they took the unusual step of asking the fifteen-year-old son of the colonel who served as military commandant of Zinder to inform his father. The colonel immediately called an alert and doubled the guard without informing the *commandant de cercle*, who represented the civilian administration. Reviewing this incident, the high commissioner of FWA (as the governor-general was called after 1945) regretted the failure to respect or consult civilian authority but found a much more disturbing element. To him it was proof that European noncoms had no understanding of, let alone sympathy for, Africans with educational or political aspirations. He called on the army to see immediately to the education of its younger cadres.[11]

This civilian reprimand brought a swift reaction from an oversensitive West African military command. General Magnan, the commander in chief, fired off an indignant letter to the high commissioner for FWA in which he claimed the army had pioneered in the evolution of African society:

> That the Army could underestimate the progress made by Africans is inconceivable, when, in fact, contrary to certain allegations, it is frequently the very originator of this progress. . . .
> Finally it cannot be denied that it is especially in contact with the western world, on the very soil of Europe that the aspirations of our Africans first developed, often at a very rapid pace. We, the cadres of the Colonial Army, were, however, the first to ascertain this evolution, to monitor it, and to channel it within the spirit of the French Community. It has been the object of our constant attention.[12]

Hyperbole aside, General Magnan's claim that the Colonial Army had accepted the new spirit of the French *communauté* did indeed reflect the modernized rhetoric of colonial language after 1945. General Seller, the second ranking officer at staff headquarters of the French army, had noted in 1946, when the draft constitution of the First Constituent Assembly had called for citizenship for overseas subjects, that the transformation of the *Tirailleurs Sénégalais* would have to begin with a new recruitment law to replace the one of 31 March 1928, which still prevailed on matters of colonial conscription. He did, however, see two exceptions which he felt would have to be made in applying the metropolitan system overseas. First, French nationals served only for eighteen months, too short a time to teach colonials the basics of military service. Second, to call up the entire annual contingent, as was the principle in France, would be impractical from a budgetary point of view, and would cause tremors in the overseas populations.[13] As events transpired, when the first constitution failed to pass, immediate pressure for drastic military reform was removed.

General Magnan's replacement as commander in chief for French forces in West Africa, General de Larminat, was prepared to initiate what would have been a truly radical modification to the *Tirailleurs Sénégalais*. Arguing that it was now time to free Africans from their "military duty" entirely, he called for a volunteer army, with autochthonous cadres quite independent of European ones.[14] His views were echoed by the civilian administration of FWA, which had long since come to regard universal conscription as a brake on colonial development.[15]

Despite this strong endorsement for reform of the *Tirailleurs Sénégalais*, no formal change in French policy occurred until 1955, when a reform in that year took a step toward a volunteer African army by allowing all men in any district to be placed in the second portion reserve whenever volunteers in that district met the quota.[16]

Recruitment

Annual military recruitment underwent a series of major changes immediately after the war. Whereas annual prewar levies had averaged approximately 12,000 men, the French army's decision in 1946 to reduce the annual contingent in FWA for that year first to 7,500 and finally to 4,320 reflected what was to become a postwar trend towards a leaner *Tirailleurs Sénégalais* contingent.[17] From then until independence, annual intakes averaged 4,000 *Tirailleurs* to supply a standing army of roughly 34,000 men, of whom 15,000 remained stationed in FWA and 19,000 were sent overseas.[18]

While the numbers of soldiers recruited declined significantly, their provenance remained remarkably constant. A brief examination of the recruitment results for 1948, the only year for which complete statistics have been found, permits a useful comparison between prewar and postwar drafts (Table 7.1).[19] It is evident that even though the two recruitments were twenty years apart, several constants remained. Côte d'Ivoire continued to be the only colony whose quota conformed precisely to census figures. Senegal moved from being slightly underrepresented in the *Tirailleurs Sénégalais* quota to conformity, while Dahomey exhibited the converse trend, from slight overrepresentation to conformity. Niger and Mauritania, always marginal to the conscription process before the war, continued to be dramatically underrepresented, although the 1948 report on recruitment commented that for only the second time since the First World War military recruitment had taken place in the *cercles* of eastern Niger.[20] Guinea and Upper Volta remained consistently overrepresented in each of the two drafts, while Soudan went from slight overrepresentation to slight underrepresentation by 1948.

Yet global recruitment figures could disguise what was a dramatic change, the percentage of volunteers in each annual cohort. Before the war volunteers could not exceed 25 percent of the annual contingent by law; after 1946, this limitation on the number of volunteers was removed.[21] The consequences were felt quickly. In the 1948 draft, 3,008 of the 8,000 recruits, or 37.6 percent, were volunteers, but this percentage was by no means equally distributed. In Guinea a remarkable 1,102 men of 1,310, or 84 percent, volunteered, whereas in Senegal, Mauritania, Soudan, and Dahomey the rates were all below 20 percent.[22]

This surge of volunteers could not be attributed to a newly discovered enthusiasm for a military career, even if service in the *Tirailleurs Sénégalais* had traditionally been popular in certain regions. Indeed, political developments elsewhere in the French empire conspired to bring about these changes. In France, the so-called Indochina amendment of 11 November 1951 modified a 1928 Law regarding overseas duty so that French conscripts could no longer be dispatched to territories which were "the theater of active operations" without their consent. Upon the

TABLE 7.1 A comparison of recruitment in FWA between 1928 and 1948

1928

Colony	Recruits	% of FWA Total	Census of 1931	% of FWA Total	Differential*
Soudan	2,781	22	2,855,658	20	+10
Guinea	2,166	17	2,234,692	15	+13
Côte d'Ivoire	1,653	13	1,866,316	13	0
Upper Volta	3,337	26	3,000,243	21	+24
Dahomey	1,084	9	1,111,906	8	+9
Niger	307	2	1,542,714	11	−82
Senegal	1,306	10	1,584,273	11	−9
Mauritania	92	1	323,498	2	−50
FWA	12,726	100	14,519,300	101	

1948

Colony	Recruits	% of FWA Total	Census of 1951	% of FWA Total	Differential*
Soudan	1,548	19	3,439,200	20	−5
Guinea	1,313	16	2,250,200	13	+23
Côte d'Ivoire	1,010	13	2,159,500	13	0
Upper Volta	1,563	20	3,113,900	18	+11
Dahomey	730	9	1,568,000	9	0
Niger	700	9	2,163,500	13	−31
Senegal	986	12	2,059,600	12	0
Mauritania	150	2	545,600	3	−33
FWA	8,000	100	17,299,500	101	

*Differential = percentage over or under quota. A + value of 23 means the colony furnished 23 percent more than its share of the quota according to population.

Sources: Census of 1931 in *Renseignements coloniaux* (1931), 487; Census of 1951 in *Service de la statistique générale: Population de l'AOF par canton et groupe ethnique* (Paris: Imprimerie Nationale, 1952); CIC General Jung, report on recruitment for 1928, Dakar, 16 August 1928, ANS, 4D70 81; Lieutenant Colonel Revol, Head of Military Cabinet, "Notes sur les opérations de recrutement concernant la classe 1948 en AOF," Dakar, 21 March 1949, ANS 4D16 81.

insistence of African politicians in the National Assembly, this same provision was extended to Africans at the same time.[23] This represented a significant victory for African conscripts, who could no longer, as a result, be sent to participate in French counterinsurgency campaigns in Indochina or Algeria against their will.

As a result of these constraints, French military authorities had to turn to other means of fulfilling their manpower needs for Indochina. One obvious step was to remove the ceiling entirely on the number of volunteers who could be accepted in the annual cohort. Another was to attempt to persuade conscripts at the end of their first year of service to convert their status from conscript to volunteer by offering a monetary bonus of 250 francs, a step which would not only make them eligible for overseas duty but also commit them to a further four years of military service.[24] There is no way of determining how many young conscripts succumbed to these sorts of pressures. What can be demonstrated is that if a hypothetical volunteer rate of 40 percent had been sustained for only a few years, the *Tirailleurs Sénégalais* could have easily reached a situation where the number of volunteers and career soldiers in the standing army would have exceeded 80 percent, versus a prewar figure of 33 percent.[25]

In the absence of formal archival records for *Tirailleurs Sénégalais* participation in the war in Indochina, very little analysis of the social and political dimensions of that campaign can be attempted. One brief discussion of the *Tirailleurs Sénégalais* by a serving officer in 1954 indicated that there were 18,500 *Tirailleurs* in Indochina in 1949 and that this number was reduced to 15,000 by 1952.[26] Another fragment of evidence is a document listing the names, ranks, and *cercles* of origin of some twenty-seven *Tirailleurs* cited to receive the Order of the Division in 1953-1954 for acts of bravery in Indochina.[27] Although entirely random, the list is consistent with the quantitative evidence suggesting that Guinea had become the heartland of *Tirailleurs Sénégalais* volunteers. No less than thirteen of the twenty-seven men cited were from that colony, with the rest more or less evenly distributed among the colonies of Dahomey, Upper Volta, Senegal, Soudan, and Côte d'Ivoire. One was from Mauritania and none from Niger. Rural and remote *cercles* such as Man and Daloa in Côte d'Ivoire, Tenkodogo in Upper Volta, and Kissidougou in Guinea were well-represented, but there were only two men from potentially urban settings, one from Bamako and the other from Porto-Novo. Another fragmentary statistic relates to African POWs liberated by the Viet Minh in October, 1954, after the defeat at Dien Bien Phu brought an end to French rule in Indochina. Of the 634 prisoners liberated, 240, or over 37 percent, were from Guinea, 106 from Soudan, 85 from Upper Volta, 67 from Senegal, 60 from Dahomey, 48 from Côte d'Ivoire, 24 from Niger, and 4 from Mauritania.[28]

The adoption of modern techniques in recruiting was another response of military recruiters to the changing postwar world. Branches of the military like the air force, hitherto particularly lacking in African personnel, now began advertising in federal newspapers like *Afrique Nouvelle*, seeking skilled applicants.[29] Posters were printed, radio appeals launched, and circulars and telephoned notices delivered to all large business firms in the Dakar area publicizing the draft and the virtues of military service.[30] Meanwhile the induction process was also being reformed. No longer were recruits subjected to debilitating marches on foot once draft boards were through with them. Movement by rail or motor transport became the rule everywhere in FWA.[31] This permitted the army to develop a wider distribution of recruits than had been possible earlier in the century. The class of 1946, for example, saw recruits from Senegal posted to regiments as far away as Bamako and Abidjan. In the Soudan, some of the contingent served locally but others were dispatched to Guinea and Upper Volta.[32] The motive, according to one observer, was clear. The army sought an ethnic mix in each regiment to avoid any charge that ethnic hostility lay behind any military intervention.[33]

The smaller postwar levies permitted recruiters to raise medical and educational standards for the *Tirailleurs Sénégalais*. In 1954, for example, the Dahomey quota demanded only 300 men from 24,000 on the recruitment lists; for Soudan, the cohort of 600 men was drawn from among 26,000 listed.[34] Although regulations stated men had to be a minimum of 172 centimeters (5 feet, 6 inches) tall to serve in the *Tirailleurs Sénégalais,* one French officer recalled that only men 175 centimeters (5 feet, 7 inches) were considered when he participated in the *Tirailleur* drafts of the 1950s.[35] The draft board in Porto-Novo, Dahomey, in 1947 offered an example of the army's concern for better educated recruits. When it was discovered that 6 graduates of the *grandes écoles* were among the 186 men declared medically fit, they were immediately incorporated into the *Tirailleurs Sénégalais* before any lottery was held.[36] Students who dropped out of school were also immediately called up in the postwar system.[37] Students could, however, by eccentric behavior or radical political expression, make themselves unattractive to the army, and some may have tried this device once inducted.[38]

Dispensations for the influential continued to be granted, but with less ease than in the prewar years. Civil servants were theoretically eligible to do one year of military service, a term rarely applied because their superiors could claim the men were indispensable.[39] Governors, meanwhile, preserved their power to dispense any individual from military service on political or social grounds.[40]

Absenteeism and flight continued to be what may be called the poor man's exemption. In fact, as Table 7.2 shows, absentee rates in 1948 had dramatically increased from a prewar average of 18.5 percent to a figure close to three men for every ten on the recruitment lists. Not only were the rates up for the federation as a whole, absentee rates had sharply increased in Côte d'Ivoire, Niger, and Soudan especially, and in Upper Volta as well. While rates remained steady in Senegal, Guinea, and Mauritania at roughly 25 percent, only in Dahomey had absentee rates fallen to a federation low of 14.5 percent. Selected *cercles* in certain colonies, however, continued to have rates far above the federation averages, perhaps in part because of their proximity to FWA borders. While such may have been the case for Bondoukou and Grand Bassam in Côte d'Ivoire, with rates of 75 percent and 54 percent, respectively, other considerations surely applied in the heartland of the Soudan, where Kita and Kayes had rates of 50 percent and Niafunké and Nioro 45

TABLE 7.2 A comparison of absentee rates between 1923–1946 and 1948

Colony	1923–1946 Absentee Rate*	1948 Absentee Rate
Soudan	14.9	26.9
Guinea	22.2	26.2
Côte d'Ivoire	11.0	35.3
Upper Volta	16.1	23.4
Dahomey	20.6	14.5
Niger	39.8	52.0
Senegal	23.4	26.5
Mauritania	27.6	27.5
FWA	18.5	28.8

*As a percentage of men on recruitment lists.

Source: Lieutenant-Colonel Revol, Head of Military Cabinet, "Notes sur les opérations de recrutement concernant la classe 1948 en AOF," Dakar, 21 March 1949, ANS, 4D16 81.

percent each. Commenting on similar results a year earlier, the governor of Soudan had attributed such high rates of absenteeism both to the "radical" political climate in FWA and to the failure of the colonial state to enforce punishment for offenders.[41] Nevertheless, with official indifference so widespread, it is impossible to avoid the conclusion that officials had more pressing concerns than absenteeism from the draft, particularly since the smaller postwar draft quotas were being easily met.

The social and ethnic composition of the *Tirailleurs Sénégalais* began to shift after the Second World War. One factor was the presence in the *Tirailleurs Sénégalais* of more second-generation career soldiers as the male offspring of men marrying in the 1920s began to come up for military duty. Growing up in military camps of larger towns, often attending the *Ecoles des Enfants de Troupe*, or EETs, the military-style academies created primarily for them by the state, many African youths came to see the army as a family occupation.[42] By the 1950s the EETs were graduating roughly sixty to seventy young men a year, all of whom were obliged to do three years of military service, usually as noncommissioned officers, upon completing their schooling.[43] French authorities occasionally expressed concern that their selectivity was creating ethnically unsuitable situations. Thus, Boisseson noted in 1956 that in the Fourth Brigade of Niger, were precautions not taken, there would have been an entirely Dahomean cadre of noncommissioned officers in command of a Djerma or Hausa rank-and-file.[44]

As the Niger Brigade example suggests, French military officials continued to practice "social engineering" in the *Tirailleurs Sénégalais* by using the variables of ethnicity and social class. Considerable variation in the practices of regional draft boards began to occur. In southern Dahomey and in urban centers generally, the emphasis was on educated youths. In rural Soudan, on the other hand, prewar practices, including the toleration of fraud, seemed to have prevailed, judging from the complaints reported in *L'Essor*, the RDA newspaper. It was reported in 1953 that far too many young boys no more than ten or fifteen years old still appeared on the draft lists. In Macina, comfortable families allowed their twenty-year-olds to hide from the recruiter or to be replaced before the boards by their younger brothers who were sure to be rejected by the medical examiners. The result was that only the poor twenty-year-olds served.[45]

Soldiers' Skills

Another new element in the *Tirailleurs Sénégalais* after 1945 was the presence of more urban Africans, often with skills the Army prized. In part, this was a natural function of the rapid urbanization taking place in FWA, with cities like Dakar and especially Abidjan growing very rapidly, and with the same phenomenon occurring in smaller cities like Bamako, Conakry, and indeed each of the provincial capitals.[46] With this growth came proletarianization and urban unemployment. When jobs became scarce, some urban youths began to look to the *Tirailleurs Sénégalais* rather than to ancestral land to which they might not any longer have easy claim. Indeed, as one military observer noted in the mid-1950s, French West Africans were now behaving like populations in Europe. When times were good, people preferred better-paying civilian employment and the freer lifestyle that went with it, but when unemployment struck, the *Tirailleurs Sénégalais* attracted a certain number of these men.[47]

Salary scales varied widely in the postwar *Tirailleurs Sénégalais*. The top salary for an African sergeant with full seniority was 1,593 francs, which was roughly ten times the monthly base pay of 150 francs for an army private.[48] By comparison, three years earlier, in 1945, a civilian farm laborer would have earned 135 francs monthly and a forest worker 195 francs.[49] Equivalences are difficult because inflation made 1948 prices much higher than in 1945. Nevertheless, while both army and civilian jobs carried with them daily food rations, the army calculated its food allowance at 5.25 francs per diem, or 157 francs a month, whereas private contractors were not obliged to reveal these values. Furthermore, army work was guaranteed for the duration of the three-year contract, while civilian laborers' jobs were clearly not. It would appear then that while base pay in the army was below civilian levels, benefits brought it up at least to civilian laborers' levels, and perhaps beyond these.

If unemployment was one motive for urban Africans to join the *Tirailleurs Sénégalais*, another reason was the acquisition or perfection of a trade. This was the case for eleven apprentice chauffeurs in the *cercle* of Abidjan who volunteered for military service in the cohort of 1948, hoping that the army would pay both for the completion of their training and the purchase of drivers' permits.[50] On the other hand, in the district of Dakar in 1946, coercion rather than voluntarism lay behind the presence of certain urban youths. The governor of Senegal attributed the high rates of absenteeism in Dakar, which had reached 64 percent that year, to the deliberate failure of employers to inquire into the military status of their employees. He confessed that to fill the Dakar quota it had been necessary to resort to a prewar tactic: "The only method which has proved effective, unfortunately, has been a round-up by the police in the vicinity of the markets on the last day [of the draft].''[51]

Partly through such reprehensible methods as these, partly as a result of wider social changes, *Tirailleurs Sénégalais* recruitment resulted in a contingent with greater skills. Of the 302 men in Dakar district who were examined by the draft board in 1946 (an untold number of whom had been shanghaied in the manner described above), 56, or 18.5 percent, were described as having skills useful to the *Tirailleurs Sénégalais*. These included men with traditional skills—seven ironworkers, three woodworkers, and one weaver—as well as modern ones—eight chauffeurs or truck drivers, three mechanics, four masons, and one or two bakers, cooks, secretaries, butchers, printers, primary school monitors, agricultural monitors, policemen, carpenters, and telephone operators.[52]

One skill among recruits that French authorities recorded both before and after the war was competence in the French language. A comparison of these data for the cohorts of 1936 and 1946 in Table 7.3 suggests that the army was successful in almost doubling the potential recruits with French language skills after the war. The problem, however, is that no data on recruits in the first portion have survived. It is therefore impossible to tell if the combined language skill rate of 3 percent was evenly distributed in both the reserve and in the military cohort for 1946. Furthermore, for purposes of training men in specialized trades, the gap between comprehension and literacy in French needs to be considered. It would seem reasonable to assume, therefore, that a shortage of recruits with French language skills con-

TABLE 7.3 Comparison of skills in French language of first and second portions of 1936 and 1946 cohorts

1936

Colony	French Speakers	Literate	Total	Total First + Second Portion	Percentage with Some French
Soudan	34	79	113	8170	1.4
Guinea	58	39	97	5600	1.7
Côte d'Ivoire	99	39	138	6368	2.2
Dahomey	28	47	75	1948	3.9
Niger	1	1	2	5097	0.1
Senegal	96	67	163	4212	3.9
Mauritania	4	0	4	70	5.7
TOTALS	320	272	592	31,465	1.9

1946

Colony	French Speakers	Literate	Total	Total First + Second Portion	Percentage with Some French
Soudan	54	64	118	7348	1.6
Guinea	43	29	72	3421	2.1
Côte d'Ivoire	150	70	220	9813	2.2
Dahomey	117	102	219	2136	10.3
Niger	37	15	52	879	5.9
Dakar district	19	13	32	302	10.6
rest of Senegal	120	149	269	8581	3.1
(Senegal total)	(139)	(162)	(301)	(8883)	(3.4)
Mauritania	8	2	10	636	1.6
TOTALS	548	444	992	33,116	3.0

Source: Compiled from recruitment reports for 1936 and 1946 in ANS, 4D167 132 and 4D90 81, respectively.

tinued to characterize the *Tirailleurs Sénégalais* after the war. If, as a speculation, three-quarters of those with French language skills actually became soldiers in 1946, this would have given the *Tirailleurs Sénégalais* only 411 youths who spoke French and another 333 who were literate out of a total intake of 4,320.[53]

While the teaching of specialized skills may have changed in the new professional *Tirailleurs Sénégalais*, basic training remained the same as the somewhat timeless process followed before the war. Upon arriving at their assigned *Tirailleur* regiment, raw recruits were issued an identity card which was coded to indicate their *cercle* of origin and depot of incorporation. An immediate measurement of aptitude was made, essentially to determine who would eventually serve overseas and who would not.[54] One critical element was how much French the recruit possessed or how quickly he managed to master the basic "petit nègre" vocabulary of the *Tirailleurs Sénégalais*. One ex-soldier recalled how raw recruits from "the bush" were so traumatized by the induction process that they literally did not know their left from their right legs, at least not when barked out as commands by a fearsome drill sergeant. Such youths would have a branch of a plant from which an herbal tea was made, called *quinquiliba* in Bambara, attached to their left leg. The marching refrain would then become "un deux [one two], quin-qui-li-*ba*, un deux, quin-qui-li-*ba*," and so on.[55] In the early years, training with live ammunition was severely limited by budget shortages, and men often were sent off to overseas combat situations to learn the job of an infantry man under fire.[56] The length of basic training could vary significantly. *Tirailleurs Sénégalais* regulations in 1918 called for nine months to a year of basic training before overseas posting,[57] while five years later the training period had been reduced, in theory, to three months.[58] In practice, however, training could be even shorter, as was the case for the First *Tirailleurs Sénégalais* Regiment of Saint-Louis in 1921, for example, where 223 of 1,812 new recruits from Upper Volta were sent off to Morocco after only forty-five days of training. As the following description of their basic training makes clear, these young men were poorly prepared for combat:

> Although their training is not complete, these young *tirailleurs* were still able to carry out marches of over 20 kilometers each with a full kit; they know how to fire rapidly in unison, which is essential for *tirailleurs* bound for war in Morocco, and most of them have thrown inert grenades with dummy firing pins.[59]

Ability to master languages was a key ingredient for the ambitious soldier. Bambara served as the unofficial African *lingua franca*, and several informants from Senegal, for example, have stated that they had to learn Bambara in order to advance their military careers.[60] Nevertheless, a mastery of French was the ultimate test. Without a solid, literate command of French, serious instruction in any army specialization was impossible.[61] In these circumstances positions of leadership among Africans in the *Tirailleurs Sénégalais* went to the more articulate, natural interpreters who emerged somewhat spontaneously from the ranks.[62]

Too often what passed for specialization in the *Tirailleurs Sénégalais* bore little relation to modern military or civilian technology. General Larminat, commander in chief of FWA forces in 1947, no doubt deliberately exaggerated the situation when he complained that the *Tirailleurs Sénégalais* had servile specialists such as "servants, orderlies, watchmen [and] cooks" rather than radio operators and mechanics.[63]

Africans in the new *Tirailleurs Sénégalais* who desired to advance their careers as specialized noncommissioned officers faced significant handicaps. The very real differences in educational background between themselves and the metropolitan counterparts with whom they ultimately competed for positions remained large, even after 1945. Inadequate schooling confined them largely to infantry and artillery units rather than the more rewarding specialized units. One partial exception were the glamorous paratroop units, where extra training was required but where willingness to take risks rather than educational level was the most important element. A training school for African "paras," as they were called, was established at Dakar in the 1950s, and by 1958 there existed an entirely new regiment of African "paras."[64]

The army may have been a school, but it was not suited to everyone. One way of illustrating this point is to examine the contrasting experiences of three *Tirailleurs Sénégalais* veterans.[65] The military career of Mamadou Pop Sané stands in marked contrast to the typical *Tirailleur* of the old Conscript Army. From a family of the Casamance region, Sané brought three skills with him when he was first conscripted into the *Tirailleurs Sénégalais* in the early 1950s. He possessed basic literacy in French after completing elementary school; he had worked briefly as an apprentice radio specialist with the Dakar police force, his first job after leaving school; and was a leading soccer player with U.S. Gorée, one of the major Senegalese clubs. The army sent him to Thiaroye to take its basic training course in communications, which he completed in three months, in contrast to many other soldiers, whose limited French often required them to take the course two or three times before being certified. Later, he qualified in such other army trades as electrician and truck driver. Sané also was able to complete his military service without ever leaving Senegal. When asked if he would accept a posting in Algeria, he knew his rights by the terms of recent legislation and refused the posting. He was even able to benefit from his soccer skills. Because he was a star player on the army team in Dakar, he avoided transfer to more remote regions of Senegal or FWA during his military stint. Upon discharge, Sané was rehired by the Dakar Police, where his career advanced steadily. As a senior police officer in 1973, he saw his military experience as the decisive moment in his life and was grateful to the army for having helped him achieve the social mobility and economic security he and his family enjoyed in the Republic of Senegal.

For M'Baye M'Bengue and Doudou N'Dao, life under French colors was an extremely unpleasant and humiliating experience. Having met as fellow students at *Ecole William Ponty*, from which they graduated as schoolteachers, both men served one year in the *Tirailleurs Sénégalais* upon completion of their studies.[66] N'Dao, because he was born in Kaolack, was a subject, while M'Bengue, from Dakar, was an *originaire*. During their military induction at Thiès, the two friends were separated, N'Dao to join other *Tirailleur* subjects from the provinces to eat "native food" from a communal bowl and to sleep on a hard board or a straw mat with only his kit bag as a pillow. M'Bengue, on the other hand, as a French citizen, was served *café au lait* each morning with other citizens, African and Europeans, was seconded to the *Tirailleurs Sénégalais*, and slept on camp cots with mosquito netting each night.

Both men share bitter memories over their treatment as educated Africans. N'Dao recalls once having criticized a young French officer for using the term *litière*, a word reserved for animals, to describe the bedding of *Tirailleurs*. The officer flared up in anger but apologized the next day, a rare event because, in his recollection, most officers were unrepentant racists. M'Bengue was asked to teach French on Saturdays to *Tirailleurs* on a voluntary basis, but was ordered to use "petit nègre" in his classes. When he objected to his officer, he was mocked the next day "as Mister artist who wants to speak French like the French," and from that point he refused to do this voluntary work. In the view of both men, the Colonial Army was a very poor school. Only a tiny handful benefited from specialized training, and then only if they brought a trade in with them, as in the case of a mechanic from Kaolack whom they both remembered.

While the views of these three individuals represent opposite poles of opinion, they serve as important reminders of a basic argument made effectively by Samuel Decalo: that all men neither enter nor leave the military with shared values.[67] The jaundiced view of the military held by these two men is not a fair reflection of the changes the *Tirailleurs Sénégalais* units underwent after 1945. Accustomed to judge people in part by the quality of their spoken French, these two assimilated schoolteachers found the *petit nègre* of the *Tirailleurs Sénégalais* an exaggerated symbol of their one-year purgatory under the colors. They even held the army responsible for what they understood to be the widespread use of *petit nègre* in Côte d'Ivoire, where they had heard that even ministers spoke this slang. Similarly, Mamadou Pop Sané's military experience may not have been as singularly successful as he was later to recall it. It also should be recalled that while many societies reward and pamper gifted athletes and performers, few people in any social formation possess such talent.

"*Promotion Africaine*": African Officers after 1945

Of all the changes in the postwar *Tirailleurs Sénégalais*, the one most heralded was the new policy of *promotion africaine*, the progressive upgrading of the quality and quantity of African officers. Yet, in reality, the first postwar decade produced dismal results. Deputy Galandou Diouf's son Moustapha had been only one of many better-educated Africans who served in the Free French units soldiering in North Africa in 1943.[68] But after demobilization in 1945, most of the best-educated Africans were lost to the *Tirailleurs Sénégalais* through demobilization. As Table 7.4 makes clear, not only did Europeans consistently outnumber Africans as officers until well after the Second World War, there were actually fewer African officers in 1954 than there had been in 1946. Not until the reforms of 1955–1956 did the numbers of African officers rise significantly.

The army blamed its failure to persuade enough men like Moustapha Diouf to make the military a permanent career on economic considerations. Tight military budgets simply did not allow for salaries comparable with the civilian sector.[69] Such an explanation was too self-serving. Part of the problem in attracting well-educated young Africans was that French military officials after 1945 had considerable difficulty in accepting a new African army based on the assimilationist principles articulated at Brazzaville. The CIC of FWA in 1946, General Magnan, acknowledged that new laws abolishing forced labor and providing for greater accessibility of Africans to citizenship would one day lead to the disappearance of the native officer, but saw no need for any changes to occur under his command.[70] The native officer understood his men's customs, and could speak their language and assess their morale. Should he become a regular French officer, he would have to instruct, command, and guide all soldiers, whether African, Asian, or European, in battle. Such responsibilities would require well-educated men with exceptional qualities of leadership, and such men in FWA were "extremely rare."[71] Indeed, General Magnan pointed out that the existing twenty-eight African officers in the *Tirailleurs Sénégalais* in 1946 were "old and faithful servants, whose intellectual abilities would

TABLE 7.4 African and European officers in *T.S.* units, 1908–1954

Year	Standing T.S. Army	Second Lieutenants	Lieutenants	Totals	European Officers
1908	10,500*	10	5	15	—
1911	11,980	3	3	6	—
1914	17,356	2	4	6	—
1924	48,000	16	0	16	437
1946	40,000*	—	—	28	558
1953	34,000	—	—	38	—
1954	34,000	—	—	23	850
1956	34,000	17	34	68†	—

*Estimated figures.
†There were also 15 African captains and 2 majors by 1956.
Sources: Maurice Abadie, *La Défense des Colonies*, 210–16; ANS, 2G12 7 and 4D143 100; *Annuaire officiel de l'armée française*, 1908, 1911, 1914; *Annuaire officiel des officiers de l'armée active*, 1924; CIC, General Piquemal, Report on Recruitment for 1924, 5 August 1924, ANS, 4D70 81; CIC, General Magnan to High Commissioner of FWA, 27 October 1946, ANS, 4D90 81; Lieutenant Colonel Villard, Head of Military Cabinet, Report, 1 October 1954, ANS, 4D175 143; Major Chailley, "L'Africanisation des cadres de l'armée" (unpublished paper, CMIDOM, Versailles, 1957).

be gradually strengthened." Among noncommissioned officers, perhaps a few with six to eight years' experience might be encouraged to sit entrance examinations for the *Ecoles d'Elèves-officiers* (as Ahmadu Fall had done), and the best of these might one day become French officers. But that would take time and none would ever advance beyond the rank of captain.

Apart from steps of a cosmetic nature, there were signs of only the most modest changes in training opportunities for African officers immediately after the war. In the fall of 1945, the old EETs, the primary schools organized as military academies in FWA, had their names changed to *Ecoles Militaires Préparatoires Africaines*, or EMPAs. The number of places was increased and the curriculum upgraded to reflect the higher aspirations of the African petty bourgeoisie. Contrary to the military's belief, African politicians and intellectuals were anxious to see more African officers trained, and they urged the transformation of these schools into modern academic *collèges*, with four years of secondary school preparation leading to the *brevet d'études du premier cycle*, or BEPC.[72] The French High Command refused, however, arguing that the goal of the schools was to accelerate the sons of soldiers to noncommissioned officer rank, not to create an African elite.[73] Nevertheless, pressure from Africans persisted, and by 1955 the *brevet* was being offered at two of the four EMPAs.

Fathers who sought a military career for their sons created considerable demand for the EMPAs. Less than one in five applicants was successful in gaining admission. In 1953, for example, only 17 percent, or 213 of 1,255 candidates, were admitted to the schools; a similar rate obtained the following year. By 1957, the number of applicants had doubled for the same number of places.[74] Parallel to the pattern in the *Tirailleurs Sénégalais* generally, a proportionately higher number of

"Promotion Africaine": African Officers after 1945

Ecole des Enfants de Troupe, French West Africa, circa 1950

students came from the interior colonies of Upper Volta and Soudan (see Table 7.5), while Senegal and especially Niger were poorly represented.[75]

The curriculum at these schools, in addition to the fundamental subjects, such as French, mathematics, physical and natural science, English, history, geography, art, and music, also included military instruction as well as what was termed *instruction civique et morale*. The military subjects covered discipline, garrison duty, infantry maneuvers, marching drill, and sports. So-called moral education contained a good deal of the Jacobin military ideology. As the CIC in 1952, General Nyo, put it, its purpose was to

> develop among the pupils sentiments of loyalty, frankness, solidarity, a sense of honour and of military duty, professional pride.
>
> The School should be one big military family . . . where all excessive severity is avoided, especially brutality of a kind which could create unfortunate complexes.[76]

By the 1950s, the EMPAs had raised the educational standards somewhat for African noncommissioned officers in the *Tirailleurs Sénégalais*. Selection procedures carried out by means of entrance exams led to low dropout rates; only nine of over two hundred cadets were dropped from the program in 1952–1953, for example.[77]

TABLE 7.5 African cadets attending EMPAs in 1953–1954

Schools	Bingerville	Saint-Louis	Ouagadougou	Kati	Total
Soudan	19	13	14	39	85
Guinea	22	34	7	8	71
Côte d'Ivoire	28	14	25	20	97
Upper Volta	26	21	51	37	135
Dahomey*	15	23	—	9	47
Niger	13	3	2	—	18
Senegal and Mauritania	17	19	6	7	49
FWA totals	140	127	105	130	502

*Totals include 17 students from Togo.
Source: Concours d'admission, EMPAs, 1953–1954, ANS, 4D174 143.

More significantly, in 1955 results of the military academy in Saint-Louis outstripped those of the civilian school there by a good margin. Whereas only 25 of 88, or 28.4 percent, of the civilian students passed the *brevet d'études du premier cycle* that year, no less than 18 of 21, or 85.7 percent of the young cadets at the EMPA of Saint-Louis were successful.[78]

More importantly for the army, these results were beginning to take effect where it counted, in the ranks of trained noncommissioned officers. As late as 1953, only 20 of 1,000, or 2 percent, of all African noncommissioned officers in the *Tirailleurs Sénégalais* were considered "first category," that is, men who had been to the same courses as French noncommissioned officers and who held equal certificates, pay, and benefits. By 1956, however, fully one-third of the Africans were of the first category.[79]

Improving the educational level of noncommissioned officers was one thing, and improving that of African officers quite another matter. French authorities acknowledged by the mid-1950s that Africanization of the officer corps, especially at middle ranks, was proceeding very poorly. Educational levels remained low, especially in comparison with the qualifications of young Africans entering the civil service and the private sector. In 1953, of the thirty-eight African officers, while twenty-one held the BEPC (representing approximately the completion of tenth grade), five had completed only six years of primary schooling while another twelve had done even less.[80] In a secret report, Lieutenant Colonel Villard, the military attaché to the high commissioner of FWA, argued that upgrading the EMPAs was an insufficient remedy since it would not attract the better-educated young Africans currently studying in the West African lycées.[81] These young African intellectuals, he argued, were spurning a military career because they did not believe they could become officers equal in stature to their French counterparts. To change this attitude, the *Tirailleurs Sénégalais* needed a complete revision of its image. As a first step, the army should modernize its barracks in FWA to make them more attractive. Second, it should launch an advertising campaign aimed at educated young Africans.

On Villard's recommendation, the *Tirailleurs Sénégalais* began a new, and revealing, recruitment campaign to attract African officer cadets. For the first time, recruiters used film screenings at schools, newspaper advertisements, posters, and pamphlets in an effort to sell the idea of a military career. While the techniques were new, the content was not. One pamphlet contained quotations from old colonial figures like Gallieni and Lyautey, the former praising an officer's career on the rather negative ground that one could always fall back to civilian life equipped with new skills if the military career did not work out. The remarks of Lyautey seemed even less likely to attract young African *évolués* of the mid-1950s. He is quoted as stressing the need to keep the Colonial Army autonomous from the French metropolitan army, an argument entirely contrary to the drive for equality and integration which had become so strong in the 1950s. The overall rhetoric of the pamphlet seemed out of tune with the times. It called for brave, intelligent, imaginative, and creative young men, willing to "devote their entire lives to the building of a greater France as once the Legions responded to the *Pax Romana*."[82]

Calling their new program *la promotion africaine*, the army issued glossy new magazines, such as *Soldat d'outre-mer* or *Frères d'armes*, showing the modern social and civil role African army officers could play in their society, in cooperation, of course, with their French "brothers-in-arms."[83] In *Soldats d'outre-mer*, for example, under the rubric of "habits of Modern life," an African officer, his wife, son, and daughter, all in European dress, were depicted eating a French meal complete with a bottle of Evian at their table. Another photograph portrayed an African sergeant teaching his son to read. Lyautey was invoked as theoretician of the army's colonial vocation as a builder of roads, markets, and dispensaries. Still another photographic article emphasized parallel opportunities for the wives of African officers, which included a variety of functions from sewing classes to adult literacy courses.[84]

Within the French military, some continued to doubt the wisdom of modernizing the African officers' corps. Major Boisseson reported in 1956 on a study of African youths' attitudes towards a military career, in which he found that African youths equated the discipline of the army with slavery, feared racial discrimination in the armed forces, and saw army officers as second class professionals since they were always subordinate to civilian authority. Boisseson maintained that the old guard African officer was a soldier with a "sacred belief" in the overall authority of the European commander. As such officers were being replaced in the later stages of the Second World War and in Indochina by newer, often hurriedly trained youths, the army was gaining slightly better educated but also less reliable cadres, men more likely to question orders from above.[85]

The highest civilian authority in FWA joined in the chorus of complaints over the slow pace of Africanization of the *Tirailleurs Sénégalais* officers' corps, and his concern was overtly political. In 1955, the high commissioner of FWA, alarmed at the news that it would take ten years before the *Tirailleurs Sénégalais* would begin to bolster the number and quality of its African officers, urged immediate action, even if this meant temporarily lowering standards through a much more rapid promotion to officer's rank of noncommissioned officers, and faster upgrading of existing African officers' ranks.[86] By 1956 there were signs the army had indeed bowed to these political pressures. The total of African officers in the *Tirailleurs Sénégalais* had risen to 68, and majors Soumaré and Fall were promoted to Lieutenant Colonel despite their limited formal education.[87] Below them stood 15 captains, 34 lieutenants, and 17 second lieutenants. In addition, the army put forward a ten-year goal to reach 410 African officers by 1966, of whom 13 would be colonels and 20 lieutenant colonels.[88]

Five Routes to Officers' Rank

In the 1950s, there were five potential routes for Africans to become officers. The first three avenues were identical to those pursued by French youths.

1. Four years of study at the prestigious *grande école* of Saint-Cyr for the ablest minority.[89]
2. Four years at one of the special schools for officer-cadets for the majority of aspirant officers.
3. Two years at the *écoles d'application* for senior noncommissioned officers wishing to take a commission. These schools were also known as the *Ecoles Spéciales Militaires Interarmes*, or ESMIA, and were located at various centers in France.
4. Two years of special training at the *Ecole Spéciale des Sous-officiers Indigènes*, at Fréjus, which, after the *Loi-cadre* law of 1956, was revived as the *Ecole de Formation des Officiers du Régime Transitoire des Territoires d'Outre-Mer*, or EFORTOM. In 1959, under pressure from Africans who resented the implied inferiority of the term "transitional regime," the same acronym came to stand for *Ecole de Formation des Officiers Ressortisants des Territoires d'Outre-Mer* [Officers' Training School for Those from the Overseas Territories].[90]
5. Promotion from the ranks to African officer for meritorious service.

The roads to Saint-Cyr and other *grandes écoles* remained virtually as impassable as ever for Africans in the 1950s. They involved sitting in competition with highly motivated and talented young metropolitan candidates, all holders of the *baccalauréat*, who had benefited from preparation at the best French *lycées*. The tiny minority of West Africans who had overcome their comparative disadvantages of opportunity to be competitive for a place at Saint-Cyr could also choose to enter competitions for other *grandes écoles* in pursuit of a liberal profession. The evidence suggests they did this, despite the army's effort to woo them.[91] In 1954, French military authorities sent majors Fall and Soumaré, the two showpiece officers of the *Tirailleurs Sénégalais*, on several visits to West African lycées to recruit young Africans for an all-expenses-paid year-long preparatory class at the *Lycée de Dakar*, established to help them prepare for the competition to enter Saint-Cyr.[92] While ten young men registered for the year in Dakar, not a single one chose to write the Saint-Cyr examinations.[93]

The route via the *écoles d'application* was only slightly more traveled by Africans in the 1950s, again because of unequal earlier educational opportunities. Educational levels for entry required at least a BEPC, and sometimes two years beyond this depending upon the type of *brevet* obtained. Again, admission was by means of a competitive examination, for which officer-cadets without military experience and French or colonial soldiers who held the rank of noncommissioned officer for two years, and who were less than twenty-eight years old, were eligible.[94] For those lacking in general education, one year of preparation at the *Péleton préparatoire* in

Strasbourg was established for those who had held noncommissioned officer rank for at least one year and who were under twenty-six years of age.[95] Fragmentary evidence suggests that very few Africans gained admission to the various ESMIA. The infantry school at Saint-Maixent had nine Africans studying there in 1955; the Armoured Corps program at Saumur counted only one African, and only one African attended the communications school at Montargis.[96]

Given the discouraging results of its efforts to create African officers on a French statute, it is no wonder that the army fell back on older paternalist solutions. One of these was by means of separate schools for native officers. One such special route, the EFORTOM program, became the most popular avenue for the creation of African officers in the years immediately preceding independence, as Table 7.6 indicates. EFORTOM opened its doors in Fréjus to its first class in October, 1956, for some 54 candidates, 34 of whom would graduate.[97] Most candidates were either African noncommissioned officers with two years in the ranks or the most promising graduates from the EMPAs.[98] Many, therefore, were second-generation soldiers, the sons of veterans who had preferential access to the EMPAs and for whom a military career was attractive. The EFORTOM process did not dramatically raise the low levels of formal education acquired by African officers. At the end of two years of study at EFORTOM, students were not brought beyond the level of the BEPC in French and in the sciences. It was hoped that in time the BEPC would become the entrance requirement and that the graduates would leave with the equivalent of the first part of the *baccalauréat* (roughly grade twelve). Military education was featured, of course, with its emphasis on molding personality through the transmission and daily application of notions of order, discipline, and duty, as one student of this subject in the interwar period had put it.[99] Subjects included were military history, instruction on French and foreign weaponry and armaments, topography, communications, motor pool, and military security. All candidates were sent for a *stage* at Pau to receive the *brevet militaire de parachutism*. The curriculum also included elements which overtly aimed at social and political control, or, in the army's phrasing, "problems of France and the world" and "social and moral preparation" for officers.

TABLE 7.6 African officers at EFORTOM, 1958–1965

	1958	1959	1960	1961	1962	1963	1964	1965	Total
Soudan	6	8	3	5	—	—	—	—	22
Guinea	1	2	—	2	1	3	1	1	11
Côte d'Ivoire	—	3	2	1	3	2	5	—	16
Upper Volta	4	5	4	3	5	3	5	5	34
Dahomey	3	2	2	4	5	2	2	2	22
Niger	1	2	1	1	—	2	3	1	11
Senegal	16	6	2	9	3	6	7	6	55
Mauritania	—	—	—	—	1	—	2	—	3
FWA totals	31	28	14	25	18	18	25	15	174

Sources: Annuaire de l'EFORTOM, 1958–65 (Saint-Raphaël: n.d.); Moshe Ammi-Oz, "Les interventions extra-militaires des forces armées nationales dans les états d'Afrique noire francophone" (thèse de 3e cycle, Université de Paris I, 1973), Annexes 8 and 9.

EFORTOM classes, circa 1955

The EFORTOM school outlived French colonialism in West Africa. It continued to graduate an average of twenty Africans a year as part of a series of bilateral military agreements negotiated with most of the new francophone African republics in the years immediately following independence, until it finally closed its doors in 1965.[100]

The list of EFORTOM graduates reads like a veritable who's who of francophone African military presidents and would-be presidents. It includes: Seyni Kountché, president of Niger, class of 1959; Mathieu Kérékou, president of Bénin, class of 1960; Moussa Traoré, president of Mali, class of 1961, and his rival for power, Yoro Diakité, class of 1958; and Seye Zerbo, president of Upper Volta until he was overthrown in a putsch in 1982.[101] Equally significantly, several of these men were the sons of serving soldiers and had acquired their primary school education in the special military academies for sons of veterans, the *Ecoles Militaires Préparatoires Africaines*, or EMPAs.[102]

The last and oldest road for African officers was through promotion from the ranks. No figures exist to measure this method, but two rather contrasting examples of men who were promoted in the immediate postwar period were Sangoulé Lamizana and Jean-Bedel Bokassa.[103] While it may be said that both men were not senior officer material because of their limited formal education, and that both would not have been promoted beyond captain in the French Colonial Army, these two "old warhorses" were very different men. While Bokassa's excesses are well known, Lamizana knew his limitations and as military president of Upper Volta, had the good sense to leave day-to-day governance to his better-qualified subordinates.

While the *Tirailleurs Sénégalais* by the 1950s had changed significantly from the Conscript Army of the interwar period, it was not yet the modern, professional army some of its planners had envisioned. Perhaps it is best to see the *Tirailleurs Sénégalais* in the last years of French colonial rule as a force divided between old and new colonial outlooks and social formations. One French officer put the question in military terms. He saw the new *Tirailleurs Sénégalais* as a compromise between skilled and educated youths who were easier to train but lacking in "social discipline," and peasant lads with their "traditional" qualities of obedience who had to be taught French before any specialized training could be undertaken.[104]

For Africans, the *Tirailleurs Sénégalais* had become symptomatic of the deeply contradictory signals that France was sending to its colonial cadres in general. On the one hand, military service had become popular with the sons of soldiers and with peasants seeking a better economic alternative. Those who avoided military service when they could continued to be youths from well-established rural families, as well as urban artisans, technicians, and intellectuals.[105] Although it was still true that underprivileged groups in FWA continued to be overrepresented in a conscription system perpetuating the myth of universality while still perpetrating many injustices, it can also be said that the military burden for FWA had become bearable, if by no means yet negligible by the 1950s.

In the period after 1945, the *Tirailleurs Sénégalais* began to reflect changes in the social formation of the larger society to which it belonged. Thus, the *Tirailleurs Sénégalais* in the postwar era consisted of a rank and file drawn from the rural peasantry and the urban proletariat, who were being commanded in part by a very diverse group of African officers. Some were old warriors promoted as a result of years of loyal and effective service. Others represented the emerging African petty bourgeoisie, products of France's limited Africanization program.

The Africanization program proved a success for France, despite its limitations. To be sure, Africans who attended such programs as EFORTOM realized they were receiving a special, separate, and unequal training compared to French counterparts. For the French military, it represented a compromise, and no doubt it brought mixed benefits from a narrow military perspective. Yet its benefits outweighed its shortcomings. First, it compared favorably with other military schemes of Africanization. The British, for example, did not launch a plan of Africanization for officers in the King's African Rifles in Kenya until 1957, and by 1961 they had created only ten so-called *effendi*, or native officers.[106] Second, in a manner not entirely foreseen even as late as 1956 when EFORTOM opened, this program yielded dramatic political dividends. Not only did it succeed in increasing the number, if not necessarily the quality, of African officers, but more importantly, it enabled the French military to transfer authority in the new states to men who shared a similar ideology. This minority of African officers were the real beneficiaries of independence. They received not only rapid promotion but, as political crises deepened, the opportunity to seize power directly.

8

French West African Veterans, 1945–1960

> *Under fire, the African veterans have proven their worth and their devotion to our cause. Despite our negligence and our mistakes in their regard, they have remained Frenchmen.*
>
> Traodec Report, 31 December 1948, Dakar

The next two chapters shift attention away from the *Tirailleurs Sénégalais* to examine the condition of French West African veterans after 1945. In this chapter, veterans are examined as a special-interest group with their own corporatist institutions, while Chapter 9 treats African veterans politically.

First, a brief comment on the quality of the sources is appropriate. Evidence is quite simply lacking to offer a complete picture of the life experiences of veterans, just as it is for the equally obscure lives of other francophone West African groups such as longshoremen or railway workers. Surviving colonial records make the political activities of veterans slightly less obscure than their social and economic lives, no doubt because the French administration appeared to have been more concerned with how veterans acted politically than with how they ran their personal lives. Unfortunately, few oral investigations have as yet been conducted, despite their urgency as more and more veterans pass away. Nancy Lawler's research on the Second World War experiences of *ivoiriens* in uniform is one significant exception. Her work, however, is focused primarily on the war rather than its aftermath.[1] Mention should also be made of my own sampling of some fifty-five veterans in Upper Volta, which I have called "the Upper Volta Survey."[2]

Unlike that for English-speaking Africa, the published literature on French West African veterans is scanty. For the former British colonies, considerable debate centers around the degree of involvement of veterans in postwar nationalist politics.[3] Interesting and important as this literature is, comparisons with francophone Africa are difficult since the terms and conditions of military service were so different. Yet only one work devotes space to French West African veterans. Virginia

Thompson and Richard Adloff offer a chapter entitled "Military Service and Veterans" in their useful but now somewhat dated *French West Africa*.[4]

While the terms "veteran" and "ex-soldier" are often used as synonyms, formal French military usage was more rigorous. The former career soldier, or *ancien militaire*, required a minimum of fifteen years of service to draw a half pension, and twenty-five years for a full pension. Members of a distinct minority of those *Tirailleurs* who chose to make the army their careers, this exclusive group should be distinguished from the great majority of French West African soldiers who served their three years as conscripts and then returned to their rural villages to resume life as peasant farmers. *Tirailleurs* who experienced the Second World War, on the other hand, whether they were careerists or conscripts, may be called war veterans, or simply veterans, the term preferred in this text. For the French military, and especially the French Treasury, however, the French phrase, *ancien combattant*, or "ex-combatant," had a more restrictive use. It referred specifically only to a soldier who had been at a war front for ninety consecutive days and had established his right to a small pension as a result. Also eligible for entitlements were those who had been taken prisoner by the enemy and incarcerated for at least six months or who had been severely disabled while in uniform. The first condition made such a soldier *un prisonier de guerre*, or prisoner of war (POW), and the second, *un grand blessé ou mutilé*, a war invalid.[5]

Not surprisingly, since thousands of West Africans had served France in two world wars, a very large number of people could be termed veterans. Indeed, in 1950, over a quarter of a million claims for some sort of military compensation had been registered with French authorities in West Africa.[6] Official numbers listing West Africans who had been recognized by the State as war veterans (see Table 8.1) were far lower. Yet even by this narrower measure, veterans constituted a significant grouping throughout the federation.

TABLE 8.1 Veterans in FWA, 1952 and 1959

Colony	1952	Percentage	1959	Percentage
Soudan	10,637	25.6	30,202	31.3
Guinea	5,435	13.1	13,443	13.9
Côte d'Ivoire	6,320	15.2	9,883	10.3
Upper Volta	8,405	20.3	17,492	18.1
Dahomey	2,526	6.1	5,987	6.2
Niger	1,264	3.0	2,677	2.8
Senegal	3,350	8.1	12,310	12.8
Dakar	2,905	7.0	3,188	3.3
Mauritania	632	1.5	1,215	1.3
FWA totals	41,474	100	96,397*	100

*The large increase in totals of veterans is a result of members of the *Tirailleurs Sénégalais* having served in Indochina and North Africa after 1952.

Sources: Lieutenant Colonel Vatinelle, Report on Veterans, Dakar, 21 December 1952, ANS, 4D178 144; 'Les Anciens Combattants dans les états," *Chroniques de la communauté*, 11–12.

Veterans' Lobbies

The earliest veterans' associations in FWA date back at least to the 1930s. One of the first for which there is any record was the *Association Fraternelle des Anciens Combattants de la Côte d'Ivoire*, whose leadership was apparently entirely European and whose main function seems to have been the distribution of *cartes des combattants*, or war veterans' cards. Only a handful of Africans were members.[7] During the period of Vichy government in FWA from 1940 to 1942, a branch of *La Légion des Combattants de Vichy* was formed, intended clearly as a political instrument.[8] The real postwar beginnings of veterans' organizations in FWA dates from 7 October 1945, with the founding of a French West African branch of a metropolitan body, the *Association des Anciens Combattants et Victimes de Guerre de l'Afrique Occidentale Française* (AACVGAOF)[9] (see Figure 8.1). The AACVGAOF was an umbrella organization grouping all veterans of the French army, European and African, citizen and subject, POWs and amputees. The secretary-general of this federal body, and its driving force until his retirement in 1957, was Papa Seck Douta. He presided both over the federal organization and its Dakar section, where he was joined by three powerful veterans' figures representing specialized groups of veterans across the entire political spectrum: Paye Dame Baye of the amputees, and Samuel Baye and Papa Guèye Fall, heading rival groupings of ex-POWs.

Papa Seck Douta was a feisty schoolteacher at *Ecole El Hadj Malick Sy* in Dakar who was born in Saint-Louis on 20 April 1887, and thus an *originaire*, or French citizen.[10] Although he ran unsuccessfully for political office in 1946 for the RDA (*Rassemblement Démocratique Africain*) in Senegal, Douta thereafter moderated his politics and by 1950 supported Lamine Guèye and the SFIO (*Section Française de l'Internationale Ouvrière*).[11] He seemed nevertheless to have maintained a politically neutral position so that he could best represent all French West African veterans.

Organization: *Association locale des prisonniers de guerre, Dakar* (ALPG) Presiding officer: Mamadou Otto Diarra Membership: 235	Organization: *Association des anciens combattants et victimes de guerre de l'Afrique occidentale française* (AACVGAOF) Presiding Officer: Papa Seck Douta Membership: ?
Organization: *Fédération nationale des prisonniers de guerre* (FNPG) Presiding Officer: Papa Guèye Fall Membership: 25	Organization: AACVGAOF, *Dakar section, Association générale des amputés et grands blessés de l'Afrique occidentale française* Presiding Officer: Samuel Baye Membership: 3,200
Organization: *Association amicale des anciens militaires coloniaux de carrière* (AAAMCC) Presiding Officer: Pierre Diagne Membership: 17	

FIGURE 8.1. Dakar-based veterans' lobbies, 1949
Source: Traodec Report of April 1949, Dakar, ANS, 4D178 144.

Over the course of his mandate he had his share of run-ins with administration officials, especially René Traodec, who ran the colonial government's *Office National des Anciens Combattants* in the late 1940s and early 1950s, a good indication that he was not prepared to adopt the government line even when socialist ministers dominated the French cabinet. Perhaps as a means of assuring that veterans were well represented in Senegalese politics, Douta chose Samuel Baye to be his right-hand man at the head of the Senegalese branch of the AACVGAOF. Samuel Baye was a close collaborator of Senghor in the BDS (*Bloc Démocratique Sénégalais*).[12]

If Douta and Samuel Baye occupied the political center in FWA, Paye Dame Baye secured the left flank and Papa Guèye Fall the right. Paye Dame Baye had helped found the *Association Générale des Amputés et Grands Blessés de l'Afrique Occidentale Française* (AGAGBAOF) in 1945 and served as its president until 1948, when he was replaced by Momar Baye. An amputee from the *Tirailleurs Sénégalais* campaign against Abdel Krim in the 1920s, Paye Dame was, like many of the war amputees in France, a political radical.[13] For Traodec, in charge of veterans' affairs in Dakar, Paye Dame was "intelligent but crafty";[14] Traodec was pleased when the war amputees voted him out of office in 1948. When Paye Dame formed a splinter group and sought to solicit funds from the private sector around the city of Dakar, French colonial officials quickly spread the word that he was unreliable.[15] His replacement, Momar Baye, was not, however, a compliant tool. He maintained cordial relations with his counterparts in France, the *Association Générale des Amputés et Grands Blessés de France*, which supported the French Communist Party.[16]

The dominant conservative force in FWA veterans' politics was Papa Guèye Fall, who, throughout the postwar era, was an ardent Gaullist.[17] Born in Saint-Louis on 15 March 1894, Fall was, like Douta, a French citizen.[18] One of Blaise Diagne's *originaire* volunteers, he enlisted in the French army during the First World War, and, at the age of forty-five, again answered the nation's call in 1939. Wounded in battle and then captured by the Germans, he returned to Senegal in 1941 to resume his career as a school principal.[19] Soon after the war's end, Fall founded a West African branch of the French *Féderation Nationale des Prisonniers de Guerre* (FNPG). Regarded by French authorities as "one of the best school principals ever trained in Africa," Fall served briefly in 1947 as attaché to the Ministry of Overseas France in Paris.

Papa Guèye Fall is well remembered in Dakar with some grudging respect for his tenacity but with little affection for his methods and manner. Two men who were his pupils and later part of his teaching staff at *Ecole Faidherbe* in Dakar describe him as an archetypical martinet, who ran his school like the army, with whistles blowing and the like, a kind man at heart but so stern that he alienated staff and students alike.[20] He took assimilation too far, in the eyes of his fellow Senegalese Muslims, converting to Catholicism, marrying a Catholic *métis* wife, and preferring to be called Jean Octave Papa Guèye Fall.[21]

Fall's FNPG was one of only three veterans' bodies on the federal level to receive state recognition, the others being Papa Seck Douta's AACVGAOF and Momar Baye's AGAGBAOF.[22] Fall was usually included in important veterans' happenings, such as the Minister of Veterans' Affairs's tour of FWA in 1953, despite the fact that his association had by far the weakest support among Africans. Indeed, membership in the FNPG remained very low, partly because Fall refused to allow his group to join the federal umbrella group, the AACVGAOF. In short, only his rigid loyalty to colonial France kept him prominent. Those ex-POWs who wanted to be part of the larger federation preferred to belong to the local association of former POWs in Dakar, presided over by Mamadou Otto Diarra, which was affiliated federally with the AACVGAOF.

Not all veterans' bodies were officially sanctioned by the colonial administration or linked to parent associations in France. In Senegal in particular, autonomous splinter groups seemed to surface more frequently than elsewhere, perhaps because proximity to the federation capital presented greater prospects for raising revenue and exercising power. Two such examples were the fringe group organized by Pierre Diagne in Dakar and by Ibra Dione in the rural *cercle* of Lambaye.

A radical opponent of Senghor in Senegal, Diagne in late 1944 founded the *Association Amicale des Anciens Militaires Coloniaux de Carrière* (AAAMCC), a body more distinguished by its longevity than its ability to attract support.[23] While Diagne was described by authorities as having only seventeen members in 1947, seven years later the AAAMCC, still alive, had moderated its tone to the point where it described African veterans as having made "sacrifices freely consented to for the expansion of French civilization in the territories of the French Empire."[24]

Lambaye *cercle* was the scene of an unusual phenomenon, a rural veterans' lobby designed by veterans for the specific purpose of challenging the power of local chiefs. There, Ibra Dione, a disabled veteran of the Second World War and a native of the region, had grouped roughly one hundred followers, each of whom paid one hundred francs' annual dues, into what he rather grandiosely called the *Association Quatrième République*. Among his closer followers was Arona Dièye, son of the village chief of Khang, and Ogo Diop, a shoemaker in Dakar who kept the association's funds and enabled the group to be registered in the colony. Colonial officials described the group as being "anti-French," a common enough label for a variety of alleged failings. In the case of Dione's association, their crime in the eyes of French officials was their ambition to replace local chiefs with their own people. Arona Dièye in fact had managed personally to take the chiefship of Khang away from his own father, and the veterans had also sought the ouster of the canton chief Lamine Dieng, an ex-policeman whom they accused of criminal acts.[25] Showing a considerable measure of political awareness, the association's man in Dakar, Ogo Diop, succeeded in persuading the radical newspaper *Réveil* to publicize their grievances.[26]

While Dakar was clearly the central focus of veterans' politics, substantial organizational work and lobbying took place in most of FWA. At the very least, each territory could count a branch of the umbrella body, the AACVGAOF, established in the territorial capital. Smaller and remote territories such as Mauritania and Niger had little else, but the Côte d'Ivoire, on the other hand, was a veritable stronghold of veterans, counting no less than twenty-two of a total of some sixty-seven veterans' lodges existing in all of FWA by 1952.[27]

Little evidence of the financial side of veterans' affairs has survived, but fragmentary evidence suggests that the various bodies generated substantial revenues. Local associations as well as the federal body regarded their revenues as private matters, and refused, with one exception, to submit their annual budgets to the scrutiny of colonial officials.[28] That one exception, the Saint-Louis branch of the

Senegalese veterans, filed a budget in 1950 of approximately 1.4 million francs, of which 760,000 francs (55%) came from their own receipts and the rest (619,000 francs or 45%) from the State.[29] Only a few details on fund-raising can be documented. Substantial annual dues, sometimes well over 100 francs, were levied; benefit evenings were held; some revenue was derived from renting of rooms in the veterans' lodges; in Niger, Hamani Diori and the RDA were said to have offered traditional-style patron-client gifts of food and animals.[30]

Diori in return reaped benefits from his cultivation of veterans. Although he himself did not have a military background, he relied on the political support of the Songhay-speaking region of southwest Niger, which had been the heartland of *Tirailleurs Sénégalais* recruitment in Niger. In contrast, Djibo Bakary, who was to be his main political rival in the late 1950s, and whom he ultimately defeated, relied on the backing of the predominantly Hausa-speaking southeast, where relatively little military recruiting occurred.[31]

Like political lobbies anywhere, the function of the veterans' associations was clear enough. They were to maximize group benefits by lobbying the government, the African politicians, and the private sector; to maintain solidarity among veterans throughout the federation; to run the local lodges and do good works on behalf of disadvantaged veterans and their families; and to raise funds to support all these activities. To facilitate its large task, the AACVGAOF ran its own newspaper, *La Voix des combattants et victimes des guerres de l'A.O.F.*[32] The paper listed executive membership and goings-on in each federal territory, reproduced parliamentary debates in which veterans' issues were discussed, and published all laws and decrees touching veterans' affairs. Deputies such as Yacine Diallo or Michel Dorange used *La Voix* to inform veterans of efforts being made on their behalf. First published out of the *Ecole El Hadj Malik Sy* where Seck Douta taught, the paper later moved to more comfortable quarters in the Dakar veterans' lodge.

Government lobbying could prove complex, given the numerous layers of political authority standing above French West African veterans. The case of Al Hajj Doudou Diallo illustrates the point.[33] Born in Mekhe, Senegal, around 1918, he was drafted into the *Tirailleurs Sénégalais* in 1938 and sent to France in 1939 in time to be taken prisoner by the Germans and held in labor camps in Brittany for two difficult years. Later, in December of 1944, he took part in the bloody soldiers' protest at Thiaroye barracks outside Dakar, and he was condemned to a long prison term as a result.[34] Released in 1946 as part of an amnesty arranged by Senghor, Diallo then joined the local Dakar branch of the West African POWs and pressed hard for the release of other ex-POWs who were Thiaroye victims, many of whom continued to serve long sentences. His continuing pressure on Lamine Guèye and especially Senghor finally resulted in a Presidential pardon for the remaining prisoners. By 1953 Diallo had become successively president of the Cap Vert local of the POWs, then federal president, and finally a member of the executive committee of the *Office National*.[35]

In many smaller centers of FWA the veterans' lodges became one of the places, beside government, mosque, or church, where Africans could turn for a helping hand.[36] In other centers, such as Abidjan, the lodges rented out rooms to war widows and their families at subsidized rates.[37]

Veterans in West Africa organized themselves into a bewildering number of groups, dictated often by definitions laid down by the metropolis. Despite their corporatist origins in France, these veterans' bodies represented an important lobby and forum for the political and economic grievances and aspirations of ex-soldiers. The discussion now turns to the efforts of individual veterans outside the corporatist structure to acquire economic benefits for themselves and their families.

Economic Benefits

In its effort to maintain ex-servicemen as a differentiated and loyal group, the colonial administration offered a variety of economic benefits to ex-servicemen. Among the privileges active servicemen and veterans alike had long sought was preferential access to chiefships for themselves and their kinsmen. As part of the reforms of the postwar era, however, French authorities implemented a new policy aimed at reducing the number of chiefships awarded to military petitioners. No longer were veterans and their kinsmen to be used to replace ineffective or uncooperative chiefs, a common earlier practice. Each inquirer was to be informed that his candidacy could not be considered unless he had "true customary rights," and if the post were vacant.[38] Of course, the fact remained that in those rare cases where chiefships were open to petitioners, soldiers and veterans had the inside track. Thus, in Côte d'Ivoire after 1945, for example, while chiefships were offered to ex-servicemen only in the remote *cercle* of Man, this was more than the general public could expect from the state.[39] Such policies caused resentment both among the nonveteran population and among the many ex-*Tirailleurs* whose hopes were unfulfilled.[40]

Postwar reforms had reduced the power of the chiefs, which should have made the position less attractive. Nevertheless, aspirations to the chiefship dominated the thoughts of many career soldiers in particular, and these men continued to petition the administration for chiefships after 1945. Three rather different cases illustrate the kinds of interventions soldiers and their families made.

The first case, in Labé *cercle*, Guinea, involved an unsuccessful bid by a serviceman to replace a sitting hereditary chief. A certain Sergeant Hassana Diallo, a Dialonké speaker originally from Yambering district, a man with an excellent war record, applied to the minister of colonies in April, 1945, for the position of canton chief. The local official recommended against Diallo's appointment, however, after his investigations revealed that the current incumbent, Tierno Mamadou Cherif Diallo, was a Fulbe descendant of Karamoko Alfa Molabé in the direct line, and had ruled effectively over forty thousand people since 1927. He concluded that it was simply impolitic to place a Dialonké in command of a Fulbe region.[41]

In the second case, a soldier intervened on behalf of his family to reverse an unfavorable French decision. In the village of Brigo, *cercle* of Kita, Soudan, the French had gone outside the Sidibé family, which had held the chiefship. The family persuaded one of its members in the military, Sergeant Moriba Sidibé, to protest this nomination, and to request that one of his brothers be chosen instead.[42] The outcome of this intervention was not indicated.

Lastly, in a third case, a soldier used his weight to help determine French policy. In the region of Yellimane in Soudan in 1946, the French planned to organize a series of autonomous villages as a canton for the first time, and to name a civilian dispatcher clerk, Mamadou Doukouré of Toubakara, as canton chief. Among the many candidates for the chiefship was the father of Adjudant Bouba Traoré. Traoré, a twelve-year career noncom, wrote a carefully crafted letter protesting the French authorities' decision. Noting that there was a legacy of hostility among the various villages since some consisted of groups of ex-slaves and others of ex-masters, Traoré recommended that the reorganization be dropped. If the administration decided to go ahead with their plans, however, then Traoré recommended his father as best qualified to be chief, as he had founded the village of Yellimane, was the oldest village chief around, and was absolutely devoted to France, having offered to the *Tirailleurs Sénégalais* the adjudant himself as a volunteer in 1934, and his youngest son in 1939. Whether Traoré's studied arguments were influential or not cannot be determined, but the records do show that the idea of a canton chief for Yellimane was quietly dropped.[43]

The absence of evidence indicating the outcome of the soldiers' interventions makes it difficult to draw firm conclusions. What does emerge is that soldiers considered themselves sufficiently influential to attempt political intervention at local levels of their societies. So, too, did their families, who had usually put them up to it.

Both in theory and practice, veterans had preferential access to reserved jobs in the private and public sectors. Private employers in France had been obliged by a law dating back to 26 April 1924 to offer preferential employment to veterans, but that law along with many others was not applied to FWA.[44] Mobilization for war in 1939, however, brought with it the need to encourage wider African participation. A decree of 21 April 1939 obliged employers, and the State, to rehire mobilized staff after the war or else show proof of why they could not do so.[45] In restricted numbers this proved a real benefit to some. One veteran stated that not only did he get back his old job as clerk but his wartime service helped him advance his career.[46] Veterans with specialized skills were a prized commodity, and they appear to have been grabbed up by employers who needed skilled workers and saw the added advantage of fulfilling a patriotic obligation. Thus, in the Upper Volta survey, the two veterans with the most marketable skills, an electrician and a mechanic, indicated that they had no difficulty in acquiring civilian jobs.

It was, however, in the public sector that veterans reaped the largest benefits. A 1948 list of reserved jobs in state service for qualified veterans included assistant police inspector; expeditionary clerk; typing clerk; forest guard; mailman; male nurse; printer's assistant; policeman; worker in public works, including foreman; interpreter; telegraphic clerk; electrician's monitor in the Post, Telegraph, and Telecommunications Department (PTT); mechanic at PTT; surveyor's assistant in Public Works (PW); draftsman at PW; veterinarian's assistant; chauffeur; and customs officer in training.[47]

Many of these specialties were filled by veterans as the economy of FWA expanded in the late 1940s and early 1950s. Bookkeeping, the PTT, and especially the police force were the most common trades for ex-servicemen. In the newly reconstituted colony of Upper Volta in 1948, for example, the entire Ouagadougou police force of some 134 men consisted of former career soldiers recruited locally. These men, moreover, were described as being of much better quality than those who had been in the force in earlier years.[48] The Upper Volta Survey included two policemen who had both been noncommissioned officers in the Second World War and who had each completed fifteen years of service in the *Tirailleurs Sénégalais*.[49] One went on to log another twenty years as a policeman before retiring in 1968.

Regional patterns were also significant. In Dahomey, where the number of literate veterans was particularly high but where the economy was less developed, veterans expressed resentment over the relatively greater opportunities of their confreres in Senegal and Côte d'Ivoire.[50] Dahomean veterans were often skilled as mechanics, chauffeurs, masons, and typists, and expressed astonishment that they could not find reserved jobs for them in southern Dahomey. Some complained that they had left well-paid jobs in France to return home, assured they would find positions.[51] Aware of these grievances and short of skilled Africans in colonies such as Côte d'Ivoire and Niger in particular, where literacy was less widespread, the French administration encouraged many Dahomean veterans to move to take up reserved jobs in these colonies in the late 1940s.[52]

The Upper Volta Survey provides a profile of veterans and the types of occupations they took up on discharge. Of the fifty-five veterans interviewed, forty-three farmed their own land, one was a farm laborer, five were salaried employees of the State, and six gave no reply. Of the five salaried veterans, there was one male nurse, one chauffeur, one mason, and two policemen.[53] Judging from this sample, it would appear that a large majority of veterans were unsuccessful in their quest for reserved jobs and had little choice but to return to peasant farming.

One reason for these poor results was that there was a narrow range to the skills of veterans. Among respondents in the Upper Volta Survey, three identified themselves as chauffeurs but only one obtained a reserved job; for male nurses, the results were even more unsatisfactory, with only one of six finding a position.[54] One interesting case involved a soldier who had been a blacksmith before beginning military service. The army had made him an armorer, but he was unable to find a salaried position on discharge and returned to village smithing after the war. The case of a respondent who listed his trade as gardener was unusual. This ex-soldier had spent twelve years in service as a grounds keeper attached to the military airport at Bamako and another three at Ouakam Air Base near Dakar, never serving overseas in his entire career as a military laborer. He returned to the land on completion of his service.

In any situation where only a small percentage of veterans received employment, grievances could be expected. Handicapped veterans, for example, lamented that their injuries made them ineligible for too many of the reserved jobs and that the State was not making an extra effort to let them fill the jobs that they could handle.[55] Other veterans, while grudgingly conceding that they lacked qualifications for many of the reserved jobs they coveted, resented the army's failure to provide them with more technical skills and education.[56]

Some resentment was also addressed to the private sector. Doudou Seck, an unemployed Senegalese ex-soldier, complained in a series of three letters to the colonial administration that he, like hundreds of others in the Dakar area, was insulted by horribly low salaries offered by Dakar merchants. As far as Seck was concerned, persistence seems to have paid off. His third letter brought him a personal interview with a senior administrator in Dakar to see what could be done for him.[57]

One form of preferential state assistance to veterans theoretically was agricultural aid. Although veterans were potential agents of modernization in their villages, the State usually missed an opportunity to enlist them as allies.[58] In Côte d'Ivoire during the 1950s veterans-turned-farmers received only token assistance from the State, although here again, these benefits, however small, were reserved for them. For example, they were issued new seeds and were gathered together for an occasional practical technical demonstration.[59] By contrast, in 1963 in the independent Côte d'Ivoire, which made important technological advances in agriculture, President Houphouet-Boigny granted a 108-hectare coffee and cocoa plantation to the Veterans' Association of Man for their collective use, a plantation which had been sold by a European to the government shortly after independence.[60] It is characteristic of veterans' appetites and unrealistic expectations, however, that the veterans of Man, while grateful for the political recognition of their loyalty to Houphouet-Boigny, could still lament, "Without Houphouet, we wouldn't have this plantation. Yes we're happy. We need another one. This isn't big enough. We're numerous now."[61]

One potential benefit that never materialized was discussed in 1945. It involved the establishment of rural farming collectives led by resettled veterans. The idea was based on an experiment undertaken in Morocco. There, in the region of Port-Lyautey, some fifty-six hectares were turned over to a cooperative of veterans to help them grow and market vegetables for Rabat; the experiment was launched in October 1945, with a subsidy of some two million francs. But West African governors were unenthusiastic, and the governor of Senegal entirely negative. It was argued on the one hand that there was no available and suitable land for this use and on the other that enough was already being done for veterans.[62]

Military pensions were without doubt the single most important area of state assistance to ex-soldiers. It provided them with a regular stipend few other West Africans could anticipate. However, pension discrimination against Africans, when compared with their French counterparts, aroused the ire of African politicians and veterans alike.[63] A decree of 16 January 1947 constituted an improvement from the French perspective when it set African military pensions at 50 percent of metropolitan ones and permitted them the same percentage increases as those for metropolitan beneficiaries.[64] Africans continued to lobby for equality in pension matters, a concession finally granted in principle by the so-called equality law of 8 August 1950.[65] Discrepancies persisted, however: first, the law was not applied to disabled veterans until 1952; second, a "temporary indemnity" was not extended to Africans; third, concessions on loans and cash advances were never granted to Africans; and lastly, throughout the entire colonial period the French treasury paid serving African soldiers and civil servants in CFA francs but issued their pensions in metropolitan francs, which had half the value.

Often, benefits sounded better on paper than they were in practice. For example, the practice of subsidizing housing seems only to have been confined to the privileged capital region around Dakar. There the SICAP Baobab housing project in the 1950s gave priority to veterans in the purchase of these state-subsidized housing units.⁶⁶ Interest-free loans were a benefit open to qualified veterans in France, and this privilege was to be extended to FWA after the 1950 reform as well. Some Africans may in fact have benefited, but the only surviving record in the colonial archive of an African loan request had a negative result. An African veteran with what were described as excellent credentials, who had become an employee of Air France, requested a loan of 300,000 francs CFA to buy a truck. His request was refused on the grounds the sum was too large.⁶⁷

French ordinances of 1945 provided for monthly cash advances for veterans unable to wait for the payment every four months (pensions in France and in francophone Africa have long been issued on the trimester system). Attempts by African lobbyists to have this benefit extended to the colonies encountered interference from the high commissioner of FWA, who argued that blanket endorsement of all metropolitan legislation was an error. Such legislation needed "adaptation" before being applied overseas, and he wished to see this preserved.⁶⁸

Veterans had varying degrees of impact on the local West African economy. Local circumstances in Dahomey and Niger, for example, created an unusual business opportunity for enterprising Africans. Because the British in 1945 had both distributed civilian clothing and allowed their demobilized African troops to keep their army issue, the French felt compelled to follow this practice in the two colonies contiguous with British Nigeria.⁶⁹ In the Dahomean towns of Porto-Novo and Cotonou, the wives of African veterans carried on a commerce in war surplus clothing some five years after the war's end.⁷⁰ There is even one recorded case of abuse in this war surplus trade. Philippe Da Silva, a minor official in the mayor's office of Cotonou, was accused by African veterans of having kept for himself most of the clothing vouchers which had been allocated to ex-soldiers soon after demobilization.⁷¹

In Upper Volta, veterans had significant impact on local economies as conspicuous consumers. In the agriculturally marginal region of Tougan, for example, veterans were responsible for considerable local price inflation. Merchants extended credit to veterans for millet purchased in advance and at a large markup, which inflated the price for other consumers, as well, and increased the profits for local traders but not for producers. Many veterans, despite their reliable incomes, were poor money managers. They were constantly broke because they tended to binge, when their thrice-yearly pensions were paid, on alcohol, gifts, and of course credit charges.⁷² The practice of money-lending to veterans in anticipation of their pension payments was widespread.⁷³ Of course, the moneylender himself might well have been an ex-soldier, like Nouma Sanguisso, a Tougan veteran who became one of the most prosperous men in the town in this manner.⁷⁴

Some ex-soldiers were able to accumulate considerable savings during their years of service. One respondent in the Upper Volta Survey, for example, stated that he had accumulated 300,000 francs of savings upon discharge.⁷⁵

One way of measuring the impact of veterans in a rural setting is to examine a particular village's experience. Greg Finnegan, in his case study of the Mossi of Tenkodogo region, Upper Volta, has ably demonstrated the role of veterans in the village of Kougsabla.⁷⁶ Four villagers went off to serve in the First World War and were therefore among the very first labor migrants in a community that was, in time, to be caught up in a regular pattern of seasonal and permanent migration south to the plantations of Côte d'Ivoire and the Gold Coast.[77] Others saw action in the Second World War and in Indochina. The most successful veteran of Kougsabla spent nine years as a clerk in the financial section of the Colonial Army. Upon discharge he was hired by a large import-export house in Ouagadougou, where he eventually became chief accountant.[78] Although he returned only three times in eighteen years to visit his village, he maintained patron-client relations by housing and supporting bright children from the village as they pursued secondary education in Ouagadougou.

What is the economic balance sheet for veterans? The vast majority did not use their savings and pensions to accumulate power or property, or to live off rents or land. Indeed, one local organizer recalled that veterans were offended by the very thought of capitalist activity. When he asked veterans on one occasion whether they had used their pensions to invest in property, veterans expressed indignation over the question and its implications.[79] Indeed, it may be argued that an important opportunity to bring about economic change in the West African interior was lost. Given the considerable economic weight of veterans as a group, the organization of cooperatives, perhaps through the investment of veterans' pensions, might have given veterans far greater local impact.[80] Yet it is also clear that veterans were considerably better off than the civilian peasantry among whom most of them lived. No ex-serviceman needed personally to till the soil if he received even a partial French pension. One respondent from the Casamance area of Senegal indicated that he was routinely able to hire a farmhand to grow his rice for him, and that this, with his pension, enabled him to live well and to enjoy prestige and respect locally.[81] It was also true, as we have seen, that for a minority of veterans some capital accumulation did take place. Certainly, for perhaps that 10 percent of veterans who were able to obtain a reserved job, military service did bring material rewards. Their entry as blue-collar workers into state employment constituted upward mobility and increased social status in West African society.

In another respect, however, the perquisites and benefits enjoyed by veterans were ambiguous. What the State offered to ex-servicemen to differentiate them and separate them from their fellow countrymen had a chimerical quality. Chiefships, reserved jobs, agricultural assistance, military pensions, cash advances on these allotments, interest-free loans, and subsidized housing—the list of fringe benefits was indeed impressive. When examined in the closer light of reality, however, each of these benefits had substantial qualifications attached. Thus, only one housing development ever materialized, the policy of cash advancements was never applied to the colonies, few if any Africans qualified for interest-free loans, farming help was minimal, and reserved jobs were limited by the narrow range of skills most veterans could bring to bear. The result was to leave veterans dissatisfied, the majority of West Africans jealous, and the Colonial State perplexed. A group of *ivoirien* veterans from the region of Man summed up the bitterness of veterans everywhere in FWA:

> —The *anciens combattants* don't have the political power to get things. So we receive no respect.
> —Already by the time we got our liberty, the RDA, the civilians, were in place. We didn't have enough money, enough power. We had no resources, even to hold meetings.

—We're not respected by the civilian population—whether they are in politics or not.
—They don't understand us. They said at that meeting those words which cut us: You have suffered for France but not for us.
—We asked for reserved jobs. No more. We asked that our children be allowed to enter directly into the military, into schools, the gendarmerie, the police, the customs. Nothing. There is nothing for us.[82]

Veterans as an Interest Group

Several factors combined after 1945 to help veterans coalesce as a distinct interest group.[83] Access to education for their children, an appetite for a more modern lifestyle, and the dispensation by the State of symbols and perquisites of higher status all combined to help veterans differentiate themselves from the rest of French West African society.

Veterans, like many other groups in West African society, placed a high value on access to education for their sons in particular. On the eve of independence in Upper Volta, for example, veterans identified education for their children as their most important need.[84] Unlike others in the larger society, however, their concrete experiences abroad, and later as returning soldiers, gave them empirical understanding of the skills necessary to rise in West African society. In the Korhogo region of Côte d'Ivoire, which had vigorously opposed European education, Senoufo-speaking veterans, having learned the importance of "paper," became strong advocates of schooling for their sons and daughters. On their return after the war, they led in the effort to consolidate villages in order to build new primary schools.[85]

Even before the war, veterans had been encouraged by the Colonial State to see access to schooling as a privilege that set their families above the mass of West African society. The State recognized this desire and provided preferential access to schooling for the sons of veterans.[86] When it is considered that in all of FWA in 1948 only 121,200 children attended primary school, the benefits of preferential access for their children was considerable.[87] Indeed, African politicians were well aware of how to appeal to ex-soldiers. Maurice Yameogo, first Minister in Upper Volta in 1958, made a promise he could hardly have afforded to keep when he assured *voltaique* veterans they could expect one school for the children of veterans at Ouagadougou, a second one for war orphans at Ouahigouya, and a night school for children past school age, in exchange for their electoral support.[88]

While many sons of soldiers did avail themselves of their opportunities, these advantages lay most clearly in the direction of a military career. While this conformed to the wishes of their fathers,[89] by the 1950s, when the expanding economy of FWA had created more civilian opportunities, not all sons expressed enthusiasm for the military life. Thus, one Senegalese veteran lamented that his son had refused entrance into the *Ecole des Enfants de Troupe*, and had also avoided studying agriculture or forestry, because he wanted to be a civil servant who did not work with his hands.[90] Somewhat similarly, a *voltaique* veteran observed that his 18-year-old son feared the discipline of the army and preferred a veterinarian's career instead; in his view, this contributed to the unfortunate situation where the military had to continue its reliance upon rural youths.[91]

One area of behavior where veterans were distinguished from the mass of the peasantry was in their desire for a modern European lifestyle. During their years in France, a few higher-ranking men of the *Tirailleurs Sénégalais* had married European women. In Dahomey, it was reported in 1947 that several noncommissioned and African officers had returned with their wives, much to the annoyance of local officials. Although administrators saw these men as pretentious "upstarts" who were especially contemptuous of African chiefs, they were forced to acknowledge that the veterans and their wives were sufficiently *évolué* to have been grudgingly accepted by the European milieu of southern Dahomey.[92]

Many veterans gravitated to the larger towns of FWA where they had greater access to consumer goods and privileges. In Ségou, Soudan, for example, the *Sureté* reported that veterans sought even the most menial laborers' jobs in town in preference to returning to rural agriculture.[93] In these urban settings, veterans often continued to maintain French practices in modern hygiene they had first encountered in military service and to adopt French tastes in dress and food. More meat in their diet than they had formerly consumed, rice, and to a lesser degree bread, became standard fare for the military, and many West Africans continued these patterns of consumption after discharge.[94] Tobacco and alcohol were also products which had their appeal among ex-servicemen, with dependency and addiction as dangerous side effects. To many opponents of westernization, especially among Muslims, the descent into alcoholism when it occurred represented a visible and deplorable consequence of the abandonment of belief.[95]

When ex-soldiers took up farming and returned to their villages, they usually became reabsorbed into local social relations.[96] Even in agriculture, however, veterans were apt to be more open to innovation than their neighbors. Veterans were frequently prepared to attempt a small experiment with a new seed or agricultural implement.[97] An *ivoirien* veteran from the Korhogo region who returned to his village made clear his unhappiness with the conservatism of local ways:

> They asked me to reenlist as a volunteer, but I said no. . . .
> We didn't stay in because our parents made fetishes so that we would stay here and work for them. That is why there was no spirit to stay in the army. Our elders had one single concern—cultivation. They only wanted us to farm for them. When we were children, they were only farmers—that's the only thing they knew—that's all they wanted. Now, if my son wants to leave here, I will give him my blessing. If you stay here, you don't do well.[98]

Peasant life for the ex-serviceman could sometimes be painful. Caught between two worlds and uncomfortable in either, the ex-serviceman might well have found the routine and ritual of village life alienating. An extreme example of this conflict is depicted by the Senegalese writer Birago Diop. One of his short stories is entitled "Sarzan," *petit nègre* for "Sergeant."[99] A veteran Bambara sergeant returns to his village after an overseas military career. Full of admiration for Western civilization and its technology, he is determined personally to bring his village up out of barbarism. He attacks local customs, dignitaries, and even the memory of the village ancestors, alienating one and all as he seeks to impose a martinet-like discipline on village affairs. His project has been to widen and improve the road linking the village to the outside world. With the passage of time, the road is widened but "Sarzan" has become a lost soul, quite mad, dressed in the tattered rags of his military uniform, and cared for by his all-forgiving kin.

French authorities made available to veterans a wide variety of lesser perquisites, none of which was important in itself, but which, collectively, served as important symbols of differentiation for veterans as a group. Ex-soldiers, for example, were among the few West Africans privileged to acquire hunting arms. In areas where hunting formed an important supplement to diet, this was of some economic benefit; elsewhere it had symbolic importance.[100] After the war the State invested considerable West African sums in the building of veterans' lodges.[101] The awarding of decorations was yet another of these symbolic gestures. The *médaille militaire* was liberally dispensed to African soldiers, and by 1952 it was held by some 8,356 veterans in FWA.[102]

The New African Military Ideology

Out of their experiences in two world wars, French West African veterans developed a pride in their accomplishments. One veteran of the First World War, Kandé Kamara of Guinea, believed that the ordeals of African soldiers had earned rights for all black people:

> If we hadn't fought, if we—the black people—hadn't fought in western wars, and been taken overseas, and demonstrated some ability of human dignity, we wouldn't have been regarded as anything.[103]

The more militant veterans of the Second World War took the argument a step further. While they continued to borrow from French republicans a faith in the virtue of French militarism, they reconstructed their beliefs around their own particular version of how events of the war had unfolded.[104] Whereas French citizens and soldiers alike were demoralized and defeatist in 1940, they held that the *Tirailleurs Sénégalais* had fought more bravely than French troops in an unsuccessful attempt to defend France from the enemy. An *ivoirien* veteran of the disaster of 1940, Laqui Konde, recalls the retreat vividly:

> Yes, we were with the French all the time. We were stronger than the whites. That bullet that hit my tooth would have killed a white. When the shooting came, the whites ran. They knew the area and we did not, so we stayed. Our officers? They were behind us. I didn't know or think about anything except life or death. We were between the two. Yes, the officers said: Stand. We will win. But it was the Germans who were stronger so we forgot what they said.[105]

After defeat, Africans felt they had saved French honor by rallying most rapidly and effectively to de Gaulle. Some even seemed to take sole credit for the victorious allied campaigns in France:

> Only the French know what we did for them. We liberated them. What greater thing could you do for them?[106]

Tirailleurs Sénégalais veterans with de Gaulle, 1958

Those Frenchmen who refused to acknowledge the enormous contributions and colossal sacrifices of African soldiers were at best old-style colonialists and at worst unrepentant racists.

However subjective they may be, these perceptions of African veterans were based on concrete experiences in the war itself, whether as front-line soldiers or as POWs of the Germans. Indeed, their familiarity with the wider world beyond FWA contributed to their awareness and political sophistication. While some veterans had experienced the degradation of Nazi POW camps, others had lived for a time in such disparate locations as Britain, the Gold Coast, and the Soviet Union, usually under harsh circumstances.[107] One official report from Dahomey stated that former soldiers there were enthusiastic about the rumor that the British and the Americans were prepared to take over the French colonies.[108] Whatever the accuracy of this report as a reflection of veterans' sentiments, it does show that ex-soldiers were well aware of the insecurity of French colonial officials in the months immediately following the war.

A second set of veterans' attitudes centered around the continuing discrimination veterans, and indeed West Africans generally, continued to experience, despite the promises of Brazzaville and after. The responses varied. One *voltaique* veteran, a corporal with 14 years in the *Tirailleurs Sénégalais*, including combat experience in Italy and Indochina, put it bluntly: "We were considered to be slaves."[109] A

General de Gaulle decorating a *Tirailleur* sergeant, circa 1944

common method of dealing with their disappointments in French behavior was to distinguish between what one Senegalese veteran called "real Frenchmen from France itself" ("les vrais Français de France"), and the unchanging racists who were still to be found living in the colonies.[110] Postal censors in Guinea commented that the main theme in letters between Guinean ex-soldiers and French families they had met during the war was their gratitude for the kindness and affection shown them, in contrast to the treatment they received from French settlers and officials in FWA.[111] The same view was loudly proclaimed on arrival home by some forty disembarking Dahomean soldiers in Cotonou.[112]

Two incidents demonstrate graphically how veterans acted on their reconstructed belief that they were better Frenchmen then their colonial overlords. Eba Aoussi recalled that as the only African schoolteacher on staff in Dabala, Côte d'Ivoire, in 1947, he alone was denied admission to the *commandant de cercle's* reception on Bastille Day. When he protested this discrimination, he was transferred to another district a month later. Furious at this insult, he decided to take political action. He thrust himself into the activities of the *ivoirien* branch of the AACVGAOF, where he soon was elected to the executive, and became a militant member of the RDA.[113]

The second incident also centered on Bastille Day festivities, and had wider ramifications. In July of 1948, the Governor of Dahomey, Cambon, imprudently decided to punish the entire Dahomean branch of the AACVGAOF because he disapproved of their sympathies for the RDA. Cambon informed the veterans' branches in both Porto-Novo and Cotonou that they would not be welcome at the annual July 14 celebrations on the grounds it was not their holiday. No doubt advised that it was a bad idea to offend the 14,000 members of the Dahomean AACVGAOF, Cambon changed his mind, but by this time the veterans of Porto-Novo had decided to boycott the entire day. Upon learning on the morning of July 14 that the veterans would not budge, Cambon ordered Montout and Sodegbé, respectively president and vice-president of the Dahomean veterans, to appear before him. Montout, a French citizen and civil servant who was head of the Dahomean Treasury, held radical political sympathies, according to the senior French civil servant in charge of veterans' matters.[114] In a fit of temper, Governor Cambon struck Montout in the face. The incident received wide publicity all over FWA, being the subject of a resolution of condemnation at the Federal Veterans Association meeting in August, 1948, and resulting in bold headlines in the opposition newspaper, *Réveil*.[115]

As for Louis Sodegbé, Montout's African vice-president, he seems to have been the calmest and wisest politician in the entire affair. The *Sureté* credited him with organizing the 1948 boycott on the grounds that this was the only way to get French authorities to deliver long-awaited combat bonuses. He had insisted, further, that these bonuses be paid directly to veterans, as was being done in Côte d'Ivoire and Soudan, and not through Cambon's administration, which was doing all it could to frustrate the legitimate political aspirations of veterans. Sodegbé was careful to caution Dahomean veterans to remain courteous and respectful to all Europeans, whether they held important positions or not, in order to maintain the good will of even the most humble potential ally.[116]

Somewhat as a group within a group, disabled veterans stood out as having a special consciousness among veterans. Often so badly handicapped that they could not return to civilian occupations, disabled veterans were perhaps for obvious reasons among the most militant professional veterans.[117] Personal victims of the war and of shameful neglect after it, disabled veterans railed against their fate and against an administration indifferent to their plight. As late as 1948 in Côte d'Ivoire, First World War amputees could be seen ambulating on their hands since the only center for artificial limbs was in Dakar, and transportation there was too costly for them.[118] A year later interim High Commissioner of FWA Chauvet was still apologizing that a lack of hospitals made it impossible to deliver and outfit artificial limbs.[119]

All veterans shared this perception that civilian society could never really grasp what it meant to have lost comrades as well as a part of themselves in battle. While veterans believed these losses could never be properly compensated, few were prepared to make the large leap over to an anti-French nationalism. Lancina Traoré, who was not himself a disabled veteran, perhaps came closest to proclaiming the veterans' universal lament: "Nevertheless, after having rendered so much service to mankind, they only ask for a little of the recognition which is their due."[120] Such shared veterans' sentiments as these do not mean that all veterans belonged to a special-interest group. After all, these West African soldiers had been recruited from many different cultural, social, and economic backgrounds, and they often possessed widely varying levels of skills before and after military service.[121] Most

Tirailleurs probably returned to their villages to be reabsorbed by ties of family and kinship. Many were no doubt disappointed that they had little of monetary value to show for their experiences but were otherwise grateful to be alive, reasonably whole, and in familiar surroundings once more.[122]

For a significant minority of veterans, particularly those congregating in the rapidly burgeoning towns and cities of FWA, it is possible to speak of their belonging to a special-interest group if not a social group. Especially striking was their membership in local associations and in the political parties of the day. Those who had suffered permanent physical damage as a result of the war and who may be described as "professional veterans," came to rely on the local lodge for social and psychological as well as economic support. Yet even for such dedicated veterans, significant obstacles prevented closer bonding. Their children did not intermarry.[123] Ex-army buddies found it daunting to maintain contact when separated by the gulf of distance, language, and skills of literacy. One informant in Senegal exchanged the odd letter with an old army buddy in Upper Volta but visits were not possible. Nor did veterans within Senegal visit each other very often.[124]

Despite such finite limits to their solidarity, veterans did stand out as a differentiated group in the immediate postwar years.[125] This was a period in which the French West African economy experienced unprecedented growth, particularly in the wage sector.[126] While the fruits of this growth were unevenly distributed, veterans had several advantages over the mass of the French West African population in the competition for these benefits. Some ex-soldiers gained preferred access to jobs in the public and private sector by virtue of their formal status as veterans, while others entered the salariat on merit by converting their acquired skills learned in the *Tirailleurs Sénégalais* into jobs. This process necessarily involved social change as well, through the growth of towns, new classes, and new lifestyles. Veterans sought and often received special privileges from the Colonial State in the spheres of education and health. Finally, the postwar era was one of heightened ideological competition, and African military men no less than others on the African continent were subjected to a wide variety of appeals for their loyalty from ethnic groups, regionalists, nationalists, and even French imperialists. The next chapter examines how African politicians and the Colonial State competed to win the political allegiance of veterans.

9

Veterans and French West African Politics, 1945–1960

> *Even in the bush, where he seems to have been reabsorbed into the ancestral milieu, the former soldier forms a class apart, escaping more and more from the traditional hold of custom. As a voter, he is the object of the careful attention of the political parties who teach him his rights, even exaggerating these, and who even take credit for fortunate decisions reached by the Government. The veterans' grievances are, unfortunately, still numerous and well founded.*
>
> Traodec Report, 31 December 1948

Of all the major postwar changes in French West Africa, the birth of African electoral politics was arguably the most dramatic. While ex-soldiers are widely held to have played some role in these political developments, the extent of that role has not been fully investigated. In English-speaking Africa, on the other hand, considerable debate centers around the importance of this involvement. In Ghana, the role of ex-servicemen in the political crisis of February, 1948, which led first to Kwame Nkrumah's imprisonment but ultimately to his political victory, has assumed a sacred place of honor in nationalist historiography.[1] Among those who have studied ex-soldiers in the Gold Coast, Eugene Schleh and Adrienne Israel support the nationalist view, while Richard Rathbone and David Killingray strongly challenge this position.[2] Killingray, in particular, maintains that while some individual soldiers may have acquired an increased political and economic understanding, their awareness may have derived from the immediate prewar crisis in the Gold Coast and not from their war experiences; the evidence is simply too weak to permit a decision on this point.[3] He argues, moreover, that "the overwhelming majority of [Gold Coast] veterans were by-standers to nationalist politics."[4] They and their ex-servicemen's associations worried more about bread-and-butter economic issues—jobs, legal aid, and assistance for unemployed veterans and the like—than about nationalist politics.

Interesting and important as this debate is, comparisons with francophone Africa are difficult since both the terms and conditions of military service and the course of nationalist politics were so different. Virginia Thompson and Richard Adloff have argued that in FWA veterans behaved more politically. A cohesive and influential group, veterans in the immediate postwar years used political means to achieve equality of treatment with their confreres in the metropolis. Despite the rejection of their quest initially, veterans responded with "unwavering loyalty" to the French administration through war, peace, and crisis. While this placed them at odds with the general political mood of FWA, by 1950 it had won them the battle for equal treatment.[5]

The Thompson and Adloff thesis requires modification in three respects. First, it misreads the political mood of veterans in the immediate postwar period. The tragic events of Thiaroye served to arouse political militancy among veterans in FWA, uniting them with the civilians who sought political changes. Second, as this chapter seeks to demonstrate, while the so-called Equality Law of 1950 was an important marker in the veterans' struggle for economic justice, the battle was by no means over by then.[6] Paternalist colonial opposition to the idea of equality for all veterans was so strong that it required formidable political pressure from African veterans in alliance with their politicians for almost an entire decade to win a series of major gains. Even then, the battle for complete equality with the metropolis could never have been won, either in the minds of veterans or substantively. Finally, and most importantly, it is *ex post facto* reasoning to argue that the innate conservatism of veterans tied them inexorably to their French rulers. To be sure, there were limits to the degree to which soldiers and veterans could be radicalized. A few men like Al Hajj Doudou Diallo, M'Baye M'Bengue, and Doudou N'Dao did transcend their earlier horizons of village, ethnic group, and colony and came to understand the contradictions in the assimilationist ideal. Others could not so easily shed the notions of discipline and hierarchy drilled into them by their French officers.[7] As this chapter seeks to demonstrate, the outcome of the struggle between the African politicians (especially the more militant RDA) and the State for the allegiance of veterans may have been weighted in favor of the State, but it was not a foregone conclusion.

It should not be surprising that politicians and officials in France and West Africa competed for the allegiance of recently demobilized soldiers. If there is one dramatic historical moment when social control over soldiers is apt to break down, it is at the moment of their demobilization. The possibility is always present that such disaffected groups might turn to banditry or even more organized forms of armed resistance to authority.[8] Indeed, after the First World War, civil administrators in FWA had worried about the potential difficulties of what they termed the *resénégalisation* of the *Tirailleurs Sénégalais*.[9] Their fears, however, proved unfounded, as most West African veterans returned quietly to their villages.[10]

African veterans of the Second World War inherited a different political context from that of their First World War predecessors. The ringing promises of the Brazzaville declaration, and on a wider basis those of the Atlantic Charter, led francophone African soldiers to anticipate a better life for themselves and their families.[11] The failure of France to keep these promises made the difficult task of returning to civilian life in West Africa even more problematic. Worse, the bloodshed at Thiaroye remained an ugly symbol of contradiction between French rhetoric and action.

Against this background of rising expectations and frustrations, France was forced to introduce a series of significant postwar changes. While falling far short of African expectations, the Fourth Republic provided for an electoral system in its colonies for the first time, for the formation of free and often radical trade unions, and for the toleration of a critical nationalist press.[12] The most immediate consequences of these changes were felt in the political arena as African political parties mushroomed in cities, towns, and villages. Competition for the allegiance of veterans became an important part of the exciting years between 1945 and 1960.

Years of Militancy, 1945–1950

Postwar political changes affecting veterans fall into three distinct phases: a time of militancy to 1950; years of accommodation from 1950 to 1956; and a period of apprehension from 1956 to 1960. The latter two phases will be discussed later in this chapter. The first period was marked by militancy both because veterans' memories of their losses in war were recent and bitter, and because the Fourth Republic seemed bent on abandoning the idea of a new and more humane French union that de Gaulle had promised at Brazzaville.

The Bamako Congress of October 1946 had set a more militant tone to West African politics. In that dusty river town in the French Soudan, a congress had been convened for all African delegates to the Second Constituent Assembly in Paris.[13] When all metropolitan parties save the *Parti Communiste Française* (PCF) chose to boycott Bamako and were able to persuade Senegalese socialist supporters Lamine Guèye and Léopold Senghor to do likewise, the great cleavage in postwar francophone Africa was born. Out of Bamako emerged an interterritorial alliance of parties that had been formed earlier in each of the colonies, commonly known as the RDA (for *Rassemblement Démocratique Africain*). Félix Houphouet-Boigny emerged as the dominant force in the RDA and its leading voice in the National Assembly, along with the more radical Gabriel d'Arboussier, who was named secretary-general.[14] Over the next four years, tensions mounted in FWA as the Colonial State turned against the RDA, using legal and extralegal means including imprisonment, banning of meetings, bribery, and even falsification of elections in a futile effort to stifle the RDA's growth.[15]

The RDA's parliamentary alliance with the PCF in the National Assembly in Paris, rather than radical African nationalism, was the root cause of this hostility from Fourth Republic governments, whether they were dominated by conservatives or socialists. African politicians were forced to choose sides. In Senegal both Lamine Guèye and Léopold Senghor remained outside the RDA, Guèye staying with the *Section Française de l'Internationale Ouvrière*, or SFIO, while Senghor eventually headed a group of moderate African politicians under the rubric of *Indépendants d'Outre-Mer*, or IOM. Though dominant in Senegal, the IOM never challenged the RDA as the leading African political group in FWA as a whole. One reason, perhaps, for the RDA's popularity was that its militancy matched the popular mood in French West African society; the trade unions, for example, launched a determined and partially successful rail strike in 1947,[16] and African deputies of all stripes were pressuring various governments of the Fourth Republic to honor the promises of Brazzaville and the First Constituent Assembly. Significantly, a common aspiration among trade unionists, civil servants, and veterans was an end to colonial discrimination, with the gradual if not immediate implementation of equal treatment under the law for metropolitan and overseas citizens.

Veterans' issues were a major element in the political agenda. Even before the war, in 1939, all men who had completed their military service had been enfranchised, though the war intervened before they could exercise this new right.[17] In the first postwar elections in 1946, only 6.97 percent, or 1,114,700 French West Africans, voted out of a total population of just under 16 million.[18] Veterans might have constituted as much as 5 percent or more of this new electorate on the first roll.[19] Even as the franchise rolls grew more extensive thereafter, it remained true that veterans constituted a core group of militants in virtually every new African political party after 1945, and beyond.[20] Nor should it be forgotten that equity for veterans was not the preserve of any one part of the political spectrum. The left saw the correction of injustice and inequality for the little man as its mandate and included veterans in its purview, while the right regarded a solution to the inequalities and neglect of veterans as part of their paternalist responsibility. The right also argued that if Africans were to be treated better in the postwar era, then the place to begin was with France's most loyal servants, the African military, rather than upstart *évolués*.[21]

In Senegal, veterans' issues lay at the heart of the political relationship between the two dominant postwar politicians, Lamine Guèye and Léopold Senghor. Guèye, the older and more established politician, had taken a calculated risk when he chose Senghor to run beside him as the second socialist candidate in the election of 1945–1946 for the Constituent Assembly. At that time, Senghor was an all but unknown émigré poet and teacher. The son of a prosperous Serère father, Senghor was the first African to receive the prestigious *aggrégation* after study at prestigious schools in France. It was during his student days in the 1930s that he came under the influence of the *négritude* movement and began to acquire a reputation as a poet of considerable merit.[22] Senghor taught grammar in lycées in Paris and Tours, became a naturalized French citizen, and was mobilized into a territorial regiment of the regular French army soon after war was declared in 1939. Captured in 1940, he endured a year of captivity as a prisoner of war, together with members of the *Tirailleurs Sénégalais*, since the Germans used race rather than nuances of French citizenship as their main criterion. It was during this incarceration that Senghor penned a series of moving poems on the theme of the *Tirailleurs Sénégalais* and their sufferings, later to be published under the title *Hosties noires*.[23] Released to Vichy in 1941 on the grounds of ill health, Senghor spent the rest of the war teaching in a lycée in Joinville, a working-class suburb of Paris.[24]

Senghor's intellectual and political development was so dramatic since those distant war years that it is perhaps difficult to picture him as the soldiers' politician. Yet it was precisely on the basis of his military background that Guèye chose him over others who had no doubt coveted the nomination.[25] Guèye, the assimilated urban lawyer, did not relish rural campaigning. In contrast, according to Ruth Morganthau:

Senghor wore khaki and sunglasses, sat on the floor of huts and ate what he was served. Some of the "citizen" Socialists disdainfully called him the "deputy in khaki," while the numerous Senegalese war veterans, who were also voters, applauded Senghor's deliberate attempt to identify with them.[26]

When an earlier political grouping known as the *Bloc Africain* split, Senghor, disliking the strict SFIO party discipline in French voting, and unhappy with its anticlericalism, left to form the *Bloc Démocratique Sénégalais*, or BDS. At the same time, in order to operate with some strength in Paris, Senghor helped found a loose grouping of middle-of-the-road African deputies known as the *Indépendants d'Outremer*, or IOM. Guèye remained content with the modest pace of reforms and closely involved with socialist colonial officials, like Wiltord, who was governor of Senegal from 1947 to 1950, or Béchard, high commissioner of FWA from 1948 to 1951.[27] Guèye had never been popular among veterans, though they occasionally attempted to use his socialist connections. He was seen as being too close to the government, especially when known socialists served as governors or as ministers of overseas France and did little to improve things.[28] In the referendum of late 1946, veterans in the Four Communes were reported to be planning a solid "no" vote on the grounds Guèye had not looked after their pensions as he had promised.[29]

While some of the military vote may have split with this rupture, it is entirely likely that Senghor took the lion's share of the veterans' vote with him over to the BDS. One authority notes that Guèye continued to hold the loyalties of the most senior African military, as well as of those citizens who were civil servants.[30] But Senghor and his right-hand man, Mamadou Dia, were able to do very well by securing the rank-and-file veterans and the rural voters. In 1949, for example, Senghor once more donned his khaki uniform to tour the countryside along with Mamadou Dia to launch the BDS as a mass-based party, appealing even to those not yet eligible to vote.[31]

The pages of the respective party newspapers make it clear that the BDS, and not the SFIO, was the Senegalese veterans' party. The SFIO's newspaper, *L'AOF*, devoted very few items to veterans' affairs. They felt obliged to give front-page coverage on 1 August 1947 to the proposed National Assembly Law to unify the status of soldiers in the French union, distinctly claiming it as their proposal since it was sponsored by Rabier, the SFIO deputy for Oran.[32] But they did not run another major veterans' item until six months later, when they ran a short article entitled "Le Délicat Problème des anciens combattants," urging the French government's line of patience on African veterans.[33]

In marked contrast, the BDS paper, *Condition humaine*, served the veterans' cause often and well. The very first issue of this paper, in February, 1948, ran a front-page story entitled "Nos Anciens Combattants," which reproduced a letter from Senghor to the minister of overseas France (the new title for the minister of colonies), pointing out that while disabled veterans from the First World War were eligible for raises granted to their French counterparts by the Law of 13 August 1947, they had not as yet received this benefit.[34] In May the paper ran another letter from Senghor, this time to the minister of veterans' affairs, asking for "franc for franc" pensions for African veterans to share with their metropolitan confreres.[35] Nor did Senghor lose an opportunity to remind Senegalese veterans of who really stood up for them. When the SFIO chose to vote with the French government to oppose equality of pensions for veterans, the BDS paper quickly reported this with the headline, "SFIO contre la démocracie en Afrique."[36] In early 1949 Mamadou Dia wrote a lead story deploring the lack of jobs for deserving veterans. He called on the government to help these men recapture their lost years by means of loans in goods, tools, or modest construction materials.[37]

Using his strongest language, Senghor himself wrote in his party's newspaper in May of 1949 that veterans' equality would remain elusive so long as the Ministry of Finance and Economic Affairs continued to follow the principles of the old *pacte coloniale*, "everything by and for the metropolis."[38] This ministry, Senghor asserted, was prepared to use any excuse, however lame, to save the French taxpayer a franc or two at the African veterans' expense.

The major difference between the IOM and its rival, the RDA, in this period was the manner in which each party was treated by the French colonial administration. Whereas the IOM was tolerated as a nuisance, officials routinely described the RDA as "the enemy."[39] Even Houphouet-Boigny's momentous split with the PCF in October, 1950 could not assuage some officials. Army intelligence, for example, went so far as to brand the RDA a subversive organization, still capable once more of embracing its former friends in Moscow.[40] The same report noted that veterans throughout FWA were "profoundly influenced" by the RDA, which was able to exploit their legitimate pension grievances. Almost all local committees of the RDA had a veteran as a member of the executive, and the RDA worked diligently to win over active as well as retired soldiers. Some RDA agents were even arrested and given six-month sentences for "subversive propaganda directed towards the morale of the troops."[41] Senior RDA politicians like D'Arboussier were well aware of this tactic of invoking military security to impede their political activities and even their freedom. Although suspected of trying to hamper military recruitment in 1947, D'Arboussier was clever enough not to be caught out in this manner.[42]

Despite official harassment, the RDA was able to attract a considerable veterans' following in the same areas where it had widespread popular support: that is, in most of the Soudan, in western Upper Volta, in parts of Niger, and in virtually all of Côte d'Ivoire.[43] An *ivoirien* ex-sergeant, Alphonse Dionkla, found himself in trouble for having been curious enough to have attended an RDA meeting while still an active soldier:

> We were informed of what was going on, but an African soldier could say nothing. At the bottom of our hearts, though, we were with them. I got fifteen days once at Bouaké because three of us went to a meeting of Houphouet at Sakasso. We didn't know we'd been seen. We got back to camp and they said: Go to the office. There they said: you were at the meeting held by the deputy Houphouet—fifteen days! [Why did you go?] To hear—to listen to what he, the deputy, had to say. It was good for us.[44]

The governor of the Côte d'Ivoire, Péchoux, launched an outright attack on the RDA. The persecution reached its peak in January of 1950, when some seventeen RDA supporters died of their wounds after two serious confrontations at Dimbokro and Bouaflé, and an RDA Senator, Biaka Boda, was murdered in that same period.[45]

Matching their strength in the interior of FWA, the RDA also maintained steady pressure in the National Assembly in Paris, the scene of its greatest triumph. No amount of French gainsaying could detract from the enormous prestige earned by Houphouet-Boigny and the RDA as sponsors in 1946 of the bill that abolished forced labor in the colonies. It came to be known colloquially throughout FWA as the Houphouet-Boigny Law.[46] The *ivoirien* veteran Namongo Ouattara attributed his political activism both to his war experiences and to Houphouet's gesture:

> After the war, General de Gaulle had said that we, the Africans, had suffered greatly. Because of this he wished to raise our pay. General Catroux, who was the commander of our regiment, the 14th R.T.S., at Alexandria, said that if Africans had bread to eat that would suffice for them. Africans don't understand money. . . .
> When I returned to the Côte d'Ivoire in 1947, I found out that Houphouet had already started to end forced labor. And I thought again about the words of General Catroux, and also of all those who had died over there and I thought I will enlist in the cause of liberty for the Africans.[47]

In Parliament the RDA spokesman on veterans' affairs was Hamani Diori, the deputy from Niger.[48] Each year he rose to speak during the debate preceding the vote on annual military credits, usually in November or December, as well as during the debate on supplementary military credits.[49] Diori would stress the inequalities experienced by African soldiers and veterans alike:

> Belonging to different races, natives of different countries, one thing alone unites them [the *Tirailleurs Sénégalais*]: the state of inferiority in which they are placed within the ranks of the French Army.[50]

Like the IOM, the RDA, represented by Diori, called for a "franc for franc" increase in pensions and explained the sleight of hand practiced by the French treasury. Metropolitan Frenchmen received pensions based on a 500 percent increase over the fixed pensions of 1929 francs as stated in the ministry of veterans' affairs manual. Overseas pensions, on the other hand, were calculated on a much lower rate by manipulation. As an illustration, Diori pointed out that while a French sergeant major received a pension of 33,592 francs annually after fifteen years of service, an African at the same rank and service could count on only 3,900 francs.[51] In April of 1949, Diori introduced an RDA-PCF sponsored amendment which would have brought in equality of pensions, but the amendment was defeated when the socialists and the conservatives voted against it.[52] Two months later, the RDA was once again defeated in Parliament in its attempt to pass a "unified statute," that is, a single set of statutes for all soldiers and former soldiers under the French flag. Adopting the veterans' slogan "mêmes sacrifices, mêmes droits" (equal sacrifices, equal rights),[53] the RDA legislation would have had immediate symbolic as well as concrete impact had it been passed. What were felt to be derogatory terms such as "*tirailleur*" would have been replaced by "*soldat*," and no distinctions in financial matters would any longer have been tolerated, whether for salaries, loans, army issue, or retirement pensions. The law would have applied retroactively to all retired career soldiers, effective 1 January 1948.[54]

Of course, the RDA did not confine itself merely to the issue of justice for veterans, and here again it differed from the IOM. While the IOM was willing to back the RDA on equality of pensions, it did not share the RDA's radical antimilitarism. Ouezzin Coulibaly of the RDA, for example, used the pension issue as a springboard for criticism of the *Tirailleurs Sénégalais*, which, he claimed, was being converted into a band of cheap mercenaries to put down French workers or to serve in the dirty war in Indochina.[55]

Back in West Africa, the RDA gave veterans' issues primacy in their agenda. At the party congress in Abidjan in January 1949, its second resolution was directed to redressing of veterans' discrimination.[56] The resolution was linked to a new bill, the so-called Ouezzin Coulibaly Law, which would have had broad benefits for overseas veterans. Not only did it include equality provisions, it proposed a new administrative office for veterans that lay outside politics and would be controlled by the democratically elected veterans' associations. The product of careful briefing from the federal veterans' lobbies, Coulibaly's proposal also contained a long list of secondary requests, ranging from job reservations and subsidized housing to the modernization of the Dakar center for the distribution of artificial limbs. So long as the RDA remained tied to the PCF, such ambitious proposals were easily defeated.

During this period of veterans' militancy, their political gains were modest. The first halting step in the direction of equality for veterans came on 16 January 1947 when a FWA decree set African military pensions at half the size of pensions in France. Though an improvement over lamentably low prewar rates, this administrative action hardly satisfied the West African constituency.[57] Not until August of 1950 was a majority of deputies in the National Assembly prepared to go at least part of the way suggested by the RDA by passing the so-called Equality Law.[58] Among its shortcomings from the perspective of the African military, the 1950 Law did not extend to Africans the temporary indemnity which was paid to European soldiers and officers on overseas duty. Nor were the increases retroactive or even immediate for all categories of payments to veterans. The disabled, for example, had to wait until 1952 until the equalization rates were applied to them. Certain benefits, such as cash advances on pensions, were, on the advice of colonial officials who felt such privileges an unnecessary indulgence for Africans, never applied to FWA.[59] Still, the 1950 Law remains a benchmark in the political struggle of West African veterans to receive their due. It conveniently marks the end of the era of militancy for veterans and ushers in a time of accommodation. To understand this partial success it is necessary to examine the activities of the bureaucratic organization constructed to control African veterans, the *Office National des Anciens Combattants* (hereafter referred to as *Office*).

The *Office National des Anciens Combattants*

Most French authorities came to understand the importance of keeping veterans as part of an alliance if a reformed French colonialism was to survive in the changed postwar world. From the early 1920s the Ministry of Veterans' Affairs had established a bureau called the *Office National des Anciens Combattants* at the Hôtel des Invalides in Paris. In 1948, a branch of the *Office* was created in Dakar.[60] Whereas in France the *Office* was expected to treat veterans' affairs bureaucratically but not politically, in FWA it was also designed to disseminate government propaganda and to monitor and, if possible, control the various veterans' lobbies which had mushroomed all over the federation.[61] The propaganda function was handled through the publication of a "Bulletin d'information," which was put together in Paris and offered the French government's perspective on its policies in North Africa and Indochina, for example.[62]

Attempted political control and manipulation of veterans was the responsibility of the presiding officer of the *Office*, René Traodec. A veteran himself and a strident Gaullist, Traodec ran the West African *Office* autocratically from its creation in 1948 until 1953, when he was forced to step down as a result of African political pressure. The *Office*, it is true, did have an Executive Council that purported to represent the various interest groups in FWA. The chairman of the Executive was the colonel in charge of the *cabinet militaire* of the CIC in Dakar, and he was joined by seven other *ex officio* members from the highest ranks of the civilian and military bureaucracy of FWA. To represent veterans some twenty-one appointees were named by the colonial government. In 1949 serving members were two grand counsellors of the FWA Federation, Robert Delmas and his African colleague, Sogui Diouf, together with Papa Seck Douta from AACVGAOF, and eight other European and ten other Africans.[63] In 1953 the Council consisted of thirty-three members, including sixteen Africans, which meant that Europeans still were in a majority.[64]

Membership aside, the Executive Council was not designed to be much more than window dressing for Traodec. The Executive met only once annually and discussed only the most routine questions. In 1953, for example, when Traodec's term was not being renewed because of African political pressure, the Executive minutes show that it dealt only with the question of when it would be possible to include soldiers from Indochina as war veterans, and whether the *Office* itself was complying with the general administrative obligation to offer veterans the first chance at reserved jobs.[65]

The *Office* received a substantial budget to cover its various functions. In 1948, for example, 12,742,400 francs CFA were apportioned, 72 percent coming from the ministry of veterans' affairs in France, and the rest, 28 percent, representing the contribution of the Federation of FWA.[66] In the following year, the overall budget had fallen to 11,625,000 francs CFA, but FWA's contribution had increased to 35 percent, or slightly over four million francs CFA. In the initial years of operation of the *Office*, most of these expenditures were for the heavy capital costs involved in building various veterans' centers across FWA, as well as in financing a series of missions into the interior to locate and register military claimants. By 1953, the primary expenses of the *Office* had been reduced to a partial annual subsidy to the territorial veterans' associations, and the overall budget had fallen to 6.5 million francs CFA, not without protest from both Seck Douta of the AACVGAOF and the high commissioner of FWA.[67] Of course, expenses in real terms were reduced even further by the rapid postwar inflation.

Table 9.1 breaks down that portion of the *Office*'s budget allocated as direct subsidies to the various territorial associations. It shows clearly that Dakar and Senegal received a substantial percentage of these funds. While this may in part reflect the gravitation toward the federal capital of veterans originally from the provinces, it probably also reflected the officials' response to the volume of political pressure being generated by Dakar-based veterans' lobbies.

TABLE 9.1 *Office National*'s budget allocated by territory, 1948–1949

	1948		1949	
Colony	CFA Fr.	Percentage	CFA Fr.	Percentage
Soudan	333,750	13	440,000	14
Guinea	263,750	10	430,000	14
Côte d'Ivoire	313,750	12	430,000	14
Upper Volta	313,750	12	485,000	15
Dahomey	283,750	11	255,000	8
Niger	176,750	7	255,000	8
Senegal and Mauritania	556,250	22	440,000	14
Dakar	335,750	13	415,000	13
FWA totals	2,577,500	100	3,150,000	100

Source: Traodec Report, Dakar, April 1949, ANS, 4D178 144.

The Struggle for Veterans' Support

During the years of veterans' militancy from 1945 to 1950, the political parties and the colonial administration engaged in a major battle for the allegiance of African veterans. On the French side, the reports and correspondence of René Traodec as secretary-general of the *Office* revealed a paternalist mind at work. Traodec candidly outlined his tactics. The executive membership of each veterans' territorial organization should be rigged in such a way that a European remained in control. An African veteran could be encouraged to take the position of president, but in an entirely honorific manner. A European veteran, preferably acting as treasurer, would assure fiscal honesty and control. Traodec went one step further. The local associations could be permitted to plead the veterans' cause in rural areas, and to lobby the private companies for jobs. Above all they were to stay out of politics and to be loyal and patriotic to France.[68]

At first glance, it might have appeared that Traodec had succeeded admirably in establishing a "shadow cabinet" of well-placed European veterans. In Senegal, Robert Delmas, member of an important French merchant family, figured prominently in veterans' affairs.[69] For a time in 1947, he actually alternated with Seck Douta as federal president of the AACVGAOF. In 1948 he headed the Dakar local, but a year later he let Samuel Baye take over the presidency while he remained on in the Executive as honorary president.[70] In Côte d'Ivoire, a French veteran named Olivier Reinach was the driving force in veterans' politics, but with his African protégé, Philippe Yacé, taking the post of president.[71] In Soudan, the French *eminence grise* was a veteran named Muller. A former major in the army and an employee of the French firm Peyrissac in Bamako, Muller served as treasurer of the Soudan veterans.[72] In Dahomey, a French civil servant named Montout, head of the finance section of the Dahomean administration, was association treasurer while a Dahomean, Louis Sodegbé, served as president.[73] In Guinea and Niger, it seemed that French veterans felt confident enough not to bother even finding an African front man. A French veteran and amputee named Durand was president of the Conakry veterans, while a French veteran, Joubert, who was *trésorier-payeur* of the territory of Niger, served as president of the Niamey veterans' local.[74]

To be sure, many Europeans in FWA may indeed have shared Traodec's preference that they continue to control veterans' affairs. Where they differed, however, is in the degree to which they allowed partisan domestic politics to operate so explicitly. Traodec proved to be an ardent and partisan Gaullist. Despite his call to keep politics outside veterans' issues, in fact, his political bias was so marked that he had great difficulty working with any Frenchman who did not share his political preferences. He constantly railed against Reinach, who was a prominent socialist.[75] He attempted unsuccessfully to have Montout, a French disabled veteran whom he labelled a "crook," but who was a political radical, voted out of office by the Dahomean veterans in favor of a Gaullist veteran, Mathie, head of the transport office in Cotonou.[76] Nor could Traodec have been pleased with the political affiliation of Robert Delmas, who attached his political future to Léopold Senghor at an early stage of the postwar era in Senegal and went on to play an important role in Senghor's BDS.[77]

The view of Traodec and others that veterans' politics was a case of French manipulation and African compliance lacked depth of understanding of the African veterans' interest. From an African perspective, letting French veterans have a prominent position in their associations made perfectly good sense. First, the Frenchmen were former officers, to whom they still felt bound as clients and ex-soldiers. Moreover, these French veterans were extremely influential people in the public and private sectors of FWA life, people very well placed to offer jobs and other favors to their former comrades-in-arms. African veterans understood better than Traodec the value of having close ties with private companies and well-placed colonial officials.

Sometimes African expectations of their fellow French veterans were too great. While it was certainly possible for the *trésorier-payeur* of Niger to make sure that his bureau observed the law and offered reserved jobs to African veterans as a priority, it was quite another matter, for example, for a judge to show favoritism. This seems to have been the expectation of African veterans in the case of a French justice of the peace named Schmitt, in Katiola *cercle*, Côte d'Ivoire.

The predominantly African veterans of Katiola had elected Schmitt president of their local of the AACVG of Côte d'Ivoire. If some veterans supposed that ties of solidarity would serve them well if they came before the bench, they were disappointed. Schmitt had sentenced an old and respected veteran, an ex-sergeant major named Maliké Keita, to a fine of 2,000 francs CFA for having killed an elephant in the classified forest of Niangbo. The law allowed a fine of from 50 to 10,000 francs CFA and a jail sentence of up to one year, so Schmitt reasoned that he had given a moderate sentence. Keita, however, thought otherwise, and took this decision to the court of appeals in Abidjan, which acquitted him. Soon after, most Africans quit the Katiola veterans' local in protest, and would not return until Schmitt was replaced.[78] Schmitt lamented that African veterans were punishing him for their failure to comprehend the French system of justice.[79]

The colonial administration occasionally did succeed in its goal of having compliant Africans fronting for controlling Europeans. In Kankan, Guinea, for example, in the elections of their local executive, veterans were offered only one slate of sixteen men headed by former Lieutenant Sidibé Sala. All these men had reserved jobs, and one was the chief of the Kankan police force. Veterans in each canton of Kankan had the option either to accept or reject the slate, which was voted in by a count of fourteen to three. The *commandant de cercle* was present at the election and offered a gift of 4,000 francs from the local budget to buy a cow in celebration of the victory of the slate.[80] It would have taken considerable courage for three cantons to have said no to all of this state pressure.

Another recourse of the colonial administration was to bribe or bully local veterans. In Dinguiray, Guinea, a former lieutenant in the *Tirailleurs Sénégalais* named Abdou Karim was persuaded to resign from the RDA and cooperate as a "loyal servant" of France in exchange for a reserved job in the local administration.[81] In Niger, where many of the veterans were in fact Sara ex-soldiers from Chad who had signed up illegally as *Tirailleurs Sénégalais* soldiers under the Niger quota, veterans were threatened with confiscation of their veterans' cards and deportation to Chad if they were known to be RDA activists.[82] It appears also that the political affiliation of veterans had considerable bearing on whether they could expect prompt and fair treatment from the *Office*. In Leo, Upper Volta, in 1949, two ex-adjutants, Bahouna Namoro and Dibo Liba, were identified as RDA organizers. Both had requested advances on their pensions but this "privilege" was being held up pending confirmation of their politics.[83]

As a last resort, the colonial administration could remove independent-minded African veterans from positions of authority. One dramatic example was Traodec's attempt to trump up charges against Souleyman Cissé, a Guinean subject living in Bobo-Dioulasso, Upper Volta.[84] An employee of the French firm SCOA and an RDA militant, Cissé was also president of the local veterans' association, much to the dismay of Traodec, who backed Ouattara Niandegué, a former lieutenant who supported de Gaulle's *Rassemblement du Peuple Français*, or RPF. With Traodec's interference, the Bobo-Dioulasso veterans did split politically, with a minority following Ouattara and breaking off to form their own independent *Amicale des Anciens Combattants*.[85] Ouattara also appealed to Seck Douta at the AACVGAOF to investigate Cissé for having personally appropriated veterans' funds. Seck Douta appointed a commission, which relieved Cissé of his functions and named Ouattara as provisional president pending an election in which only "legal" veterans, that is, those against whom there was no litigation, could vote. Unfortunately for Traodec, the year-end elections in 1949 saw Cissé triumph over Ouattara by the resounding vote of 1,800 to 17.[86] A new investigation by Seck Douta this time exonerated Cissé. A humbled Traodec was forced to admit that Cissé's "crime" had been to refuse to turn over the association's books (which was his right) rather than any hard evidence of fraud.[87] The dispute ended formally in 1951 when Ouattara issued a formal apology to Cissé for having brought a false accusation forward.[88] It was no doubt also significant that by 1951 the RDA had made its peace with the colonial administration and the harassment of its militants was diminishing.

The Cissé affair shows the power of the RDA in the struggle to maintain control of veterans. While unable to draw on the resources that the *Office* could use, the RDA did have some advantages. In Côte d'Ivoire, Houphouet-Boigny was able to integrate veterans as militants into his local RDA branches to such an extent that it was often difficult to tell where one organization ended and the other began.[89] For example, the head of the ex-POWs of Côte d'Ivoire, Vamoutoré Touré, was also the president of the RDA branch in Tiébissou. In 1947 the RDA secretary-general, Denise, provided Touré with a large sum of money from the party treasury to distribute as a gift to ex-POWs. Denise told party organizers throughout Côte d'Ivoire to pass the names of all ex-POWs to Touré. Not only was this not illegal, the RDA was demonstrating that it could deliver benefits to veterans more efficiently than the State.[90] Similarly, when Commandant Liger visited the *cercle* of Toumodi, Côte d'Ivoire, in 1949 to register veterans, he discovered that he simply could not get his work done without RDA support. As he made his formal address to the assembled veterans, an RDA militant declared: "Mr. Houphouet is the only one in charge here. If you expect to get the job done, you'd better get hold of the RDA Party Secretary who looks after us."[91] When Liger replied that he would register veterans only with the help of a government clerk and the interpreter of the subdivision, two-thirds of his audience of one hundred veterans walked out on him.[92] He was forced to make a return visit in 1951 to RDA strongholds in Côte d'Ivoire and Upper Volta to complete his task.[93]

The Soudan was another territory where RDA strength among veterans was significant. As early as December of 1946, intelligence reports had commented on the rising star of one Modibo Keita, a schoolteacher and "RDA agitator" in the Sikasso region who was said to be attracting a significant following among veterans because the local *commandant de cercle* was from the "old colonial school" and refused to moderate his ways.[94] While Keita's political success can certainly be attributed to more than the fortuitous opposition of a bungling colonial official, it is nevertheless interesting to see how he, like all ambitious West African politicians of the day, saw the importance of getting the veterans on his side at an early stage of his career.[95]

A third RDA stronghold relying on veterans was Niger. Hamani Diori, the RDA spokesman on veterans' issues in the National Assembly, was from Niger and took some pains to maintain this base of support. He was said to have used party funds to "bribe" veterans with cola, sheep, and cash payments.[96] It would seem that in Niger, as in Côte d'Ivoire, the RDA could communicate with its members who were veterans quickly and effectively. Liger noted that the RDA could transmit news of pension changes from Paris to Niger faster than the colonial administration could, and that Diori took credit for all improvements in veterans' benefits that were occurring.[97]

While these case studies do not by themselves indicate the outcome of the struggle between Traodec and the RDA for the loyalty of veterans, it would seem that by 1950 African veterans all over FWA had grown tired of the *Office* secretary-general's high-handed ways.[98] Through the good offices of Lamine Guèye, Seck Douta, on behalf of the AACVGAOF, pushed hard for Traodec's replacement, together with the nomination of an African as assistant secretary-general. Lamine Guèye, in turn, dutifully wrote his fellow socialist, François Mitterand, who had just become minister of overseas France, noting that Traodec refused ever to consult African political representatives, that under him the *Office* used "authoritarian colonial methods," and that he did not consult his executive committee either.[99] While Mitterand backed his official publicly, he queried the FWA administration as to whether Traodec should be kept on. The high commissioner of FWA defended

Traodec vehemently as an invaluable expert, a veteran who had himself fought alongside Africans and knew their habits well, a man who had done wonders on a small budget, and who should be renewed, but perhaps only for two years.[100]

Traodec completed his term in 1952. He was replaced by M. Maylie, another European veteran, with an African scheduled to be named at some later date as assistant secretary-general.[101] By 1957, when Africanization was making significant strides in the military generally, the first African secretary-general of the *Office* was named when retired Lieutenant Colonel Soumaré, by then the most senior African officer in the French army, was appointed.[102]

The Liger Campaign: Veterans Accommodated, 1950–1956

In the competition for the allegiance of French West African veterans, the issue of financial compensation was perhaps the most important element of all. The RDA had recognized this and taken pains to offer gifts to veterans from time to time. The *Office*, on the other hand, had concentrated its efforts on the building of veterans' lodges and the inducement of reserved jobs, but had done precious little as late as 1948 to facilitate the payment of various forms of veterans' compensation.

In the new atmosphere of the French union, the State's failure to honor its commitments to African veterans had become a scandal. To be sure, the "old" colonialism had neglected its obligations to African veterans of the First World War. Yet the same tradition of cynicism and parsimony appeared to be continuing, despite the pressure of African politicians and veterans alike. It was estimated in 1948 that over 12,000 career veterans had been waiting ten years for full pensions and were having to make do with sporadic and politically manipulated cash advances.[103] Bureaucratic delays were interminable. A senior official cited a case in Guinea, which he described as being one of the more efficient areas of military administration, in which it had taken three years from the time the military intendant at Kindia first wrote Paris until a decision was actually made to grant an African career soldier his pension, and then another two years for the soldier to get paid, after some twenty letters had been exchanged.[104]

Realizing a crisis was at hand, French authorities finally took steps to bring about a change on 27 March 1948 by naming Henri Liger to head a major registration campaign.[105] Liger's purpose was to conduct a census of veterans of both wars, to provide them with official certificates if possible, and in general to see that justice was done them. By the end of his mission in 1950, he would have traveled the length and breadth of FWA and opened no less than 250,000 dossiers, registering the claims of widows, orphans, and ex-soldiers alike for back pay and various other forms of compensation.

The Liger campaign was a large operation which grew even more substantial as it proceeded. Although the campaign was originally designed to address the claims of Second World War veterans, it soon became clear that because little had ever been done to compensate veterans of earlier campaigns, these men or their surviving dependents could not be refused attention when they came forward.[106] Thus, whereas Liger began in October of 1948 with a staff of fifteen French noncommissioned officers as secretaries, their number had doubled by early 1949. The task was prodigious, as Liger discovered upon completion of his pilot project in Guinea in the opening months of 1949. It had taken a trained army secretary two hours to open a file and prepare a duplicate *livret*, or record book, for each veteran. The 3,200 outstanding cases in Guinea alone represented thirty months' work, the entire tour of duty for a single noncom. As a result, Liger recommended that the one specialized noncom be assisted by three civilian secretaries, to be lent by the colonial administration in each territory, or the campaign would fail.[107] Typically of joint military-civilian operations in the past, however, Liger would later complain that the civilian authorities never gave his campaign the high priority it deserved.[108]

By any measure, the indefatigable Henri Liger achieved an enormous success. In slightly over one year he logged no less than 48,000 kilometers on the ground and 180 hours in the air, and deposited at least fifteen detailed reports on his findings.[109] In three months alone, from January to March, 1949, he visited every *cercle* in Côte d'Ivoire, Upper Volta, Dahomey, and western Niger, traveling some fourteen thousand kilometers by military truck over sometimes rudimentary roads.[110] As Table 9.2 indicates, by July of 1950 Liger had settled 162,692 cases, or slightly over 60 percent, with another 30 percent pending and roughly 10 percent still to be contacted. Despite his prodigious efforts, it remained true that five years after the war 40 percent of the veterans and their families were still awaiting resolution of their claims. By the end of 1952 it was reported that while cases for Second World War veterans had finally been processed, 1,586 pre-1939 cases were still pending, held up because they lacked full dossiers.[111]

By the mid-1950s African veterans no longer felt grossly mistreated. Even if the French treasury had saved millions of francs by delaying justice to thousands of African veterans, years of colonialism had left few Africans so naive as to have expected full equality with French nationals. A somewhat obscure figure in the history of French colonialism in West Africa, Henri Liger certainly deserves more recognition than he has received in general histories. The success of his campaign did far more than the machinations of Traodec at the *Office* to turn the battle for

TABLE 9.2 Results of the Liger campaign as of July 1950

	Vets of 1914–38	Vets of 1939–45	Families of Deceased	Total Seen (A)	Cases Pending (B)	Not Yet Seen* (C)	A+B+C
Soudan	10,297	30,767	1,726	42,790	17,215	6,000	66,005
Guinea	4,259	16,222	1,374	21,855	10,016	1,594	33,465
Côte d'Ivoire	2,441	22,144	712	25,297	9,609	6,981	41,887
Upper Volta	5,010	26,031	1,524	32,565	14,833	9,480	56,878
Dahomey	1,325	8,782	468	10,575	4,601	759	15,935
Niger	1,181	1,979	188	3,348	1,527	975	5,850
Senegal	4,424	18,528	2,072	25,024	16,449	—	41,473
Mauritania	443	686	109	1,238	1,011	450	2,699
FWA	29,380	125,139	8,173	162,692	81,261	20,720	264,673

*Based on Liger's estimates.
Source: Liger, "Fin de mission en A.O.F.," Dakar, 13 July 1950, ANS, 4D178 144.

veterans' support back in favor of the Colonial State. Political decisions in Paris also helped, of course. The passage of the Equality Law in August, 1950, even if it fell short of truly even-handed treatment for African as well as French veterans, was another major element in the winning back of veterans' support. Thus, by the end of 1950, as African veterans found themselves registered and, more importantly, receiving cash payments from the treasury, their collective interest changed. The period of accommodation was at hand.

In the wider arena of African politics, a similar and related period of accommodation occurred at roughly the same time. While official harassment of the RDA in Côte d'Ivoire in particular remained intense, in 1950 Félix Houphouet-Boigny began a major shift in his and the RDA's political orientation. He removed the pro-communist Gabriel d'Arboussier as secretary-general and in October of that same year ended the RDA's parliamentary alliance with the PCF. Although it took a year for the effects of this shift to be felt entirely in FWA, by the end of 1951 it was clear that a new period of accommodation applied to FWA as a whole.[112]

One result of the changed climate was that the RDA no longer felt it necessary or politically profitable to champion the cause of veterans. A studied RDA silence occurred in the National Assembly and in the territorial press. In the period of accommodation from 1950 to 1956, in Soudan, for example, *L'Essor*, the newspaper of the *Union Soudanaise du RDA*, ran only minor stories touching on veterans' affairs. In 1952 there appeared, straight from the military intendant's bureau, a list of all the conditions of eligibility required to be the beneficiary of a now deceased former POW.[113] Similarly, the next year, *L'Essor* ran a letter from the RDA deputy from Soudan, Mamadou Konaté, to the minister of defense, inquiring about pension rights for illegitimate children.[114]

The moderate grouping of African parliamentarians, the IOM, adopted a similar policy. In February, 1950 Senghor used the pages of his party's newspaper, *Condition humaine*, to applaud the major improvement that occurred in the fall of 1949 when the French government raised African military pensions to one half those of their French brothers. He urged the government to take the last step to full equality.[115] This proved to be the last feature article run by Senghor's newspaper on veterans' affairs. Attention shifted to the fight for a labor code, to the so-called "war of the peanuts," and to issues of concern to the wider African society. Senghor now left veterans' issues to others in the National Assembly, a sign no doubt that in his view veterans were no longer an abused group.[116]

The new spokesman for the IOM in Paris during the years of accommodation was Mamba Sano, the Deputy from Guinea. In the early 1950s he rose on three occasions in the National Assembly to raise veterans' issues.[117] Beginning his remarks by noting that "the enemy's bullet never discriminated on the basis of race," Sano stressed how his party's efforts, and those of other men of good will, had suppressed the more flagrant anomalies, and that the principle of equality of pensions had been formally recognized by France. He worried, however, about the erosion in value of these pensions because of the exchange rate of the CFA franc. After a decree of 4 December 1954 introduced an index of correction for African and other overseas pensions to cover losses in exchange, Sano confined himself to the relatively modest complaints over the acute shortage of centers to outfit artificial limbs for disabled veterans and the slow pace at which veterans' lodges were being built.

Thus, the first half of the 1950s resembled a brief golden age for African veterans. While it is an error to conclude that "by 1950 the battle for equal treatment was substantially won,"[118] five years later this statement would be valid. The Equality Law of 1950 was followed two years later by a FWA decree definitively introducing equal pensions to French West African servicemen.[119] In December of 1954, the French government introduced indexation of pensions paid in CFA francs to protect veterans against any future devaluation.[120] Of course, these significant gains were achieved at a political and social cost. Once again the African military came to be viewed by the general population as conservative intermediaries tied too closely to French rather than African interests.[121] No doubt this accounted for the sudden lack of enthusiasm for veterans displayed by their erstwhile allies, the African politicians.

Years of Apprehension, 1956–1960

While African veterans experienced postwar years of militancy to 1950, and then years of accommodation until 1956, the remaining years of the decade marked a time of apprehension. This third period of the postwar era saw the triumph of regionalism, and with it a real danger to the collective interests of veterans. The instrument of this decentralization in FWA was the 1956 *Loi-cadre*, or Framework Law, introduced by Gaston Defferre, then minister of overseas France, with the active backing of Félix Houphouet-Boigny.[122] The passage of this legislation in the waning days of the Fourth Republic signaled the end of the federal structure which had existed in FWA for fifty years. While the old federation with its capital at Dakar had the advantage of scale, most African politicians believed, as did Houphouet-Boigny, that the colonial administration at Dakar was too closely tied to Paris and difficult for them to penetrate. The new framework of eight territorial governments created in 1956 was a weaker structure, but easier to control. African politicians recognized that French political authorities in each territory could be more easily influenced, and, more importantly, that each unit could be an effective springboard for permanent sovereignty should independence become an available option. With France forced to cede independence to Morocco and Tunisia in 1956, and with the British in the same position in the Gold Coast in 1957, pressures in this direction were growing stronger.

Not all politicians saw the *Loi-cadre* as a positive step. Senghor, who vehemently opposed the legislation, correctly predicted it would lead to the "balkanization" of FWA. Indeed, by the end of 1960 each and every one of the former territories had become independent.[123]

By 1957, major political tensions had returned to FWA, this time brought on by France's unpopular war in Algeria and the approaching of independence in the nearby British colony of Gold Coast. Once again, veterans' issues began to be taken up by the major political parties, a sign that even if veterans were no longer at the forefront of political events, the military vote was nevertheless still valuable, and, perhaps, volatile. As official administrative reports made clear, the veterans' propensity to vote and to help the various parties convince others to do likewise in rural areas continued to make them a desirable element in any African political party with serious electoral aspirations.[124]

As a leading force in FWA, the RDA was quick to respond to veterans' issues once again. In the Soudan, the RDA newspaper, *L'Essor*, together with its sister organ for youth, *Tribune des jeunes*, were cited by military intelligence as being Communist mouthpieces for daring to protest against the use of the *Tirailleurs Sénégalais* in the Middle East and in Algeria.[125] Another RDA paper, *L'Action*, listed a series of veterans' demands that might have been taken directly from the AACVGAOF portfolio: "franc for franc" payment of pensions; wider allocation of reserved jobs; free hospitalization for veterans and their families; faster, and even monthly, payment of pensions as was permitted in France.[126] At the Bamako Congress of the RDA in 1957, veterans succeeded in adding their demands to the party platform. They requested and received RDA backing for formal veterans' representation on the councils of each territorial government, together with the party's promise to reserve 10 percent of all college and university bursaries for sons of veterans and war victims.[127]

These were difficult years politically for Senghor and the BDS. He had strenuously opposed the *Loi-cadre*, not only because he feared the consequences of regionalism but also because he recognized that Senegal had the most to lose in a decentralized FWA. Nor did he find any satisfaction in the dying Fourth Republic's agony in Algeria. To maintain solidarity with African veterans, Senghor and the BPS reluctantly supported a boycott against the Bastille Day celebrations of 14 July 1957, as well as the decision of veterans in Saint-Louis to boycott the French centenary celebrations of Faidherbe's founding of the *Tirailleurs Sénégalais*, which were planned for 24 and 25 August 1957.[128]

African soldiers and veterans alike had several reasons to be apprehensive about the breakup of the federation of FWA. First, members of the African military may have been the strongest federalists in FWA. For them the federation, like the army itself, had been a unifying element, an attractive substitute for ties of ethnicity and regionalism. The *ivoirien* veteran Daouda Tuo Donatoho put it this way:

> There were Africans from all over, but we all wore the same uniforms. Some had different tribal scars—a bit different, but we all wore the same uniform. If someone asked where I was from I'd say: the *Côte d'Ivoire*. Before, at home, I'd have said: Korhogo.[129]

Second, the specter of independence posed a financial and moral threat. For soldiers and veterans, their military and, in a sense, social contract had been with France, but independence might very well mean the transfer of pension obligations and control to what they believed were often venal local politicians. Also, with no African deputies to represent them any longer in Paris, they feared, with some justification as it turned out, the loss of their political leverage.[130]

The French counterinsurgency war in Algeria did not make the choices of the African military any easier. While career soldiers continued to serve under the French flag in what by now had become an unpopular repression, many ex-soldiers expressed their disapproval in a vivid manner by refusing what for them had always been an important annual ritual, the honoring of their fallen comrades in conjunction with Bastille Day ceremonies. For the military, then, the conjuncture of the war in Algeria and the *Loi-cadre* heightened a conflict in loyalties. The dreaded day was fast approaching when they would have to choose between France and Africa.

10

Epilogue: Collective Memory and the African Military

> *France is our mother.*
> *It is she who feeds us*
> *With potatoes*
> *And rotten beans!*
> Tirailleur marching song of the Second World War[1]

This marching song effectively summarized the contradictory experiences of Africans under French colors. In the early years, France used the coerced to coerce others by employing what was essentially a slave army of *Tirailleurs* to impose its rule in West Africa. Africans recognized this feature of the early *Tirailleurs Sénégalais* forces. Soldiers who took on engagements were called "children of the government,"[2] a phrase which could equally cover servile origins as well as the substitution of government service for the extended family. Indeed, in the cases where masters turned over slaves to the army to receive the enlistment bonus as compensation, it could literally be said that the army had in fact become the new master.

As the *Tirailleurs Sénégalais* evolved with the changing nature of French colonialism, patron-client ties came to require ideological reinforcement. The essential ingredients in the dominant Republican ideology came to be extended first to African cadres, and then even to the rank and file. African soldiers, like their French brothers-in-arms, were bound by nonclass elements such as honor, duty, and patriotism to defend France, the home of the enlightenment and of the *mission civilisatrice*. Yet this universalistic national and imperial ideology was often combined with the crude racism of white officers and of civilian society. In France, the depiction of a grinning African *Tirailleur* in the billboard advertisements of "Banania," a popular breakfast food, symbolized the contempt with which France's military "servants" were viewed. The caricatured soldier was made to proclaim in *petit nègre* French, "Y a bon banania" (an approximate colloquial translation might be "it be good, Banania").[3]

Such contradictions were particularly painful to the assimilated African elite. Because Léopold Senghor shared part of the African soldiers' experience in war and captivity, he could not contain his anger and frustration at this treatment. He penned these lines at the end of the Second World War:

> You are not poor men, with nothing in your pockets, without honor.
> I will tear down the *banania* smiles from every wall in France.[4]

A distinguished grammarian and a soldier himself, Senghor was doubly provoked to think that France did not sufficiently appreciate either the strides achieved in modernization of French Africa or the enormous sacrifices African soldiers had made on behalf of the metropolis.

Still another contradiction occurred in the contrast between French and African experiences upon discharge from the army. Whereas all soldiers were obliged to accept a command structure while in the army, Frenchmen, once their military obligation was fulfilled, could return to a society with democratic norms. Colonial subjects, on the other hand, upon returning to civilian life, continued to encounter a command structure ruling over their daily lives. Indeed, it might be said that of the two arbitrary worlds of colonialism and militarism, the latter was preferable in that it had more clearly defined rules and traditions.

Reactions like those of Senghor highlighted the need for modifications in colonial military rhetoric after 1945. The French army, however, was unable to articulate a military equivalent of the new colonial ideology of Brazzaville and the French union. No doubt their difficulty was a function, in part, of their unwillingness to understand postwar changes and their unhappiness over military setbacks in Indochina and North Africa. In the end, French officers stumbled towards an integrationist ideology of *"promotion africaine,"* the Africanization of African officers, without believing themselves that the French Fourth Republic and its empire could be saved in this way. Africans, for their part, could juxtapose the contradictory rhetoric of universal citizenship and professional advancement with the meager evidence that military service had done much to improve education and job prospects. In reality, postwar society was not capable of absorbing ex-soldiers readily in ways that could make even partial use of their past experiences. African soldiers and veterans could not help but be frustrated over such deeply contradictory signals.

African civilian and military figures in the postwar era were not merely passive recipients of a received ideology. They stood to gain from fuller imperial integration, and, therefore they modified Republican military ideology to suit their interests. The African military, in particular, developed its own peculiar variant of the Jacobin military ideal. Their version of how the Second World War unfolded was certainly at odds with the dominant Gaullist view. African solders, veterans, and civilians argued that their fathers and brothers had fought and died (and with more enthusiasm than did their French brothers) to save France and humanity from Nazi totalitarianism, but also to win equality and brotherhood for all in the French union, thus completing the Revolutionary ideals of liberty, fraternity, and equality.

It is important to recognize that these African modifications to military ideology had new overtones and implications. As articulated by politicians, there was a potential threat. The new French family or union might have to face separation and even divorce if the debt to colonials was not honored. Thus Doudou Guèye, a Senegalese radical nationalist who was imprisoned for a time for his politics while editor of *Réveil* in 1950, could later write angrily in *L'Action* about the shameful neglect of African military heroes. Protesting French veneration of the explorer Savorgnan de Brazza while neglecting the accomplishments of his Senegalese sergeant, Malamine, Guèye cried out for commemoration of those African soldiers "who gave their lives on battlefields all over the world to save the honor of the French nation."[5]

The vision of African military accomplishment in French uniform became a powerful symbol of political and even cultural consciousness. The exploits of the *Tirailleurs Sénégalais* in a sense took on a life of their own, and those who questioned this aspect of ideology did so at their peril. The Senegalese historian and politician Abdoulaye Ly, for example, dared to belittle the achievements of the *Tirailleurs Sénégalais* in a speech, later expanded and published as a pamphlet by *Présence Africaine*, entitled *Mercenaires noirs*.[6] Failing to distinguish between the hundreds of thousands of conscripts who were obliged to serve in the *Tirailleurs Sénégalais* and those who volunteered to become career soldiers, Ly drew analogies between the *Tirailleurs* and those "barbarian" Scythian archers who ravaged cities of classical Greece. Sparing neither the left nor the right, he attacked all African politicians for having hypocritically championed the cause of soldiers and veterans in the National Assembly, and chastised the African press for following the French lead in celebrating the black soldier along with the jazz musician as the archetype "nigger." Ly's attack on the *Tirailleurs Sénégalais* was denounced at the time and long remembered as a grave political blunder.[7]

Abdoulaye Ly failed to recognize that the *Tirailleurs Sénégalais* had become incorporated as part of the African collective memory. What Africans chose to preserve of the history of the *Tirailleurs Sénégalais* was not precisely the same as what was remembered in France, or what French officials wished to see remembered. The best illustration of this is the attempt to commemorate three tragic events of the Second World War, the seedbed of the new African military ideology.

Two of these events are obscure today, despite their magnitude. One was the murder of Captain Charles N'Tchoréré on the lower Somme near Airaines on 7 June 1940; the second was the massacre of African *Tirailleurs* at Chasselay-Montluzin, outside Lyons on 17 June 1940. The third disaster, and the one best remembered by Africans, was the uprising at Thiaroye, and its bloody aftermath, described in Chapter 6.

Captain Charles N'Tchoréré was one of the few African officers to have served in the *Tirailleurs Sénégalais* in both World Wars.[8] By one of the terrible ironies of war, he led a company of the Fifty-third RICMS in defense of Airaines, a village on the Somme not far from Amiens, and was killed on 7 June, the very same day his son, Corporal Jean-Baptiste N'Tchoréré, fell, also on the lower Somme. In a letter he had written to his son, N'Tchoréré revealed his deep commitment to the assimilationist ideology: "I have here in front of me your last letter; how proud I am to find in it the phrase, 'Whatever happens papa, I will always be ready to defend our dear country France.'"[9] N'Tchoréré's honor as a proud French officer was to cost him his life. Whereas his son was killed in combat, Captain Charles N'Tchoréré was taken prisoner. With French officers as eyewitnesses, N'Tchoréré was ordered by a young

Memorial at Airaines to the Fifty-third *Régiment d'Infanterie Coloniale Mixte Sénégalais*
Text reads in translation:
"To 1200 glorious dead African *Tirailleurs* of the Fifty-third Régiment d'Infanterie Coloniale Mixte Sénégalais (RICMS) who fought on the strong points of the Somme held by the Fifth Division of the Infanterie Coloniale, on 5, 6, and 7 June 1940.
In dying they fell united fraternally so you could remain French.
Homage of the town of Airaines."

Panzer officer to place his hands over his head and to stand with the African rank-and-file troops.[10] When N'Tchoréré refused, insisting on his right to be treated equally with the other French officers, his German captor shot him on the spot.[11]

Chasselay-Montluzin is an obscure village located above the Rhone Valley, some twenty kilometers northwest of Lyons. In 1940 it lay close to the main road artery to Lyons, the *Nationale 7*.[12] On 17 June, the Twenty-fifth Regiment of the *Tirailleurs Sénégalais* was ordered to defend this main entry to Lyons and was deployed throughout the village. Some one hundred Africans, with a 75 mm

Memorial at Airaines to Captain Charles N'Tchoréré.
Text reads in translation:
"To Captain N'Tchoréré who died a hero on 7 June 1940 and to all the soldiers of Black Africa who shed their blood for France."

battery, occupied a Roman Catholic convent at Montluzin, while other companies in the regiment took up positions elsewhere in the area. The convent became the focus of a bitter five-hour battle. When it fell, a German officer reprimanded Mother Superior Sister Clothilde for allowing her convent to be used as a defensive strong point. She replied, "We and our goods belong to France."[13] Awarded the *Croix de Guerre* after the Liberation, the Mother Superior was then able to indicate publicly what could not be revealed during Vichy rule. She stated that after the Panzers had stormed the convent, they angrily lined up the 212 surviving African wounded against the convent wall and executed them by firing squad. The men were buried later by local villagers very near to Chasselay.

These two bloody incidents have subsequently become monuments to collective memory in both a figurative and literal sense. The French government has taken steps to preserve the memory of these African victims of German brutality who "died for France" in such a tragic manner. In the small town of Airaines in northern France, the Ministry of Veterans' Affairs, in collaboration with the local population, erected a statue to the fallen Gabonese captain, Charles N'Tchoréré, as well as a monument to other Africans of the Fifty-third RICMS who fought to defend the Somme.[14] At Chasselay-Montluzin after the war, the cemetery for the fallen African soldiers was redesigned to resemble the architecture of the western Sudan. The ministry of veterans' affairs and the National Association of French Veterans also brought in an African Muslim cleric and the Moro Naba, traditional emperor of the Mossi people of Upper Volta, to help officiate at the ceremonies.[15]

Despite these efforts to preserve collective memory, neither memorial has meaning either for Africans or for the local French population living in those regions today. To be sure, both sites are carefully maintained by the ministry of veterans' affairs, and each locale is included during such annual veterans' occasions as Bastille Day. But African dignitaries are no longer brought in to participate in those events, and even the names of the two battles are remembered in Africa by only a tiny handful of aging African veterans.

The tragic events of Thiaroye, on the other hand, are quite another matter. The perpetrators of the brutality at this military base on the outskirts of Dakar on 1 December 1944 were members of the French Colonial Army, not an invading enemy force. French officials hoped that with the passage of time this bungled attempt to restore military discipline to African soldiers pushed beyond the limits of their patience would disappear from the African collective memory.

News of a political disaster of this magnitude could not be suppressed. Word spread like wildfire throughout FWA, aided in part by French officials' actions, some of which were ill-considered. They made a public example of the Thiaroye prisoners by parading the alleged ringleaders through the streets of Dakar and then conducting a court-martial that became something of a show trial.[16] They ordered the immediate dispersal of over 1,000 remaining ex-POWs who were at Thiaroye to their colonies of origin. As a result, Thiaroye was on everyone's lips for months afterward. Friends wrote about it to African soldiers still in France and awaiting demobilization.[17] Others wrote to French families that had befriended African soldiers in the hopes of enlisting French public support for a pardon for the men held in prison.[18] Always the response was the same: shock and indignation over a brutal act of repression against soldiers whose only crime was to claim money that was rightfully theirs.

Thiaroye remained deeply embedded in the collective memory of African veterans. As far away as Upper Volta and Côte d'Ivoire some thirty years after the event, veterans were able not only to recall the event but also to give explicit and accurate summaries. A *voltaique* veteran put it this way: "It concerned prisoners-of-war who were returning to their homes. When they arrived in Dakar, they insisted upon their rights, but that was badly received by the white military establishment."[19] One veteran in Côte d'Ivoire was so convinced of Thiaroye's impact that he stated rather overconfidently that "everyone" had heard of these events.[20]

In Senegal, as might be expected, popular memory of Thiaroye burned vividly, despite French officials' efforts to thwart this. For militant parties like the RDA, Thiaroye became an important political symbol of colonial oppression, and they made every effort to keep its memory before the public. On 19 February 1950, as part of an International Day of Struggle against Colonialism, the RDA, its youth wing, the newspaper *Réveil*, the Communist trade unions, and a series of veterans' associations combined to organize a public pilgrimage to the cemetery at Thiaroye where the fallen soldiers were buried.[21] When the colonial officials banned the march and deployed armed European soldiers around the cemetery, the demonstrators were forced instead to proceed to the war memorial in the center of Dakar, where they discreetly placed their wreaths "in homage to the massacred of

Thiaroye, victims of Colonialism." Sporadically, later in the 1950s, ceremonies at Thiaroye were organized. During the period of the de Gaulle referendum in 1958, the *Parti Africain de l'Indépendance*, campaigning in favor of a "no" vote, which would mean immediate independence, organized a pilgrimage of its militants, along with African students and trade unionists, to Thiaroye cemetery, and afterwards clashed with the police.[22]

The events of Thiaroye were a political embarrassment not only to French officials but to African assimilationists. Himself a former POW, Léopold Senghor wrote a moving poem entitled "Tyaroye" which condemned French actions as those of "a nation forgetful of its mission of yesterday." Still optimistic about the future, he concluded, however, that the deaths of Thiaroye had not been in vain. The fallen men were martyrs to "a new world which will be tomorrow."[23] In 1946 Senghor remembered his responsibilities to the victims of Thiaroye when he lobbied actively for a general pardon for those still in prison.

After independence, however, Senghor found Thiaroye too painful and embarrassing a moment in the history of Franco-Senegalese relations to be commemorated. Perhaps he resented the manner in which the symbol of Thiaroye had been appropriated by his more radical political opponents. Despite occasional requests from Senegalese veterans, Senghor did not permit a monument to the fallen at Thiaroye and allowed the cemetery to remain in shabby disrepair.[24]

Nevertheless, Thiaroye's place in the collective popular memory seems assured. For an older generation of Africans who lived through the events and the political awakening that followed the war, Thiaroye represents the ingratitude of a colonial system. For a younger generation of francophone Africans today, a monument to Thiaroye has been shaped in a form more touching than bronze or cold granite. *"Le Camp de Thiaroye"* is the title of the latest major film of the noted Senegalese film maker Ousmane Sembène.[25] Destined for a wide audience in Europe and especially in Africa, this film seems an entirely appropriate way for Africans to preserve a set of collective memories meaningful to them, if not to their former colonial rulers.

APPENDIX

Questionnaire for Ex–*Tirailleurs Sénégalais*

I. CIVIL STATUS

Date and place of birth (village and *cercle*) _____

Current place of residence _____

Did your father perform his military service? Yes _____ No _____

If yes, was he a career soldier? Yes _____ No _____

Date and locale of your first mobilization _____

Length of service _____

Status of service: conscript _____ volunteer _____

Did you re-enlist after your first tour of duty? Yes _____ No _____

Locale and duration of overseas postings _____

Were you ever a prisoner of war? Yes _____ No _____

If yes, where and when? _____

Dates of ranks achieved in service _____

Date and place of your final demobilization _____

Civilian career after service _____

Decorations received _____

II. RECOLLECTIONS OF MILITARY SERVICE

1. Who was responsible for your incorporation in the army?

    your family _____ your *chef de canton* _____

    your village chief _____

    your *commandant de cercle* _____ others _____

2. If you were a volunteer, why did you choose the army as a career? _____

3. Did you have a trade before entering the army? Yes _____ No _____

4. If yes, what was it? _____

5. Did you acquire or improve a trade in the army? Yes _____ No _____

6. If yes, what was it? _____

7. Did you continue this trade after discharge? Yes _____ No _____

    I adopted another trade, that of _____

III. RECOLLECTIONS OF DEMOBILIZATION

1. When you returned home after demobilization, did you encounter difficulties in readapting to civilian life? _____

2. Did you ever hear of the mutiny at Thiaroye in 1944? Yes _____ No _____

    It involved _____

3. Are you a member of your local branch of the *Anciens Combattants et Victimes de Guerre*? Yes _____ No _____

4. Did you ever receive your *carte de combattant*? Yes _____ No _____

5. Did you vote in the first elections after the Second World War? Yes _____ No _____

6. Were you a member of a political party in that period? Yes _____ No _____

    If yes, it was the _____

Questionnaire for Ex-Tirailleurs Sénégalais

7. Did the political parties help veterans in their struggle for recognition of their rights?

    Yes _____ No _____

8. Did they finally grant equal rights to veterans in Africa as those obtained by veterans in France? Yes _____ No _____

9. If no, why not? _____

10. After your demobilization did you try to find a job outside your native village?

    Yes _____ No _____

11. If yes, was it in a large city? _____

    in the capital of the *cercle*? _____

12. Were you successful? Yes _____ No _____

13. If not, what were the reasons? _____

14. Did you have a son who served in the army? Yes _____ No _____

15. If yes, did you encourage him in the direction of army service? Yes _____ No _____

16. If you have a young son today, would you encourage him to undertake a military career? Yes _____ No _____

NOTES

CHAPTER 1
Four *Tirailleur* Armies

1. See section titled "A Note on Sources," at the end of the book for a full discussion of the variety of written and oral records dealing with the topic.
2. After long neglect, important historical studies of some of these groups are beginning to appear. See, for example: Catherine Coquery-Vidrovitch and Paul E. Lovejoy, "The Workers of Trade in Precolonial Africa," in *The Workers of African Trade*, ed. Catherine Coquery-Vidrovitch and Paul E. Lovejoy (Beverly Hills: Sage, 1985), 9–24. Also see two articles by Luise White: "Prostitution, Identity, and Class Consciousness in Nairobi During World War II," *Signs* 11 (1986): 255–73; and "A Colonial State and an African Petty Bourgeoisie: Prostitution, Property and Class Struggle in Nairobi, 1936–1940," in *Struggle for the City: Migrant Labor, Capital, and the State in Urban Africa*, ed. Frederick Cooper (Beverly Hills: Sage, 1983), 167–94. Two social and labor histories of FWA are: Monique Lakroum, *Le Travail inéqal: Paysans et salariés sénégalais face à la crise des années trente* (Paris: L'Harmattan, 1982); and Nicole Bernard-Duquenet, *Le Sénégal et le Front Populaire* (Paris: L'Harmattan, 1985).
3. See Clayton, *France, Soldiers and Africa*, 357–59, for an overview of the limited role of *Tirailleurs Sénégalais* in these conflicts.
4. Adolf Hitler, *Mein Kampf*, trans. Ralph Manheim (Boston: Houghton Mifflin, 1943; original ed. 1925), 644–45. The subject of German attitudes toward black African troops is treated in more detail below in Chapter 6.
5. Racial prejudice against nonwhite troops was by no means confined to Germany. For British and American examples, see Keith Nelson, " 'The Black Horror on the Rhine': Race as a Factor in Post–World War I Diplomacy," *Journal of Modern History*, 42 (December 1976), 606–27.
6. David Irving, *The Trail of the Fox* (New York: Avon, 1977), 72.
7. Charles J. Balesi, *From Adversaries to Comrades-in-Arms: West Africans and the French Military, 1885–1918* (Waltham, Mass.: Crossroads Press, 1979), 112–13.
8. The phrase is from Senghor's poem entitled "Prayer for Peace," in Senghor: *Prose and Poetry*, trans. and ed. J. Reed and C. Wake (London: Heinemann, 1976), 136.
9. Frantz Fanon, *Black Skin, White Masks*, trans. Charles Lam Markmann (New York: Grove Press, 1968; original ed. 1952), 104–05, note 32.
10. For the role of a dominant ideology as a powerful instrument of control, see Antonio Gramsci, *Letter from Prison*, trans. Lynne Lawner (New York: Harper & Row, 1975). The best single study of French colonial ideology is Raymond F. Betts, *Assimilation and Association in French Colonial Theory, 1890–1914* (New York: Columbia University Press, 1961).
11. Three somewhat dated biographies of Senghor are Jacques L. Hymans, *Léopold Sédar Senghor, an Intellectual Biography* (Edinburgh: Edinburgh University Press, 1971); Irving L. Markovitz, *Léopold Sédar Senghor and the Politics of Negritude* (New York: Atheneum, 1969) and Ernest Milcent and Monique Sordet, *Léopold Sédar Senghor et la naissance de l'Afrique moderne* (Paris: Seghers, 1969). See also Jean Filipovich's excellent "Léopold Senghor's

Approach to Opposition: A Policy of Reconciliation" (unpublished master's thesis, McGill University, 1981).
12. Albert Memmi, *The Colonizer and the Colonized*, trans. Howard Greenfeld (Boston: Beacon Press, 1967), 16.
13. Trotsky's views are described by Teodor Shanin, *The Awkward Class: Political Sociology of Peasantry in a Developing Society: Russia, 1910-1925* (Oxford: Clarendon Press, 1972), 184-92.
14. Nicos Poulantzas, "The New Petty Bourgeoisie," in *Classes and Class Structure*, ed. A. Hunt (London: Lawrence & Wishart, 1977), 113-24. The term "petty bourgeoisie" has been subject to abuse as Gavin Kitching and Colin Leys have pointed out. See Gavin Kitching, *Class and Economic Change in Kenya: The Making of an African Petite Bourgeoisie, 1905-1970* (New Haven: Yale University Press, 1980), especially 443-45; Colin Leys, "The 'Overdeveloped' Post Colonial State: A Re-evaluation," *Review of African Political Economy*, 5 (1976): 45-46.
15. For helpful treatments of the views of Marx and his followers on this subject, and for guidance on its utility in the Third World, see Leys, "Post Colonial State," 45-46, and especially Carol A. Smith, "The Petty Bourgeoisie as a 'Fundamental' Revolutionary Class in Nicaragua," *Labour, Capital and Society*, 19 (1986): 10-11.
16. Byron Farwell, *The Great War in Africa (1914-1918)* (New York: Norton, 1986), 122-23.
17. David Killingray, "The Idea of a British Imperial African Army," *Journal of African History*, 19 (1979): 421-36. Killingray points out that the British overcame this reluctance and were planning to send West African troops to Palestine and even to the western front had the war lasted to 1919. They came to view the deployment of Africans against Europeans as a calculated risk that had to be taken at some point rather than as a breach of an inviolable principle.

CHAPTER 2
Slaves into Soldiers

1. Yves de Boisboissel, "Contribution à la connaissance de l'histoire militaire de l'Afrique occidentale française: Tirailleurs et spahis sénégalais et soudanais," *Revue internationale d'histoire militaire* 4 (1956): 48.
2. Ibid., 47-48.
3. Shelby C. Davis, *Reservoirs of Men: A History of the Black Troops of French West Africa* (1934; reprint, Westport, Conn.: Negro Universities Press, 1970), 28-29.
4. For a definitive treatment of the etymology of the term *laptot*, and for conditions of employment for African soldiers and sailors in nineteenth century Senegal, see J. Malcolm Thompson, "In Dubious Service: The Recruitment and Stabilization of West African Maritime Labor by the French Colonial Military, 1659-1900" (Ph.D. dissertation, University of Minnesota, 1989).
5. Schmaltz to Minister of the Navy, 4 September 1819, ANS, 2B 4.
6. Minister of the Navy Portal to Schmaltz, 22 March 1820, ANS, 2B 5.
7. F. Zuccarelli, "Le Régime des engagés à temps au Sénégal (1817-1848)," *Cahiers d'études africaines* 2 (1962): 420-61; and François Renault, "L'Abolition de l'esclavage au Sénégal: L'Attitude de l'administration française (1848-1905)," *Revue française d'histoire d'outre-mer* 58 (1971): 5-80.
8. Boisboissel, "Tirailleurs et spahis," 52.
9. Schmaltz to Minister of the Navy, 27 March 1820, ANS, 2B 5.
10. Minister of the Navy Comte de Chabrol to Governor of Senegal, 31 July 1827, ANS, 1B 17; and Minister of the Navy Baron d'Haussez to Governor of Senegal, 5 March 1830, ANS, 1B 14.

11. Zuccarelli, "Régime des engagés," 438-39.
12. Claude Faure, "La Garrison européenne du Sénégal (1779-1858)," *Revue d'histoire des colonies* 8 (1920): 82-83; ANS, 2B 18, 1B 31.
13. Renault, "L'Abolition," 23.
14. Governor of Senegal to Minister of the Navy, 17 July 1853, ANS, 2B 28.
15. Faure, "Garrison européenne," 95.
16. J. D. Hargreaves, ed., *France and West Africa: An Anthology of Historical Documents* (London: Macmillan, 1969), 99-102.
17. Faure, "Garrison européenne," 98.
18. Davis, *Reservoirs*, 40. For differing views of Faidherbe's military and political activities, see Leland C. Barrows, "The Merchants and General Faidherbe: Aspects of French Expansion in Senegal in the 1850's," *Revue française d'histoire d'outre-mer* 61 (1974): 236-83; and Margaret O. McLane, "Commercial Rivalries and French Policy on the Senegal River, 1831-58," *African Economic History* 15 (1987): 39-67.
19. Boisboissel, "Tirailleurs et spahis," 54.
20. Faidherbe to Minister of the Navy, 16 December 1857, and same to same, 25 August 1858, ANS, 2B 32.
21. Faidherbe to Minister of the Navy, 16 December 1857, ANS, 2B 32.
22. Faidherbe to Minister of the Navy, 25 August 1858, ANS, 2B 32.
23. Ibid.
24. Renault, "L'Abolition," 21, 24.
25. For a stimulating account of the complexities brought on by the winding down of slavery in Africa, see the collection of studies in Suzanne Miers and Richard Roberts, eds., *The End of Slavery in Africa* (Madison: University of Wisconsin Press, 1988). What the authors term "the ambiguities of freedom" are explored in the introduction (pages 3-68), which also contains an extensive bibliography.
26. J-L. Boutillier, "Les captifs en A.O.F. (1903-1905)," *Bulletin de l'I.F.A.N.*, sér. B. 30, 2 (1968): 515.
27. Renault, "L'Abolition," 21-22.
28. Governor Pinet-Laprade to post commandants at Dagana, Podor, Aéré, Soldé, Matam, Bakel, and Médine, 11 April 1868, ANS, 3B 80.
29. At that time West African battalions numbered approximately six hundred men. Hippolyte-Victor Marceau, *Le Tirailleur soudanais* (Paris: Berger Levrault, 1911), 13.
30. Philip D. Curtin, *Economic Change in Precolonial Africa: Senegambia in the Era of the Slave Trade* (Madison: University of Wisconsin Press, 1975), Vol. 1, 35-36; and Jean Bazin, "Guerre et servitude à Ségou," in *L'Esclavage en Afrique precoloniale*, ed. Claude Meillassoux (Paris: Maspero, 1975), 135-181.
31. Edouard Guillaumet, *Le Soudan en 1894: la vérité sur Tombouctou, l'esclavage au Soudan* (Paris: Albert Savine, 1895), 156.
32. Ibid.
33. Anne-Marie Duperray, *Les Gourounsi de Haute Volta* (Stuttgart: Franz Steiner, 1984), 63-66.
34. Myron Echenberg, "Jihad and State-building in Late Nineteenth Century Upper Volta: The Rise and Fall of the Marka State of Al-Kari of Boussé," *Canadian Journal of African Studies* 3 (1969): 531-61.
35. Interviews with Adama Diara of Warou, Burkina Faso, 31 March and 5 April 1967, cited in Echenberg, "Al-Kari," 557.
36. Lieutenant Governor Poiret to Governor-General of FWA, 7 February 1920, ANS, 17G 43. This letter and others in the dossier recount the services to the French cause rendered by the family of Samori Touré.
37. Faure, "Garrison européenne," 85-87.
38. Cited and translated by J. D. Hargreaves in his *France and West Africa*, 99.

39. Marceau, *Le Tirailleur soudanais*, 3. All subsequent references to Marceau's views on the various ethnic groups are from pages 2 and 3 of this same work.
40. For strong parallels in Mozambique where Portuguese colonial officials also applied biological determinism to their judgments about suitable ethnic recruits for their colonial army, see Allen Isaacman and Anton Rosenthal, "Slaves, Soldiers, and Police: Power and Dependency among the Chikunda of Mozambique, ca. 1825-1920," in Miers and Roberts, *The End of Slavery in Africa*, 244.
41. Gilbert de Boisseson, *Le Recrutement en AOF et AEF* (Paris: CMIDOM, 1956).
42. Charles Mangin, *La Mission des troupes noires, compte-rendu fait devant le Comité de l'Afrique française* (Paris: Comité de l'Afrique française 1911), 29.
43. See Chapter 4 below.
44. Boutillier, "Captifs en A.O.F.," 513-35; Richard Roberts and Martin Klein, "The Banamba Slave Exodus and the Decline of Slavery in the Western Sudan," *Journal of African History* 21 (1981): 375-94; Martin Klein and Paul E. Lovejoy, "Slavery in West Africa," in *The Uncommon Market: Essays in the Economic History of the Atlantic Slave Trade*, ed. H. A. Gemery and J. S. Hogendorn (New York: Academic Press, 1979), 181-212.
45. Lieutenant Governor of Guinea to Governor-General of FWA, 23 October 1907, ANS, 4D 30; Boutillier, "Captifs en A.O.F.," 518-20, 527-29.
46. Lieutenant Governor of Guinea to Governor-General of FWA, 7 February 1920, ANS, 17G 43.
47. Charles Mangin, *Regards sur la France d'Afrique* (Paris: Plon, 1924), 220.
48. Abdel-Kader Mademba, *Au Sénégal et au Soudan français: Le Fama Mademba* (Paris: Larose, 1931), 51.
49. Ibid.
50. All from Lieutenant Governor Poiret to Governor-General of FWA, Conakry, 7 February 1920, ANS, 17G 43.
51. Mangin, *Regards*, 220.
52. Curtin, *Economic Change*, Vol. 1, 35-36.
53. Henri Frey, *Campagne dans le Haut-Sénégal et dans le Haut-Niger (1885-1886)* (Paris: Plon, 1888), 86; Mme. Paul Bonnetain, *Une Française au Soudan: Sur la route de Tombouctou (Du Sénégal au Niger)* (Paris: Librairies Imprimeries réunies, 1894), 67.
54. Under Secretary of State for Colonies to Governor of Senegal, 4 July 1890, ANSOM Senegal, XVI, 66a.
55. *Projet de rapport sur le recrutement des Tirailleurs*, Saint-Louis, 15 July 1882, ANS, 4D 29.
56. Ibid.
57. Lieutenant Governor of Guinea to Governor-General of FWA, 28 October 1907, ANS, 4D 30.
58. Ibid.
59. See Chapter 2 above.
60. "La France et les cadres africains et malgaches: 1, les cadres avant l'indépendance," *Frères d'armes* 10 (1964): 16.
61. "La France et les cadres africains," 16.
62. General Louis Archinard, "Les Troupes indigènes," in *L'Empire colonial français* (Paris: Plon, 1929), 225.
63. H. O. Idowu, "Assimilation in Nineteenth Century Senegal," *Bulletin d'Institut Fondamental de l'Afrique Noire*, sér. B, 30 (1968), 1422-47.
64. "La France et les cadres africains," 18. No trace of Adekempi-Thompson can be found in the *Annuaire de L'Armée Française*'s listing of African officers in 1908, the first *Annuaire* to offer such information.
65. See page 15 above.
66. General Louis Faidherbe, *Le Sénégal: La France dans l'Afrique occidentale* (Paris: Hachette, 1899), 366 for founding, and 369 for results by 1871. He notes that among the

dropouts were 6 "traitors," that is, graduates of the school who went over to France's enemies.
67. Beyond the mention of its opening, no details of this early school have survived. General Nyo, "L'Evolution des cadres militaires africains de 1914 à 1956," *Tropiques* (June, 1956): 33.
68. In Lamine Guèye's socialist newspaper, *L'AOF*, 5 March 1948. Official gazettes list M'Baye as the most senior of five "native" officers in 1908, of four in 1911, 1912, and 1914. See *Annuaire officiel de l'armée française* (Paris: Imprimerie Nationale, 1908, 1911, 1912, and 1914).
69. See Chapter 3 below.
70. The current research of J. Malcolm Thompson promises to add considerably to this subject. See also Thompson, "In Dubious Service."
71. Faidherbe to Minister of the Navy, 16 November 1857, ANS, 2B 32.
72. Governor of Senegal to Minister of the Navy, 2 May 1896, ANSOM Senegal XVI, 36.
73. Rapports d'ensemble, troupes de groupe, 1912, 1913, ANS, 2G12 7.
74. Orsat, in François Descostes, ed., *Au Soudan (1890-1891): Souvenirs d'un Tirailleur Sénégalais d'après sa correspondance intime* (Paris: Picard, 1893), 32.
75. Ibid., 34. Ironically, Orsat himself was killed in the Soudan in 1890 when he was shot in the heart at very close range.
76. Ibid., 25.
77. Frey, *Campagne*, 58.
78. Balesi, *Adversaries*, 13, 18.
79. Minister of the Navy Comte de Chabrol to Governor of Senegal, 31 July 1827, ANS, 1B 14.
80. Faure, "Garrison européenne," 51.
81. H. Gatelet, *Histoire de la conquête du Soudan Français, 1878-1899* (Paris: Berger Levrault, 1901), 488-89.
82. Ministère des Colonies, *Annuaire du Territoire du Haut-Sénégal et Moyen Niger, 1900-03* (Paris: Imprimerie Nationale, 1904), 3.
83. Marceau, *Tirailleur soudanais*, 37.
84. Faure, "Garrison européenne," 86.
85. Faidherbe to Minister of the Navy, 16 December 1857, ANS, 2B 32.
86. Faure, "Garrison européenne," 90.
87. Paul Vigné D'Octon, *Au pays des fétiches* (Paris: Alphonse Lemerre, 1890), 112.
88. Orsat in Descostes, *Souvenirs*, 29-30.
89. Marceau, *Tirailleur soudanais*, 36.
90. Marceau, *Tirailleur soudanais*, 36; Vigné D'Octon, *Au pays des fétiches*, 111.
91. For the shift in French thinking about a "stabilized" working class see Frederick Cooper, "From Free Labor to Family Allowances: Labor and African Society in Colonial Discourse," *American Ethnologist* 16 (1989), 745-65.
92. Jean Ferrandi, *L'Officier colonial* (Paris: Larose, 1930), 90.
93. Ibid., 91-92.
94. Minister of the Navy Comte de Chabrol to Governor of Senegal, 31 July 1827, ANS, 1B 14.
95. Faure, "Garrison européenne," 12.
96. Marceau, *Tirailleur soudanais*, 37.
97. Ibid., 37.
98. R. Lassalle-Séré, *Le Recrutement de l'armée noire* (Paris: Librairie moderne de droit et jurisprudence, 1929), 97.
99. *Historique du 1er Régiment des Tirailleurs Sénégalais* (Saint-Louis: n.d.), 11, 15.
100. General Commanding Caudrelier to Lieutenant Colonel Mangin, Dakar, 6 July 1910, ANS, 149 AP 4.

CHAPTER 3
The *Tirailleurs Sénégalais* and the First World War

1. Roberts and Klein, "Banamba," 393.
2. Marceau, *Tirailleur soudanais*, 81.
3. Marc Michel, *L'Appel à l'Afrique: Contributions et réactions à l'effort de guerre en A.O.F., 1914-1949* (Paris: Editions de la Sorbonne, 1982), 30.
4. Ibid., 404.
5. For one illuminating case study, see Timothy C. Weiskel, "Labor in the Emergent Periphery: From Slavery to Migrant Labor among the Baoule Peoples, c. 1880-1925," in *The World System of Capitalism: Past and Present*, ed. Walter L. Goldfrank (Beverly Hills: Sage, 1979), 207-33.
6. Abadie, *Défense*, 47.
7. Michel, *l'Appel*, 6-7.
8. Ibid., 15-39.
9. Denise Bouche, *Les Villages de liberté en Afrique Noire, 1887-1910* (Paris: Mouton, 1968), 108; Roberts and Klein, "Banamba," 375-94.
10. See Chapter 2 above.
11. The only full-length biography of Mangin is the considerably dated Charles Bugnet, *Mangin* (Paris: Plon, 1934). A useful discussion of Mangin's activities in the First World War may be found in Leland C. Barrows, "The Impact of Empire on the French Armed Forces, 1830-1920," in *Double Impact: France and Africa in the Age of Imperialism*, ed. G. Wesley Johnson (Westport, Conn.: Greenwood Press, 1985), 53-91.
12. By both birth and marriage (his second wife was a daughter of the French general and anti-Dreyfusard politician Godefroy Cavaignac), Mangin closely fits the general view that a majority of French turn-of-the-century officers came from former landowning families, or the upper bourgeoisie, and often held antirepublican sentiments. See Clayton, *France, Soldiers and Africa*, 9.
13. Marc Michel, *La Mission Marchand, 1895-1899* (The Hague: Mouton, 1972).
14. His book by that title was published in 1910 by Hachette, and enjoyed wide circulation and debate.
15. De Vogüé, writing in *Le Figaro*, 11 October 1909.
16. Count Eugène Melchior de Vogüé, *Les Morts qui parlent* (Paris: Plon, 1899), cited in Davis, *Reservoirs*, 120.
17. Ibid.
18. Led by General Archinard, the group also included Gouraud and Audéoud. Michel, *L'Appel*, 2-3.
19. One writer claims that between 1909 and 1911 over 4,000 articles were published for and against the idea of a Black Army. See Charles Balesi, "West African Influence on the French Army of World War I," in *Double Impact: France and Africa in the Age of Imperialism*, ed. G. Wesley Johnson (Westport, Conn.: Greenwood Press, 1985), 99 and n.
20. See below for more on the critics of the *Tirailleurs Sénégalais*.
21. Two treatments of this opposition are Abdoulaye Ly, *Mercenaires noirs: Notes sur une forme de l'exploitation des Africains* (Paris: Présence Africaine, 1957); and Marc Michel, "Un Mythe: La 'Force Noire' avant 1914," *Relations intérnationales* 2 (1974), 83-90.
22. An additional reason for Mangin's fame, not at all apparent at the time, was that he was one of the few French officers to predict a long and costly war of attrition with Germany. In such a protracted struggle, he argued that colonial troops might very well turn the war in France's favor. Charles Mangin, *La Force Noire* (Paris: Hachette, 1910), 88-89.
23. Mangin's precise term was "un reservoir intarissible d'hommes." *La Force Noire*, 81.
24. Mangin, *La Force Noire*, 276.

25. Charles Mangin, "Caractères physiques et moraux du soldat nègre," *La Revue anthropologique* 10 (1911) 1–16.
26. Ibid., 10.
27. Ministère des Colonies, *Annuaire de l'A.O.F., 1917–1921* (Paris: Imprimerie Nationale, 1922).
28. For more on Ponty's ideology and policies, see G. Wesley Johnson, "William Ponty and Republican Paternalism in French West Africa (1866–1915)," in *African Pro-Consuls: European Governors in Africa*, ed. Lewis Gann and P. Duignan (Stanford: Hoover Institute, 1978), 127–56.
29. Governor-General Ponty to Minister of Colonies, 23 September 1909, ANS, 5D37 89.
30. Davis, *Reservoirs*, 134–36; Lassalle-Séré, *L'Armée noire*, 100.
31. Mangin, *La Force Noire*, 285.
32. His preference for force over negotiation at Marrakech earned Mangin the life-long enmity of Lyautey. Bugnet, *Mangin*, 119.
33. C. M. Andrew and A. S. Kanya-Forstner, "France, Africa, and the First World War," *Journal of African History* 19 (1978), 15.
34. Charles Mangin, *Comment finit la Guerre* (Paris: Plon, 1920), 257.
35. *La Démocratie du Sénégal*, 4 December 1913, as cited in Michel, *L'Appel*, 34. As Michel points out, this ironic statement came from a newspaper that was soon to become the voice of Blaise Diagne, after his election as mayor of Dakar and deputy to the French National Assembly in 1914.
36. Mangin, *Comment finit la Guerre*, 258.
37. General Gassouin Report, 8 June 1920, ANS, 4D143 100. Also see Chapter 4 below.
38. General Gassouin Report, Dakar, 8 June 1920, ANS, 4D143 100.
39. See Chapter 4 below for the results of this survey.
40. Mangin in letter dated Baronville, 12 August 1914, in Charles Mangin, *Lettres de guerre 1914–18* (Paris: Fayard, 1950), 13–14.
41. Léon Bocquet and Ernest Hosten, *Un fragment de l'Epopée sénégalaise: Les tirailleurs sur l'Yser* (Brussels: G. Van Oost, 1918), 21; Maurice Dutreb, *Nos Sénégalais pendant la Grande Guerre* (Metz: Editions des Voix Lorraines, 1922), 42–44.
42. Balesi, *Adversaries*, 98.
43. Ibid., 99.
44. Captain Eugène Devaux, quoted in Balesi, *Adversaries*, 111.
45. Balesi, *Adversaries*, 110.
46. Ibid.
47. Davis, *Reservoirs*, 126.
48. Alistair Horne, *The Price of Glory* (London: Macmillan, 1962), 100.
49. George F. G. Stanley, "The Fighting Forces," in J. M. S. Careless and R. Craig Brown, eds., *The Canadians, 1867–1967* (Toronto: Macmillan, 1967), 772–73.
50. Horne, *Glory*, 309.
51. Ibid., 102.
52. Réginald Kann, "Les Troupes Indigènes en France," *Le Temps*, 26 April 1923.
53. Leon Wolff, *In Flanders Fields: the 1917 Campaign* (New York: Ballantine Books, 1958), 66.
54. P. Painlevé, "La Politique de guerre de 1917," *Revue de Paris*, 15 March 1922, 316.
55. Davis, *Reservoirs*, 128.
56. For an interesting but biased view of the affair, see Nelson, "The 'Black Horror on the Rhine'," 606–27. See also Jacques Nobécourt, *Une Histoire politique de l'armée: De Pétain à Pétain, 1919–1942* (Paris: Editions du Seuil, 1967), 91–94 and 108–09; and Balesi, *Adversaries*, 123–24.
57. Paul Azan, *L'Armée Indigène Nord-Africaine* (Paris: Lavauzelle, 1925); Mangin/de Torcy debate in *Le Mois Coloniale et Maritime*, no. 97 (1911), 78–86.

58. For an interesting discussion of the medical problems of African soldiers in France during the First World War, see Balesi, *Adversaries*, 104–07.
59. Kann, "Les Troupes indigènes."
60. John Keegan, *The Face of Battle* (Harmondsworth, England: Penguin, 1976).
61. The general reference for the Thirty-seventh at Verdun is Horne, *Glory*, 101–02.
62. Ibid.
63. See Chapter 6 below.
64. Stanley, "The Fighting Forces," 773.
65. Ibid.
66. James L. McWilliams and R. James Steel, *The Suicide Battalion* (Edmonton: Hurtig, 1978).
67. Horne, *Glory*, 100.
68. F. Ingold, *Les Troupes noires au combat* (Paris: Berger Levrault, 1940), 70, cited in Balesi, *Adversaries*, 110.
69. Bakary Diallo, *Force-Bonté* (Paris: F. Rieder, 1926) 123, translated and cited by Balesi, *Adversaries*, 110.
70. Diallo, *Force-Bonté*, 84.
71. Ibid., 123.
72. Keegan, *Face of Battle*, 52, citing S. L. A. Marshall's classic study, *Men Under Fire*.
73. Abdel-Kader Mademba, *Au Sénégal et au Soudan français: Le Fama Mademba* (Paris: Larose, 1931), 8–9. This book is a history of the Sy family's, and especially Mademba Sy's service under the French.
74. He is called Mamadou Racine in French accounts. See, for example, "La France et ses cadres africaines."
75. Mademba, *Le Fama Mademba*, 109.
76. Ibid., iii, frontispiece dedication.
77. Ibid.
78. Abdel-Kader Mademba to Mangin, 19 July 1917, in Mangin Papers, *Archives Nationales*, 149AP11.
79. Second Lieutenant Abdel-Kader Mademba to General Mangin, from hospital camp 69, Menton, 19 July 1917; and same to CIC of FWA, from Rufisque, Senegal, reporting on his mission, 4 June 1917; both in Mangin Papers, *Archives Nationales*, 149 AP 11.
80. Second Lieutenant Abdel-Kader Mademba to CIC of FWA, from Rufisque, Senegal, reporting on his mission, 4 June 1917; in Mangin Papers, *Archives Nationales*, 149 AP 11.
81. Ibid.
82. Comments in Adrien-Edgar Allègre and Amadou N'Diaye Duguay-Clédor, "Déces de Commandant Abd el Kader Mademba," *La France coloniale*, no. 274, 1 September 1932.
83. CIC General Bonnier to Governor-General Angoulvant, Dakar, 28 October 1918, ANS, 4D53 81.
84. Allègre and Duguay-Clédor, "Déces."
85. Abdel-Kader Mademba, *Recrutement de Tirailleurs et main d'oeuvre au Soudan Français* (Tananarive: n.d., circa 1919).
86. Abdel-Kader Mademba to Mangin, 5 December 1917, Mangin Papers, *Archives Nationales*, 149 AP 11.
87. For more on the mutiny see Guy Pedroncini, *Les Mutineries de 1917* (Paris: Presses Universitaires de France, 1967).
88. Abdel-Kader Mademba to Mangin, 5 December 1917, Mangin Papers, *Archives Nationales*, 149 AP 11.
89. Allègre and Duguay-Clédor, "Déces." The military traditions of the Sy family did not end with Abdel-Kader Mademba's death. His son, Cheikh-Claude Mademba Sy, later to be known by the assimilated name of Claude Mademba Sy, also became an African officer in the *Tirailleurs Sénégalais*. When Senegal became independent in 1960, he, like

hundreds of other military and civilian leaders, maintained dual French and Senegalese citizenship. Promoted to major after accepting a transfer to the Senegalese National Army, he was placed in command of the Senegalese battalion sent as part of the United Nations peacekeeping force in the Congo later in 1960. After his retirement from the military, Claude Mademba Sy became a diplomat, serving as Senegal's ambassador to various European capitals before leaving diplomatic service to live in Paris.

90. Governor-General of FWA J. van Vollenhoven to Minister of Colonies, "Rapport sur le recrutement des troupes noires," Dakar, 25 September 1917 (hereafter as Van Vollenhoven Report), ANS, 4D143 100.
91. The law governing conscription was passed on 30 July 1919; an addition governing deferments for students was dated 20 October 1920, and another reducing service for French citizens to eighteen months was passed on 23 March 1926. These texts, together with decrees amending procedures for West Africa, can be found in the appropriate numbers of the *Journal officiel de l'Afrique occidentale française*.
92. Richard D. Challener, *The French Theory of the Nation-in-Arms, 1866–1939* (New York: Russell & Russell, 1965), 174–75.
93. Rapport d'ensemble, 1920, Dakar, ANS, 4D143 100.
94. Van Vollenhoven Report, ANS, 4D143 100.
95. Ibid.
96. Van Vollenhoven Report, ANS, 4D 143 100; Colonel J. Molard, "Rapport sur la colonne de Dédougou," Dédougou, 1 September 1916, ANS, 1D 212.
97. Marc Michel, "La Genèse du recrutement de 1918 en Afrique noire française," *Revue française d'histoire d'outre-mer* 58 (1971): 441–48.
98. A. Prévaudeau, *Joost van Vollenhoven (1877–1918)* (Paris: Larose, 1953).
99. The Four Communes were Gorée, Saint-Louis, Rufisque, and Dakar.
100. As quoted and translated by G. Wesley Johnson, *The Emergence of Black Politics in Senegal* (Stanford: Stanford University Press, 1974), 184.
101. Johnson, *Emergence*, 183–91.
102. See Chapter 9 below.
103. Johnson, *Emergence*, 194–95.
104. Johnson, *Emergence*, 195; and the following reports on the Diagne mission: Governor-General Angoulvant, Paris, 18 March 1919, ANS, 4D143 100; General Bonnier, Dakar, 9 October 1918, ANS, 4D53 81; Blaise Diagne, December, 1918, ANS, 4D143 100.
105. As quoted in Hélène D'Almeida-Topor, "Les Populations dahoméennes et le recrutement militaire pendant la première guerre mondiale," *Revue française d'histoire d'outre-mer* 60 (1973): 238.
106. D'Almeida-Topor, "Les Populations dahoméennes," 196–241; Patrick Manning, *Slavery, Colonialism and Economic Growth in Dahomey, 1640–1960* (Cambridge: Cambridge University Press, 1982), 264.
107. Patrick Manning, "L'Affaire Adjovi: La bourgeoisie foncière naissante au Dahomey, face à l'administration," in *Entreprises et entrepreneurs en Afrique (XIXè et XXè siècles)*, II, ed. Laboratoire "Connaissance du Tiers-Monde" (Paris: L'Harmattan, 1983), 256–57.
108. D'Almeida-Topor, "Les Populations dahoméennes," 233.
109. There were, however, occasional mutinies of new recruits. For one such mutiny in the forest zone of southern Côte d'Ivoire in 1927, see Military Commander of Côte d'Ivoire to CIC of FWA, Bingerville, 14 January 1928, ANS, 4D42 28.
110. M. Delafosse, "Les points sombres de l'horizon en Afrique occidentale," *Bulletin du Comité de l'Afrique française* (June, 1922): 274–77; Lieutenant Governor Antonetti, Abidjan, to Governor-General, Dakar, 6 July 1920, ANS, 4D54 81.
111. Michel, *L'Appel*, 404.
112. Ibid., 406.
113. Ibid., 405–08.

114. F. de Kersaint-Gilly, "L'Essai sur l'évolution de l'esclavage en Afrique occidentale française: Son dernier stade au Soudan française," *Bulletin du comité d'études historiques et scientifiques de l'A. O. F.*, 9 (1924): 474.

CHAPTER 4
Military Conscription in Theory and Practice

1. The CIC, General Gassouin, Report of 8 June 1920, ANS, 4D143 100.
2. Ibid.
3. Military Cabinet of the Governor-General of FWA, "Etude sur le recrutement en AOF," Dakar, 6 January 1922 (hereafter as Preliminary Report, 1922), ANS, 4D143 100.
4. Military Cabinet of the Governor-General of FWA, "Etudes sur le recrutement en AOF. Résultats des sondages effectués dans les colonies du Groupe en 1922," Dakar, 5 July 1923 (hereafter as Final Report, 1923), ANS, 4D72 81.
5. Ibid.
6. Ibid.
7. Samir Amin, ed., *Modern Migration in Western Africa* (London: Oxford University Press, 1974), 74–77.
8. Patrick Manning has noted that low male-to-female ratios similar to these were prevalent in the era of the slave trade. See Patrick Manning, "The Enslavement of Africans: A Demographic Model," *Canadian Journal of African Studies* 15 (1981): 499–526.
9. Michel, *L'Appel*, 403–04.
10. See below for more on these high unfit ratios.
11. Final Report, 1923, ANS, 4D72 81.
12. Albert Sarraut, *La Mise en valeur des colonies françaises* (Paris: Payot, 1923), 65. Sarraut added that in case of a general mobilization a much more extensive call-up of trained native reserves would be required.
13. Fred Cooper has demonstrated that, especially from the International Labour Organization Convention of 1930 until after the Second World War, the British and French avoided explicit labor policies in favor of tacit understandings. See Cooper, "From Free Labor," 745–65.
14. Davis, *Reservoirs*, 179–182. The Law of 1919 applied also to FEA, Madagascar, the Somali Coast, and the Pacific colonies. In 1923 a further decree extended conscription to French Indochina. Captain Houdry, "Le Recrutement en AOF," *Renseignements Coloniaux* (1929), 374.
15. The CIC, General Peyregne, Report on Recruitment for 1927, 29 July 1927, ANS, 4D70 81.
16. Ibid.
17. Lieutenant Colonel Mordrelle, Head of Military Cabinet, "Etude sur le recrutement," Dakar, 27 April 1920, ANS, 4D54 81.
18. Médecin-Capitaine Lacaze, Medical Report on Class of 1933, Saint-Louis, 31 October 1933, ANS, 4D167 132.
19. See the especially angry letter of the CIC, General Benoit, to the Minister of Colonies, Dakar to Paris, 6 November 1930, ANS, 4D84 81.
20. Report on Recruitment for Upper Volta in 1931, Ouagadougou, 28 April 1931, ANS, 4D84 81.
21. Preliminary Report, 1922, ANS, 4D143 100.
22. See Table 7.1 in Chapter 7 below for an attempt to show the correlation between quotas and population censuses. It must be said that such estimates cannot be precise in any statistical sense since they are predicated on scientifically weak census data. What can be stated is that French authorities made an effort to be equitable with the data they were given.
23. For a detailed examination of the demographics of Sine-Saloum, see Charles Becker, Mamadou Diouf, and Mohamed Mbodj, "L'évolution démographique régionale du Séné-

Notes

gal et du bassin arachidier (Sine-Saloum) au vingtième siécle, 1904–1976," in *African Population and Capitalism: Historical Perspectives*, ed. Dennis D. Cordell and Joel W. Gregory, 76–94 (Boulder, Colo.: Westview Press, 1987).

24. Final Report, 1923, ANS, 4D72 81.
25. Johnson, *Emergence*, 190; Marc Michel, "Citoyenneté et service militaire dans les quatres Communes du Sénégal au cours de la Première Guerre mondiale," in *Perspectives nouvelles sur le passè de L'Afrique Noire et de Madagascar: Mélanges offerts à Hubert Deschamps* (Paris: Editions de la Sorbonne, 1974), 299–314.
26. The CIC, General Doré, "Rapport sur le recrutement des troupes indigènes en A.O.F.," Paris, 6 May 1931, ANS, 4D42 28.
27. Governor-General's circular letter to all Lieutenant Governors, Dakar, 17 September 1931, ANS, 4D143 100. Senegal's underrepresentation applies only to the interwar period. It should be recalled that in the First World War, Senegal proportionately may have offered up the largest share of soldiers for the *Tirailleurs Sénégalais*. See Michel, *L'Appel*, 406.
28. For example, the Annual Report on Recruitment for Guinea, Conakry, 13 November 1921, ANS, 4D56 81.
29. Inspector of Colonies Picanon, extract from his "Rapport de mission en Haute-Volta, 1924–25," ANS, 4D79 81.
30. The term *bons absents* was an abbreviated form of *bon pour le service armé mais absent*. Men in this category are discussed at length in Chapter 5 below.
31. For comparative purposes, Table 4.2 also highlights the *cercle* of Bougouni, a largely Bambara-speaking region of southeastern Soudan. Bougouni *cercle* had an estimated population of 212,000 according to the 1951 census for Soudan, placing it very close to the average for the colony at that time. It was typical of the densely populated farming savanna of West Africa from which most *Tirailleurs* were recruited.
32. Médecin-Capitaine Lacaze, Medical Report on Class of 1933, Saint-Louis, 31 October 1933, ANS, 4D167 132.
33. Cooper, "From Free Labor," 751.
34. An officer who served on one such mobile draft board has left a revealing account of the recommended techniques. Captain Houdry, "Le recrutement en Afrique occidentale française," *Renseignements coloniaux* (1929), 373–77.
35. The CIC, General Bonnier, Report on Recruitment for 1918, 9 October 1918, ANS, 4D53 81.
36. Governor-General Boisson, circular letter to all Lieutenant Governors, Dakar, 6 December 1938, ANS, 4D42 28.
37. The CIC, General Jung, Report on Recruitment for 1928, Dakar, 16 August 1928; the CIC, General Benoit, Report on Recruitment for 1929, Dakar, 14 August 1929, both in ANS, 4D70 81.
38. For a case study of African military labor on this project, see Myron Echenberg and Jean Filipovich, "African Military Labour and the Building of the *Office du Niger* Installations, 1925–1950," *Journal of African History* 27 (1986): 533–51. For more on forced labor in the history of FWA, see Babacar Fall, "Le Travail forcé en Afrique occidentale française, 1900–1946: Cas du Sénégal, de la Guinée et du Soudan," thèse du 3ème cycle, Université de Dakar (Dakar: 1984); Lakroum, *Le Travail inégal*, 59–63; and Bernard-Duquenet, *Le Sénégal et le Front Populaire*, 91–104.
39. The Lieutenant Governor of Soudan blamed the unpopularity of the labor brigades for the dramatic increase in the number of volunteers in his colony, where volunteers went from 29 men in 1927 to 625 in 1929; included in Report of the CIC, General Benoit, 14 August 1929, ANS, 4D70 81.
40. Interview with Dramane Sarambé, Tougan, Burkina Faso, 15 June 1973.
41. Nancy E. Lawler, "Soldiers of Misfortune: The *Tirailleurs Sénégalais* of the Côte d'Ivoire in World War Two," 3 volumes (Ph.D. dissertation, Northwestern University, Evanston, Ill., 1988) III, 887–88.

42. The annual reports on recruitment from 1927 to 1939 gave incomplete information on recruits able to speak, read, or write French, and on such specified trades and skills as those of wood-carvers, blacksmiths, masons, bakers, tailors, and leather workers. ANS, 4D70 81.
43. The CIC, General Jung, Report on Recruitment for 1929, Dakar, 16 August 1928, ANS, 4D70 81.
44. Lassalle-Séré, *L'Armée noire*, 148.
45. Ibid., 149.
46. Ibid., 150.
47. The CIC, General Benoit, Report on Recruitment for 1929, Dakar, 12 July 1929, ANS, 4D42 28.
48. Ibid.
49. The CIC, General Freydenberg, Report on Recruitment for 1932, Dakar, 15 September 1932, ANS, 4D70 81.
50. Abou Digu'en, *Notre empire africain noir: Ses problèmes politiques et militaires* (Paris: Lavauzelle, 1928), 65–66. See Chapter 8 below for further discussion of the army and civil trades.
51. Lieutenant Governor's Report on Recruitment in Upper Volta for 1932, Ouagadougou, 28 April 1931, ANS, 4D84, 81.
52. General A. Duboc, *Les Sénégalais au service de la France* (Paris: Edgar Malfère, 1939), 11.
53. Ibid. The point is confirmed by veterans from forest zones of Senegal and Côte d'Ivoire. Interviews with Pierre Diémé, 15 April 1973, Ziguinchor; and with Eba Aoussi, 24 May 1973, Abidjan.
54. Lieutenant Governor of Niger, Report on Recruitment for 1947, Niamey, 12 July 1947, ANS, 4D90 81.
55. A. I. Asiwaju, *Western Yorubaland under European rule, 1889–1945: A Comparative Analysis of French and British Colonialism* (Atlantic Highlands, N.J.: Humanities Press, 1976), 124.
56. See, for example, Stephen Baier and Paul E. Lovejoy, "The Tuareg of the Central Sudan: Gradations in Servility at the Desert Edge (Niger and Nigeria)," in *Slavery in Africa: Historical and Anthropological Perspectives*, ed. Suzanne Miers and Igor Kopytoff (Madison: University of Wisconsin Press, 1977), 391–444; Dennis D. Cordell, *Dar al-Kuti and the Last Years of the Trans-Saharan Slave Trade* (Madison: University of Wisconsin Press, 1985); and E. Anne McDougall, "A Topsy-Turvy World: Slaves and Freed Slaves in the Mauritanian Adrar, 1910–1950," in Miers and Roberts, *The End of Slavery in Africa*, 362–88.
57. Mangin's full title was *Membre du Conseil Supérieur de la Guerre et Président de la Commission Interministerielle des Troupes indigènes*; ANS, 4D10 89.
58. General Mangin on behalf of the Commission, Report of 24 May 1921 (hereafter as Mangin 1921 report), ANS, 4D10 89.
59. Ibid.
60. Minister of Colonies to Governor-General of FWA, 30 July 1920, ANS, 4D42 28.
61. Procès-verbal de la commission interministerielle des troupes indigènes, 9 June 1922, Mangin Papers, *Archives Nationales*, 149AP12. Some indication of the considerable stature of the commission comes from its membership. Mangin was president; the vice-president was General Mazillier; there were in addition two representatives of the ministries of Justice and Finance, and twelve from the Ministry of War, including generals Hergault and Jung, one admiral from the navy, and three representatives from the Ministry of Colonies (Governor-General Angoulvant, M. Duchêne, and the CIC of FWA, General Benoit).
62. Ibid.
63. Ibid.
64. CIC General Gadel, Dakar, 16 October 1920, replying to a questionnaire relating to the formation of native officers, ANS, 4D166 132.
65. Minister of Colonies to CIC of FWA, 16 March 1923, ANS, 4D10 89.

66. Mangin 1921 report, ANS, 4D10 89; and Minister of Colonies Sarraut to Governor-General of FWA, 20 June 1921, ANS, 4D42 28.
67. Potential officers from Morocco did not attend the Fréjus center. Instead, Lyautey created a separate and more substantial training program for them in the palace at Meknès, where some 43 students were enrolled in a four-year program in 1931. Paul Catrice, "L'Emploi des troupes indigènes et leur séjour en France," *Etudes: Revue catholique d'intérêt général* (20 November 1931): 399–400.
68. Ibid., 400.
69. General Nyo, "L'Evolution des cadres," 35.
70. In August of 1951 a fourth EET was opened at Ouagadougou. Report of Lieutenant Colonel Villard, Head of Military Cabinet, to Director General of Finances, Dakar, 12 September 1953, ANS, 4D170 143.
71. For a useful guide through the complexities of the French and French colonial system of education, see David E. Gardinier, "The French Impact on Education in Africa, 1817–1960," in *Double Impact: France and Africa in the Age of Imperialism*, ed. G. Wesley Johnson (Wesport, Conn.: Greenwood Press, 1985), 333–44.
72. CIC General Magnan, Instruction to all Lieutenant Governors of FWA, Dakar, 10 September 1948, ANS, 4D170 143.
73. Interview with Gen. Fall, Dakar, 18 May 1973.
74. See below, page 69.
75. General Guignard to Mangin, Saint-Raphael, 14 May 1924, Mangin Papers, *Archives Nationales*, 149AP12.
76. Ibid.
77. *Historique du 1er Régiment*, 10.
78. Minister of Colonies to CIC of FWA, 16 March 1923, ANS, 4D10 89.
79. CIC General Benoit, Report on Recruitment of Career Soldiers and African Cadres, Dakar, 12 July 1929, ANS, 4D42 28.
80. Ibid.
81. CIC General Benoit to Governor-General of FWA, 12 July 1929; General Benoit received an affirmative reply from General Peltier, at the Directorate of Military Services in Paris on 9 October 1929; all in ANS, 4D42 28.
82. Biographical sketch in *Annuaire Mémoriale d'EFORTOM, 1956–1965* (St. Raphaël: n.d.).
83. David Gardinier, *Historical Dictionary of Gabon* (Metuchen, N.J.: Scarecrow Press, 1981), 148–49.
84. These fragments of information about his military career are from the *Annuaire Mémoriale d'EFORTOM, 1956–1965*.
85. See conclusion below.
86. Interview with Pierre Diémé, Ziguinchor, Senegal, 15 April 1973.
87. *Afrique en guerre*, no. 126, 8 June 1944, 1–3.
88. *Paris-Dakar*, 2 December 1960. A year later he was ousted by Senghor for having sided with Mamadou Dia in an alleged coup attempt.
89. Interview with General Fall, Dakar, 18 May 1973.
90. His cohort of 1931 at Saint-Maixent formed an *amicale*, or old boys' association, and kept in touch through a bulletin which they called *Epaulets*. Interview, 18 May 1973. While it is clear that General Fall formed lifelong ties of friendship with European officers as a peer, it should be remembered that this was rare—first, in the sense that he was a citizen, and second, in that he was able to qualify in a competition to gain access to the ESMIA.
91. *Paris-Dakar*, 2 December 1960, on occasion of his promotion to general.
92. Ibid., for his participation in the landings in Provence; interview of 18 May 1973 for other details.
93. *Paris-Dakar*, 2 December 1960. General Fall visited part of Mali Federation units in Léopoldville as part of the UN peacekeeping force there.

94. Interview with General Fall, Dakar, 18 May 1973.
95. Ironically, one such individual was Stanislas Mangin, a son of the general, who felt the excessive paternalism of the *Tirailleurs Sénégalais* was unfaithful to his father's assimilationist dream of the army as a school for true African modernization. Interview with Stanislas Mangin, Paris, 16 February 1977.

CHAPTER 5
Consequences of Conscription: Military Migration

1. I am grateful to Professor A.I. Asiwaju for the idea. In his article on this subject, he points out that these refugees not only sought shelter from the military recruiter but were also fleeing the *commandant de cercle* and his police who frequently demanded work gangs for forced labor. A.I. Asiwaju, "Migrations as Revolt: The Example of the Ivory Coast and the Upper Volta before 1945," *Journal of African History* 17 (1976), 577–94.
2. The term in French was *broyeurs des noirs* and literally means "crushers" of blacks. It was originally ascribed to General Charles Mangin. See Horne, *Glory*, 232.
3. Lieutenant Governor Antonetti, Abidjan, to Governor-General, Dakar, 6 July 1920, ANS, 4D54 81.
4. Final Report, 1923, ANS, 4D72 81.
5. He was so quoted in Final Report, 1923, ANS, 4D72 81.
6. Philippe David, *Les Navétanes: Histoire des migrants saisonniers de l'arachide en Sénégambie des origines à nos jours* (Dakar: Nouvelles Editions Africaines, 1980).
7. The CIC, General Benoit, Report on Recruitment for 1929, Dakar, 14 August 1929, ANS, 4D42 28.
8. The CIC, General Freydenberg, Report on Recruitment for 1932, Dakar, 15 September 1932, ANS, 4D70 81.
9. Whenever conscription has been imposed as a tribute by an alien state, rather than as a duty owed the larger community, attempts to avoid service have usually met with applause rather than opprobrium. In his illuminating study of rural France in the nineteenth century, Eugen Weber describes very similar practices to avoid service that were prevalent in rural France until the 1880s. See Eugen Weber, *Peasants into Frenchmen: The Modernization of Rural France, 1876–1914* (Stanford: Stanford University Press, 1976), 292–95.
10. Lieutenant Governor of Senegal, Report on Recruitment for 1932, Saint-Louis, 28 May 1932, ANS, 4D84 81.
11. Lassalle-Séré, *L'armée noire*, 150–51.
12. The CIC, general Jung, Report on Recruitment for 1928, 16 August 1928, ANS, 4D70 81.
13. Ibid.
14. Manning, *Slavery, Colonialism and Economic Growth*, 265–67; James S. Spiegler, "Aspects of Nationalist Thought among French-speaking West Africans, 1921–1939" (Ph.D. dissertation, Nuffield College, Oxford, 1968).
15. CIC, General Benoit, to Governor-General of FWA, 1 July 1929, ANS, 4D42 28.
16. Ibid.
17. For more on smuggling between southern Dahomey and Nigeria, see Ogunsola John Igué, "Evolution du commerce clandestin entre le Dahomey et le Nigeria depuis la guerre du 'Biafra'," *Canadian Journal of African Studies* 10 (1976): 235–58.
18. Lieutenant Governor of Guinea, Report on Recruitment for 1931, Conakry, 23 May 1931, ANS, 4D84 81.
19. Lieutenant Governor of Niger, Report on Recruitment for 1934, Niamey, 8 March 1934, ANS, 4D82 81. For an excellent discussion of the military recruitment of Sara youths see Chapter 2 of Mario Azevedo, "Sara Demographic Instability as a Consequence of French Colonial Policy in Chad (1890–1940)" (Ph.D. dissertation, Duke University, Durham,

N.C., 1975). For a general discussion of the burden of military recruitment on the demographically besieged populations of FEA, see Dennis D. Cordell, "Extracting People from Precapitalist Production: French Equatorial Africa from the 1890s to 1930s," in *African Population and Capitalism: Historical Perspectives*, ed. Dennis D. Cordell and Joel W. Gregory (Boulder, Colo.: Westview Press, 1987), 144–45.

20. Challener, *The French Theory of the Nation-in-Arms*, passim.
21. The CIC, General Piquemal, to Governor-General of FWA, Dakar, 15 May 1925, ANS, 4D167 132.
22. Lieutenant Governor of Niger, Report on Recruitment for 1931, Niamey, 15 May 1931, ANS, 4D69 81.
23. Lieutenant Governor of Niger, Report on Recruitment for 1934, Niamey, 8 March 1934, ANS, 4D82 81.
24. The CIC, General Benoit, to Governor-General of FWA, 12 July 1929, ANS, 4D42 28.
25. The CIC, General Villain, Report on Recruitment for 1935, Dakar, 9 August 1935, ANS, 4D167 132.
26. Médecin-Capitaine Lacaze, Medical Report on Class of 1933, Saint-Louis, 31 October 1933, ANS, 4D167 132.
27. Abadie, *Défense*, 75.
28. Governor-General of FWA to Minister of Colonies, 3 June 1945, ANS, 4D36 14.
29. Clayton, *France, Soldiers and Empire*, 359.
30. Lawler, "Soldiers of Misfortune," III, 968.
31. Files entitled "Marriages with Europeans, 1946" in ANS, 4D36 14.
32. Joe Harris Lunn, "Kande Kamara Speaks: An Oral History of the West African Experience in France, 1914–18," in *Africa and the First World War*, ed. Melvin E. Page (London: Macmillan, 1987), 38.
33. Minister of Overseas France to High Commissioner for FWA, Paris, 4 July 1946, ANS, 4D36 14.
34. Interview with Medical Colonel Macé, St. Raphaël, France, 2 March 1977. Dr. Macé added that while he felt French and African soldiers may have been important agents in the spread of venereal disease in Africa, he did not find that sexually transmitted diseases constituted a severe medical problem for African soldiers, and certainly not in comparison to pulmonary infections.
35. Jean Bazal, *Marseille galante* (Marseilles: P. Tacussel, 1980), 82.
36. When colonial officials in FWA balked at this surcharge on the population of that colony, it was agreed to place half the costs in the budget of the Ministry of Veterans' Affairs in Paris. Director General of Political Affairs of FWA to Inspector General of Colonies, Dakar, 30 August 1946, ANS, 4D36 14.
37. Governor-General of FWA to Minister of Colonies, Dakar, 3 June 1945, ANS, 4D33 14. What the Governor was protesting unsuccessfully was the army's insistence on demobilizing all Africans back in FWA, leaving it to ex-servicemen and their families to find a way to fund their travel back to Europe.
38. All cases are from the file "Marriages with Europeans, 1946" in ANS, 4D36 14.
39. This is precisely the theme of Ousmane Sembène's second novel, *O pays, mon beau peuple!* (London: Methuen, 1986, first published 1957). The story describes the difficulties faced by a young African in France who returns to Senegal with a French wife.
40. The general question of veterans' reintegration in FWA is addressed in Chapters 8 and 9 below.
41. As indicated in "A Note on Sources" at the end of the book, the subject of veterans in FWA still awaits its historian.
42. It should be noted that no *Tirailleurs Sénégalais* units were stationed on French soil until the 1930s.

43. Myron Echenberg, "Les Migrations militaires en Afrique occidentale française, 1900–1945," *Canadian Journal of African Studies* 14 (1980): 449–50.
44. Ibid.
45. See Figure 5.4.
46. Echenberg, "Migrations," 449–50.
47. Delafosse, "Les Points sombres," 275.
48. See Monique Lakroum, *Le Travail inégal*, and Nicole Bernard-Duquenet, *Le Sénégal et le Front populaire*.
49. Sidiki P. Coulibaly, "Les Migrations voltaiques: Les origines, les motifs et les perceptions des politiques" (Ph.D. dissertation, Université de Montréal, 1978), 72–75; Dennis D. Cordell and Joel W. Gregory, "Labour Reservoirs and Population: French Colonial Strategies in Koudougou, Upper Volta, 1914 to 1939," *Journal of African History* 23 (1982): 220.
50. An excellent discussion of the Jacobin principles of the draft can be found in Challener, *The French Theory of the Nation-in-Arms*.
51. Weber, *Peasants into Frenchmen*, 293–95.
52. Ibid., 80.
53. Preliminary Report, 1922, ANS, 4D143 100.
54. Challener, *The French Theory of the Nation-in-Arms*.
55. Criticisms of African military service can be found throughout the annual reports on recruitment in the archives of FWA. Well-educated men such as M'Baye M'Bengue, Doudou N'Dao, and Al-Hajj Doudou Diallo expressed in interviews their indignation over what they correctly regarded as patently inferior treatment. Diallo, in particular, referred to the humiliation of the *Tirailleur* uniform and to his embarassment over having to march barefooted and in the *Tirailleur* uniform in front of family and friends, many of whom were *originaires* and could serve in the regular army and wear its less garish dress. Interviews with M'Baye M'Bengue and Doudou N'Dao, Dakar, 4 May 1973, and with Al-Hajj Doudou Diallo, Dakar, 18 May 1973. For a critical French view, see Catrice, "L'emploi des troupes indigènes," 388–409.
56. Ministère de la Guerre, *Statistique médicale de l'armée métropolitaine et de l'armée coloniale*, 19 vols. (Paris: Imprimerie Nationale, 1925–1942).
57. If it can be demonstrated that large increases in pulmonary diseases occurred, for example, at times of large-scale demobilization, such as 1919 and 1945, then the correlation would be obvious. This research has yet to be done. As for venereal diseases, a potentially significant variable not only for mortality but also for fertility, soldiers may have introduced new strains of infectious venereal disease, though here again, little research has been done. One article that does investigate venereal disease for its impact on fertility is A. Retel-Laurentin, "Fécondité et syphilis dans la région de la Volta noire," *Population* 28 (1973): 793–815.
58. See Joel W. Gregory and Victor Piché, "Mode de production et régime démographique," *Canadian Journal of African Studies* 19 (1985): 73–79; and Dennis D. Cordell, Joel W. Gregory, and Victor Piché, "African Historical Demography: The Search for a Theoretical Framework," in *African Population and Capitalism: Historical Perspectives*, ed. Dennis D. Cordell and Joel W. Gregory (Boulder, Colo.: Westview Press, 1987), 14–32.
59. Cordell and Gregory, eds., *African Population and Capitalism*, 27.
60. See Table 4.2.
61. As reported by CIC, General Jung, Report on Recruitment for 1928, Dakar, 16 August 1928, ANS, 4D70 81.
62. Ibid.
63. The CIC, General Piquemal, to Governor-General of FWA, 15 May 1925, ANS, 4D69 81.
64. Ibid.

CHAPTER 6
"Morts pour la France":
The *Tirailleurs Sénégalais* and the Second World War

1. A few general overviews exist, the best of which is Rita Headrick, "African soldiers in World War II," *Armed Forces and Society* 4 (1978): 502–526; see also Michael Crowder, "The 1939–1945 war and West Africa," in *History of West Africa*, vol. 2, ed. J.F.A. Ajayi and Michael Crowder (London: Longman, 1971) 596–621. Two recent studies of the war's impact on Côte d'Ivoire are Jean-Noel Loucou, "La deuxième guerre mondiale et ses effets en Côte d'Ivoire," *Annales de l'Université d'Abidjan*, série I, *Histoire* 8 (1980): 183–207; and Nancy E. Lawler, "Soldiers of Misfortune."
2. There are important gaps, especially of divisions virtually destroyed in 1940, but the reports and regimental diaries (*journaux de marche*) that do survive are invaluable. They may be consulted at the Archives Nationales, Service Historique de l'Armée, Château de Vincennes, Paris (hereafter ANSHA).
3. Roger Bruge, *Juin 1940: Le Mois maudit* (Paris: Fayard, 1980); and *Les Combattants du 18 juin* (Paris: Fayard, 1982), vol. 1, *Le Sang versé*.
4. Michel, *L'Appel*, 405–08.
5. The order of battle in May and June 1940 was as follows. First and Sixth *Divisions d'Infanterie Coloniale* or DIC: both were on the Aisne and the Argonne, 15 May to 11 June, and then made an orderly retreat to the Vosges foothills by 22 and 23 June; Fourth and Fifth DIC: these two divisions bore the brunt of the German attack on the Somme from 23 May on, and were virtually destroyed by the Panzer attack of 5 June; Second, Seventh, and Eighth DIC: several regimental units from these divisions were detached from the *Armée des Alpes* and rushed to the Seine on 5 June, and the regiments fought in the major river valleys of central France.
6. Among other considerations, First World War losses and war weariness among the French population had made it politically attractive to substitute large numbers of colonial troops for metropolitan ones. See Challener, *The French Theory of the Nation-in-Arms*.
7. "Observations sur les rapports sur le recrutement pour 1940," Director of Political and Administrative Affairs to Head of Military Cabinet, Dakar, 11 June 1940, ANS, 4D4 14.
8. The first lower figures are from Colonel Crevecoeur, "L'Union française," conférence prononcée au N.A.T.O. collège, 30 January 1952, CMIDOM, Versailles, 1952; using archival sources, Lawler cites the higher figures. She also provides an invaluable table with the dates for the disbanding or disappearance of all African units after the armistice. See Lawler, "Soldiers of Misfortune," vol. 1, 264, 268, and 289.
9. John S. Ambler, *The French Army in Politics, 1945–62* (Columbus, Ohio: Ohio State University Press, 1966), 74.
10. Report by Colonel Le Masle, Chef d'Etat Major for CIC, General Magnan, Dakar, 13 May 1946, ANS, 2D2 14. For accounts of Africans serving in the Free French Forces see Denise Bouche, "Le retour de l'Afrique occidentale française dans la lutte contre l'ennemi aux côtés des Alliés," *Revue d'histoire de la deuxième guerre mondiale* 29 (1979): 41–68; and Anthony Clayton, *France, Soldiers and Africa*, 137–52.
11. As was the case in the First World War, this French recruitment placed a heavier burden on French subjects than on other colonial peoples. To be sure, British military recruiting was much heavier in 1939–1945 than in 1914–1918, but it probably did not very much exceed half a million men, many if not most of whom were porters and members of other labor units. These men would have been drawn from population in excess of 100 million, compared to 200,000 French African soldiers from a population of roughly 18 million, a ratio of roughly two to one.
12. Hélène de Gobineau, *Noblesse d'Afrique* (Paris: Fasquelle, 1946), 12.

13. Diallo, *Force-Bonté*, 109.
14. As Chapter 3 has shown, educated Africans in the coastal towns of Senegal and Dahomey asserted their rights to citizenship and were prepared to volunteer for First World War duty to make good their claims.
15. For a work heavily emphasizing patron-client ties see Balesi, *Adversaries*.
16. Mangin's paternalism is vividly demonstrated throughout his prolific writings. See, for example, Charles Mangin, *Regards sur la France*.
17. *La Gazette du Tirailleur* (Dakar, 1940).
18. For a British equivalent of this strip, see *A Spear for Freedom* (Nairobi, 1943?), Great Britain, Ministry of Information, recruitment booklet for East African Military Labour Service, as cited by Headrick, "African Soldiers," 504. This was a picture-book with short captions in English, Swahili, Chinyanja, and Luganda depicting the transformation of an East African cattle herder into a British soldier. The emphasis was on good food and the order, discipline, and cleanliness of army life, which may say something about British preoccupations but much less about their ideological concerns, compared with the *Mamadou* material.
19. No issue for 15 June seems ever to have appeared. Very few copies of this periodical now survive. One series can be found in the library of the Institut Fondamentale d'Afrique Noire (IFAN), Dakar, Sénégal.
20. Houdry, "Recrutement," 373.
21. Anonymous report on the death of Captain Charles N'Tchoréré (n.d.; 1940?), ANSHA, 34N/1081.
22. Fourth RTS, Année 1938: Annual report by Colonel Blaizot, Toulon, 15 September 1938, ANSHA, 34 N/1-81.
23. Interview with Pierre Diémé, Ziguinchor, Sénégal, 15 April 1973.
24. General Yves de Boisboissel, "Un siècle d'héroisme au service de la France: Le centenaire des Tirailleurs Sénégal," *Tropiques, Revue des Troupes Coloniales* 55, June (1957), 22.
25. Rapport de sondage du contrôle postal militaire, 11 May 1940, ANSHA, 34N/1081.
26. Boisboissel, "Le Centenaire," 23-31; Captain G. Bonnet, *Mémorial de l'Empire: A la gloire des Troupes Coloniales* (Paris: Sequaria, 1941). The best general accounts in English of the fall of France and the subsequent Vichy years until liberation are Jean-Pierre Azéma, *From Munich to the Liberation, 1938-1944*, trans. Janet Lloyd (Cambridge: Cambridge University Press, 1984); and Robert O. Paxton, *Parades and Politicians at Vichy* (Princeton: Princeton University Press, 1966).
27. See Chapter 8 below. This view was widely current among several informants: interviews with Léopold Basse, Dakar, 11 April 1973; with Jean Ahui, Abidjan, 24 May 1973; with Paul Vicens, Ouagadougou, 8 June 1973.
28. Lawler, "Soldiers of Misfortune," vol. 2, 697, 699-700.
29. Ibid., vol. 1, 151, and 3, 871.
30. Ibid., 1, 156-57.
31. Bruge, *Le Sang versé*.
32. Bruge, *Le Mois maudit*, 118.
33. Entry by Captain Campos-Hugueney in *Journal de marche* of the Forty-fourth RICMS relating to events of 5 June 1940; entry is dated 22 August 1940, ANSHA, 34N/1-81. For two well-documented Nazi atrocities against African soldiers, see Chapter 10 below.
34. Balesi, *Adversaires*, 110.
35. Nelson, "The 'Black Horror,' " 606-27.
36. For Hitler's rantings on the subject of black African troops, see Hitler, *Mein Kampf*, 644-45; for more on the subject of Nazi views of Africa, see Alexandre Kum'a N'dumbé III, *Hitler voulait l'Afrique: Les plans secrets pour une Afrique fasciste, 1933-1945* (Paris: L'Harmattan, 1980).
37. The *Pommersche Zeitung* of 28 July 1940 reported thus on the resistance of black African

Notes

troops at Condé-Folie, near Amiens, lower Somme; as quoted in Rapport du Lieutenant-Colonel Polidori, Fifty-third RICMS, in captivity, 3 July 1940, on the operations of 4-5-6 and 7 June, ANSHA, 34N/1081.

38. Bruge, *Le Mois maudit*, 118.
39. Lawler, "Soldiers of Misfortune," vol. 2, 729–30 and 1, 162 n. 157. According to Nancy Lawler, it was possible that the German, who was described by several informants, may have been either an old colonist or a visitor to the Côte d'Ivoire.
40. Gobineau, *Noblesse*, 12. Hélène de Gobineau estimates there were 20,000 African POWs, of whom perhaps half died in captivity. See above, page 88 for official French estimates placing the total number of African POWs at around 15,000.
41. Irving, *The Trail of the Fox*, 72. Irving's tone suggests very little sensitivity to the plight of the African soldiers. Rather than disapproving of Rommel's behavior, he treats the incident as illustrative of Rommel's passion for detail. The filming, incidentally, was for Goebbels's propaganda film *Victory in the West*, which depicts African soldiers dancing and cavorting in what is suggested to be an uncivilized manner.
42. Mail from West Africa could be channeled through the neutral offices of the neighboring colony of Portuguese Guinea to prisoners in Europe. It is unlikely that very much contact actually took place through such a cumbersome route. Governor-General of FWA to Minister of Colonies, 26 February 1944, ANS, 4D31, 14.
43. Gobineau, *Noblesse*, 12–5.
44. Ruth Schachter Morgenthau, *Political Parties in French-Speaking West Africa* (Oxford: Clarendon Press, 1964), 137–38.
45. Gobineau, *Noblesse*.
46. Ibid., 99–102.
47. Minister of Colonies to Governor-General of FWA, 31 October 1944, ANS, 4D31 14.
48. An excellent study of the hostility POWs feel for their liberators can be gleaned from James Clavell's novel, *King Rat*, based upon the experiences of British and American prisoners of the Japanese in Singapore.
49. Charles de Gaulle, *Mémoires de Guerre* (Paris: Plon, 1959), vol. 3, *Le Salut, 1944–1946*, 30–32; Jean de Lattre de Tassigny, *History of the French First Army*, trans. Malcolm Barnes (London: Allen & Unwin, 1952), 173–75; John Keegan, *Six Armies in Normandy* (Harmondsworth: Penguin, 1982), 302–03; Raoul Salan, *Mémoires: Fin d'un empire. I: Le sens d'un engagement, Juin 1899 Septembre 1946* (Paris: Presses de la Cité, 1970), 132–33.
50. De Gaulle, *Le Salut*, 32.
51. Jérôme Ollandet. *Brazzaville, capital de la France Libre: Histoire de la résistance française en Afrique (1940–1944)* (Brazzaville: Editions de la Savane, 1983).
52. De Gaulle, *Le Salut*, 32.
53. Ibid.; de Lattre, *French First Army*, 172–73.
54. De Lattre, *French First Army*, 177; Salan, *Le Sens d'un engagement*, 133.
55. Michel, *L'Appel*, chap. XVII.
56. As a French citizen from outside the Four Communes of Senegal, Yacé was a tiny minority within a minority in FWA. The quotation is in Lawler, "Soldiers of Misfortune," vol. 2, 458. Yacé's perception recalls the position of Blaise Diagne and the Senegalese citizen volunteers of the First World War.
57. Paul Ladhuie, "L'Etat d'esprit des troupes noires consécutif à la guerre 1939-1944," mémoire inédit de *l'Ecole Nationale d'Administration de la France d'Outre-Mer* (ENFOM), 1946.
58. Ladhuie, "L'Etat d'esprit"; Fernand Poujoulat, "Evolution de la mentalité des tirailleurs sénégalais au cours de la guerre 1939–1945," mémoire inédit de *l'Ecole Nationale d'Administration de la France d'Outre-Mer*, Paris, 1946, no. 124. All following references to the incidents of 1944–1945 are from these two works unless otherwise stated.
59. Ladhuie, "L'Etat d'esprit," 13–16.

60. For an examination in greater detail of the uprising at Thiaroye see Myron Echenberg, "Tragedy at Thiaroye: The Senegalese Soldiers' Uprising of 1944," in *African Labor History*, ed. Robin Cohen, Jean Copans, and Peter Gutkind (Beverly Hills: Sage, 1978), 109-28.
61. Minister of Colonies to Governor-General of FWA, 31 October 1944, ANS, 4D31 14.
62. Department of Political Affairs, Government General of FWA, ANS, 4D178 144.
63. "Sur les incidents de Thiaroye," ANS, 4D178 144.
64. Lawler, "Soldiers of Misfortune," vol. 2, 472.
65. "Renseignements," no date, contains a letter translated from Arabic and written by one Ibrahima of Dakar to his friend, Abdurahmane Traoré, stationed in a military camp in the Bouches-du-Rhone *département*, ANS, 4D31 14.
66. See, for example, *Réveil*, 25 August 1946, 2.
67. "Renseignements," reports of 2 December and 6 December 1944, ANS, 2D3 1.
68. Translation into French of Seydou Nourou Tall's statement to the *Tirailleurs Sénégalais*, 25 June 1941, ANS 4D31 14.
69. See Chapter 9 below for more on Senghor's postwar activities.
70. "Renseignements," 6 December 1944, ANS, 2D3 1.
71. Lamine Guèye, *Itinéraire africain* (Paris: Présence Africaine, 1966).
72. Administrator, District of Dakar and region to Governor-General of FWA, 4 December 1944, ANS, 4D31 14.
73. Minister of Colonies to Governor-General of FWA, 31 October 1944, ANS, 4D31 14.
74. Governor-General of FWA, circular letter to all Lieutenant Governors, 14 December 1944, ANS, 4D31 14.
75. See Chapter 7 below.
76. Governor-General of FWA to Minister of Colonies, 7 December 1944, ANS, 4D31 14.
77. The ANS archives are silent on this subject, and newspaper accounts vary considerably, but see George Padmore in the *West African Pilot*, 30 August 1945; the socialist *L'Espoir de Nice*, and the conservative *Le Patriote de Nice et du Sud-Est* for 20-22 August for the Saint-Raphaël disturbances; and both newspapers again on 6 September 1945 for the Antibes incident.

CHAPTER 7
The New Army, 1945-1960

1. Catherine Coquery-Vidrovitch has contributed significantly to our understanding of these economic changes. She has demonstrated, for example, that they owe their origins not to the Gaullists but to planners under the short-lived Vichy administration from 1940 to 1942. See her article, "Vichy et l'industrialisation aux colonies," *Revue d'histoire de la deuxième guerre mondiale* 29 (April, 1979), 69-94. For a sharp criticism of the unequal distribution of the fruits of this growth, see Jean Suret-Canale, *Afrique noire occidentale et centrale: III, 1. De la Colonisation aux indépendances (1945-1960)* (Paris: Editions Sociales, 1972); and Jacques Marseille, *Empire colonial et capitalisme français: Histoire d'un divorce* (Paris: Albin Michel, 1984). Political changes have been examined by at least two generations of historians and political scientists. The bibliography is extensive, but the pioneering study by Morganthau, *Political Parties*, still remains an indispensable guide. Also useful is Edward Mortimer, *France and the Africans, 1944-1960: A Political History* (London: Faber & Faber, 1969). More recently, two valuable works have appeared: Joseph Roger de Benoist, *L'Afrique occidentale française de 1944 à 1960* (Dakar: Nouvelles Editions Africaines, 1983); and Patrick Manning, *Francophone Sub-Saharan Africa, 1880-1985* (New York: Cambridge University Press, 1988).
2. Alistair Horne, *A Savage War of Peace: Algeria, 1954-62* (Harmondsworth: Penguin, 1979); Jacques Tronchon, *L'Insurrection malgache de 1947* (Paris: Maspero, 1974). For a brief

account of the role of the *Tirailleurs Sénégalais* in Indochina and Algeria, see Clayton, *France, Soldiers, and Africa*, 357–59.
3. Myron Echenberg, "Tragedy at Thiaroye," 109–28.
4. Paul-Marie de la Gorce, *The French Army: A Military-Political History*, trans. Kenneth Douglas (New York: George Braziller, 1963), 352–54.
5. "L'Armée au service de route," *Tropiques* 44 (1946), 30–32.
6. Director of Political, Administrative and Social Affairs of FWA to Head of Military Cabinet, Dakar, 2 October 1946, ANS, 5D31 14.
7. Manning, *Francophone Sub-Saharan Africa*, 112.
8. Charles de Gaulle, *Vers l'armée de métier* (Paris: Berger-Levrault, 1934).
9. See Abdou Wany N'Diaye's review in *Réveil*, 4 November 1947, 1.
10. Horne, *Savage War of Peace*, 23–28.
11. Governor of Niger to Governor-General of FWA, Niamey, 11 September 1945, reported on the incident; Governor-General of FWA to CIC General Magnan, Dakar, 1 October 1945, expressed his indignation; all in ANS, 5D31 14.
12. CIC General Magnan to Governor-General of FWA, 4 October 1945, Dakar, ANS, 5D31 14.
13. General Seller, Deputy Head of l'Etat—Major of the Army, 1er Bureau, Paris, 24 June 1946, "Note pour la direction des troupes coloniales," enclosed in Minister of Overseas France to High Commissioner of FWA, 14 August 1946, ANS, 5D31 14.
14. Report of CIC General de Larminat, Dakar, April, 1947, ANS, 5D16 14.
15. Director of Political, Administrative and Social Affairs of FWA to Head of Military Cabinet, Dakar, 2 October 1946, in ANS, 5D31 14.
16. Boisseson, *Le Recrutement*, 7.
17. CIC's Memorandum, Dakar, 30 April 1946, ANS, 4D52 28. No archival repertory for recruitment statistics is available for the period after 1948. For 1954–1957, overall figures are available at CMIDOM. They indicate the following annual levies : 1954 3,000; 1955 4,000; 1956 3,000; 1957 5,350.
18. Lieutenant Colonel Villard, "Valorisation des Cadres," 15 October 1954, ANS, 4D175 143.
19. Lieutenant Colonel Revol, Head of Military Cabinet, "Notes sur les opérations de recrutement concernant la classe 1948 en A.O.F.," Dakar, 21 March 1949, ANS, 4D16 81 (hereafter as Revol, 1948 report).
20. Revol, 1948 report, ANS, 4D16 81.
21. Ibid.
22. Ibid. In Guinea, the following eleven *cercles* reported quotas filled entirely with volunteers: Kindia, Télémélé, Younkounkoun, Gaoual, Kankan, Siguiri, Faranah, N'Zérékoré, Kissidougou, Gueckédou, and Macenta.
23. Major Duval, "La Legislation militaire actuelle en Afrique noire" (unpublished paper, CMIDOM, Versailles, 1956).
24. Major Chailley, "L'Africanisation des cadres de l'armée" (unpublished paper, CMIDOM, Versailles, 1957), 24.
25. The calculation is as follows: 40% of 4,000 for 4-year terms which volunteers served = 6400 men; 60% of 4,000 for 3-year terms as conscripts = 7200 men. That provides 13,600 men for tours of duty. This leaves 20,400 men in for second and further tours of duty, for a total *Tirailleur* force of 34,000. These careerists total 59% of the army, but added to the 6,400 volunteers, the percentage of volunteers and careerists reaches 79%. A situation in which a smaller standing army becomes largely based on volunteers is entirely consistent with general tendencies in armies everywhere. As Jacques Van Doorn states, "the size of the army is an inverse function of the volunteer/conscript ratio." Van Doorn, *The Soldier and Social Change* (Beverly Hills: Sage, 1975) 56.
26. Villard, "Valorisation des Cadres," ANS, 4D175 143. The French commander of the ill-

fated counterinsurgency war in Indochina, General Navarre, wrote in 1956 that the French Army in Indochina consisted of 54,000 French troops, 20,000 Legionnaires (mostly Germans), 20,000 North Africans and 18,000 black Africans. General Navarre, "Les Données de la défense de l'Indochine," *Revue de défense nationale* (March, 1956), 271-79.
27. Enclosed in the same dossier as Villard's report, ANS, 4D175 143.
28. Benoist, *L'Afrique occidentale française*, 177. Many prisoners died in captivity in Indochina, and these rates of attrition are unknown. It is therefore unwise to read very much into such partial information.
29. *Afrique Nouvelle*, 19 November 1955.
30. Governor of Senegal, Report on Recruitment in the District of Dakar, 1946, 17 April 1946, ANS, 4D90 81.
31. CIC of FWA, Report on Recruitment for 1946, Dakar, 30 December 1946, ANS, 4D52 28.
32. CIC of FWA, Instructions for Recruitment of Cohort of 1946, Dakar, 9 November 1945, ANS, 4D90 81.
33. Lieutenant Jean Haeckel, "Guide à l'usage des officiers de renseignements en Afrique" (unpublished paper, CMIDOM, Versailles, 1973).
34. Recruitment for 1954 in FWA, CMIDOM archives.
35. Interview with a retired French colonel who requested anonymity, Fréjus, 17 February 1972, who noted somewhat wryly that he himself was only 172 centimeters tall.
36. Report on Recruitment in Dahomey for 1947, Porto-Novo, 6 November 1947, ANS, 4D90 81.
37. Information conveyed in 1973 by a major in the Senegalese National Army who requested anonymity. He stated that he had dropped out of medical studies in 1957 and was about to be called up for military service. Rather than serve as a conscript, he chose to volunteer, entered officers' training school for fourteen months, and went on to a very successful career.
38. Report on Recruitment in Dahomey for 1947, Porto-Novo, 6 November 1947, ANS, 4D90 81. The report adds that seven of the twelve student graduates were later discharged after passing through the draft board but before full incorporation into a regiment. Whether this is because they could assert political influence or because they betrayed a hostile attitude to the military is not clear.
39. Major Duval, "La Législation militaire."
40. Ibid.
41. Report on Recruitment in Soudan for 1946, Koulouba, Soudan, 24 April 1947, 4D90 81. The penalty for absenteeism for those who evaded their responsibility and those who assisted them was one year in prison by the terms of the decree of 23 March 1933, title 7, but it was rarely applied. Boisseson, *Le Recrutement*, 5-6.
42. See pp. 118-19 below for more on the EETs.
43. EMPA files, 1953-1956, ANS, 4D174 143.
44. Boisseson, *Le Recrutement*, 15.
45. "A la Commission de Recrutement," editorial, *L'Essor*, 26 January 1953.
46. Morganthau, *Political Parties*, 412-13, illustrates this urban growth in a series of tables constructed from census data.
47. Chailley, "L'Africanisation," 26. See also Boisseson, *Le Recrutement*, 8-9, who describes these veterans as "detribalized town dwellers."
48. *Réveil*, 1 March 1948. This newspaper published a series of soldiers' salaries to show how poorly they compared with French equivalents. They revealed discrepancies as high as 8 to 1 between senior noncommissioned officers who were French and those who were African.
49. F. J. Amon D'Aby, *La Côte d'Ivoire dans la cité africaine* (Paris: Larose, 1951), 108.

50. Retel, 1948 report, ANS, 4D16 81. He indicates that there were seventy volunteers in the Abidjan quota of eighty-five recruits.
51. Governor of Senegal, Report on Recruitment in the District of Dakar for 1946, 17 April 1946, ANS, 4D90 81.
52. Ibid.
53. For the total recruited in 1946, CIC of FWA, "Memorandum," Dakar, 30 April 1946, ANS, 4D52 28. Unfortunately, this source is entirely silent on the question of how many recruits with French language skills actually were incorporated. Thus there is no basis for assuming that the army would have been able to incorporate 75 percent of its potential with language skills. The hypothetical exercise is introduced to illustrate the continuing handicaps faced by recruiters and recruited alike in the new military era after 1945.
54. CIC, General Bonnier, Report on Recruitment for 1918, 9 October 1918, ANS, 4D53 81.
55. Interview with Mamadou Pop Sané, Dakar, 4 April 1973.
56. CIC of FWA, Report on Recruitment for 1923, Dakar, 11 August 1923, 4D70 81.
57. CIC General Bonnier, Report, 9 October 1918, ANS, 4D53 81.
58. CIC of FWA, Report on Recruitment for 1923, Dakar, 11 August 1923, 4D70 81.
59. *Historique du 1er Régiment*, 101-02.
60. Interviews with General Ahmadu Fall, Dakar, 18 May 1973; and with Pierre Diémé, Ziguinchor, Senegal, 15 April 1973. Diémé recalls using a notebook to write down Bambara phrases, and states that some knowledge of Bambara was a must for noncommissioned officers.
61. This was the view of a careful student of recruitment, R. Lassalle-Séré, *L'armée noire*, 202.
62. The same process obtains today among migrant Sarakole workers employed in the auto industry in France, for example. Those most gifted in language emerge in leadership positions, and become strike and union leaders. See the article on immigrant workers in France by Jean Benoit in *Le Monde*, 6 December 1982.
63. CIC General de Larminat, Report of April, 1947, ANS, 5D16 14.
64. Chailley, "L'Africanisation."
65. The following accounts are based on interviews held at Dakar with Mamadou Pop Sané on 4 April 1973; and with both M'Baye M'Bengue and Doudou N'Dao on 4 May 1973.
66. For an interesting discussion of the Ponty school, see Peggy Sabatier, "Did Africans Really Learn to Be French? The Francophone Elite of the Ecole William Ponty," in *Double Impact: France and Africa in the Age of Imperialism*, ed. G. Wesley Johnson (Westport, Conn.: Greenwood Press, 1985), 179-90.
67. Samuel Decalo, *Coups and Army Rule in Africa: Studies in Military Style* (New Haven: Yale University Press, 1976), 3.
68. General Nyo, "L'Evolution des cadres," 36.
69. Ibid. It is worth noting that in France, the army was losing many talented young officers because of poor salaries and low morale. De la Gorce, *The French Army*: 353-54.
70. CIC General Magnan to Minister of Overseas France, Dakar, 19 June 1946, ANS, 4D90 81.
71. Ibid.
72. See *Réveil*, 14 March 1949, for example.
73. CIC General Borgnis Desbordes to Minister of Overseas France, Dakar, 25 May 1949, commenting on a request he had received from Mamadou Dia, then a senator from Senegal in the French National Assembly; ANS, 4D 170 143.
74. Chailley, "L'Africanisation," 21.
75. For statistics on EMPAs for both years, ANS, 4D174 143.
76. CIC General Nyo to Governor-General of FWA, Dakar, 10 July 1952, in ANS, 4D170 143.
77. EMPAs, ANS, 4D 170 143.
78. Chailley, "L'Africanisation," 21.

79. Nyo, "L'Evolution des cadres," 37–38.
80. CIC General Nyo to Minister of Overseas France, 22 July 1953, ANS, 4D175 143.
81. Lieutenant Colonel Villard, Head of Military Cabinet, secret report, "Le Problème de la valorisation des cadres militaires africaines en AOF," Dakar, 1 October 1954, ANS 4D175 143.
82. Pamphlet in ANS, 4D175 143.
83. "Promotion sociale dans les armées," *Frères d'armes*, 1962, 37–41; Georges Chaffard, "Le Rôle social de l'armée en Afrique," *Soldats d'outre-mer*, June, 1958, 15.
84. Ibid.
85. Boisseson, *Le Recrutement*, 11–12.
86. Governor-General of FWA to Minister of Overseas France, 20 August 1955, ANS, 4D175 143.
87. Little is known about Soumaré's formal education, but he seems originally to have been promoted from the ranks. For details of Amadou Fall's career, see pp. 68–69.
88. Chailley, "L'Africanisation."
89. After the Second World War, "Saint-Cyr" refers to the prestigious officers' training college rather than to the town of this name near Versailles. In 1945 bombardments destroyed the military academy's buildings, and the school was moved to the small town of Coëtquidan, some forty kilometers from Rennes, where Saint-Cyr has been located ever since. See de la Gorce, *The French Army*, 131–32.

In French army parlance, Saint-Cyr and Coëtquidan are used interchangeably. Coetquidan also hosted other military schools such as a special training center for French noncommissioned officers, which Africans occasionally attended.
90. Chailley, "L'Africanisation."
91. Between 1950 and 1953 the number of African students studying for a professional civilian degree in France rose from 171 to 327. In 1953 medicine led the way with some 76 students, science followed with 56 and law was third with 50 students. See table in Benoist, *L'Afrique occidentale française*, 273. In marked contrast, not a single African was studying at Saint-Cyr.
92. Lieutenant Colonel Villard, Secret report, 1 October 1954, ANS, 4D175 143.
93. Chailley, "L'Africanisation."
94. Ibid.
95. Document entitled "Advancement and Training of African Noncommissioned Officers," no date (probably 1955), ANS, 4D175 143.
96. Ibid. Not a single African applied for or was admitted to the Air Force Officer Training program for its noncommissioned officers, located at Salon-en-Provence. As the anonymous army officer who wrote this report put it rather indignantly, the air force had made "no provision for subjects or citizens of the French union."
97. Chailley, "L'Africanisation"; also see article written by "Jah," a nom de plume of a French officer, *Tropiques*, June, 1957; and *Annuaire mémoriale d'EFORTOM, 1956–1965*.
98. Article by "Jah," 34.
99. Lassalle-Séré, *L'Armée noire*, 209.
100. For more on this transitional period see Edward M. Corbett, *The French Presence in Black Africa* (Washington: Black Orpheus Press, 1972), and Moshe Ammi-Oz, "Les interventions extra-militaires des forces armées nationales dans les états d'Afrique noire francophone" (thèse de 3e cycle, Université de Paris I, 1973).
101. *Annuaire mémoriale d'EFORTOM, 1956–1965*.
102. Diakité and Traoré, for example, were old acquaintances in this respect. Diakité studied at the Kati EMPA until 1951 when he became a corporal at the age of nineteen. Traoré also studied there at roughly the same time. *Souvenir et Devenir*, 8e année, n.s., no. 15 (1971).

The EMPA traced its origin back to the nineteenth century Ecoles des Enfants de

Troupe or EETs, and were military-style academies created by the state as a special benefit for soldiers' sons, and as a way of perpetuating a military career as a family occupation for loyal servants of France. At age eighteen, youths left the EMPAs and entered the army as noncommissioned officers, whether or not they had completed their studies. Contrary to the military's belief, African politicians and intellectuals were anxious to see more African officers trained and urged the transformation of the EMPAs into modern academic *collèges*, with four years of secondary school preparation leading to the BEPC. By 1956, an indeterminant number of EMPA graduates did complete the BEPC. See *Réveil*, 14 March 1949, for African intellectuals' opinions, and Nyo, "L'Evolution des cadres," 37-38, for gradual upgrading of the EMPA curriculum.

103. Lamizana, born in 1916, entered the army in 1936 and was promoted to lieutenant from the ranks in 1954, after serving with distinction in Indochinese combat operations. Bokassa, born in 1921, began military service in 1939, served in the Second World War and in Indochina, but was not promoted to second lieutenant until 1956. A captain at independence in 1960, he transferred his commission in that same year to become chief of staff in the Central African Republic army. Since he was a kinsman of both Boganda and Dacko, family ties as well as the dearth of trained officers for the CAR—rather than his own competence—help explain his rapid but belated rise. See Mark R. Lipschutz and R. Kent Rasmussen, eds., *Dictionary of African Historical Biography*, 2nd ed. (Berkeley: University of California Press, 1986), 261 and 270-71; and Pierre Kalck, *Central African Republic: A Failure in Decolonization* (New York: Praeger, 1971).
104. Chailley, "L'Africanisation," 26.
105. Boisseson, *Le Recrutement*, 8-9.
106. Anthony Clayton and David Killingray, *Khaki and Blue: Military and Police in British Colonial Africa* (Athens, Ohio: Ohio University Center for International Studies, 1989), 261-62.

CHAPTER 8
French West African Veterans, 1945-1960

1. Of the 109 interviews of *ivoirien* ex-soldiers, a few took place in the Bouna and Man regions, while the large majority were held in the Korhogo region. The entire texts of these valuable oral testimonies, covering over 600 pages, are reproduced in Lawler, "Soldiers of Misfortune," vol. 2, 550-804, and vol. 3, 805-1202.
2. Respondents for the survey were chosen on a random basis in two regions of the country, the Tougan area and the capital city of Ouagadougou. Of the fifty-five subjects, one had served in the First World War, eight in the interwar period, twenty-eight in the Second World War, and eighteen after 1945. Of the latter group, all were retired in 1962, whether or not they had completed fifteen years of service. The questionnaire is reproduced in the Appendix.
3. For a spirited summary of the literature on veterans of the Gold Coast and Nigeria, see David Killingray, "Soldiers, Ex-Servicemen, and Politics in the Gold Coast, 1939-50," *Journal of Modern African Studies* 21 (1983): 523-34. See also Chapter 9 below for a brief discussion of the issues in contention.
4. Virginia Thompson and Richard Adloff, *French West Africa* (Stanford: Stanford University Press, 1957). While carefully argued, the chapter on veterans was based on published sources only.
5. Eligibility was a complicated issue, often requiring legal training in the interpretation of the *code de pensions*. The standard guide was the *Code de pensions militaires d'invalidité et des victimes de guerre*, printed in the *Journal officiel de la république française*, 1953.
6. See Table 9.2 reporting the results of the Liger mission to register West African veterans.

7. Anciens Combattants dossier, 1934, ANSOM, XIII-22-176/1408.
8. Jim Giblin, "The Vichy Years in French Africa, A Period of African Resistance to Capitalism" (Master's Thesis, McGill University, Montreal, 1978), 53–54.
9. The AACVGAOF received the blessing of de Gaulle and was allowed to inherit the property of the Legion in the federal capital of Dakar. Pamphlet of AACVGAOF, n.d., ANS, 2D23 14.
10. *Voix des combattants*, no. 126, 20 July 1957.
11. For his SFIO connection, Lamine Guèye to Minister of Overseas France François Mitterand, 28 August 1950, ANS, 4D178 144.
12. Traodec Report, 31 December 1948, ANS, 4D178 144.
13. High Commissioner of FWA to all Lieutenant Governors, 20 October 1953, ANS, 17G312 143.
14. Traodec to President of Dakar Chamber of Commerce, 29 March 1949, ANS, 4D178 144.
15. Traodec to President of Dakar Chamber of Commerce, 29 March 1949, Dakar, ANS, 4D178 144. Léopold Basse, who was President of the Dakar amputees in the 1970s, believes that Paye Dame Baye was replaced because he was "autocratic and lazy," and that his successor, Momar Baye, was a much better representative of amputee interests. Interview with Léopold Basse, Dakar, 11 April 1973.
16. Governor-General of FWA to all Lieutenant Governors, 20 October 1953, ANS, 17G312 143.
17. Traodec, who shared Fall's affiliation with the RPF (*Rassemblement du Peuple* Français), was a solid supporter of Fall, whom he described thus: "He is very French and he is honest. His stiff manner and speech do not immediately convey sympathy, but he is a man of considerable merit who is well worth watching." Traodec Report, 31 December 1948, ANS, 4D178 144.
18. All biographical information on Papa Guèye Fall, unless otherwise stated, comes from his Dossier du Personnel, ANS, I.C. 3053.
19. Fall's military career was distinguished. He received the *Croix de Guerre* for 1914–1918 and again for 1939–1940. Though he began his military service as a common soldier, he rose through the ranks to captain, and was promoted to major in the reserves in 1949. Dossier du Personnel, ANS, I.C. 3053.
20. Interview with M'Baye M'Bengue and Doudou N'Dao, Dakar, 4 May 1973.
21. Fall came from a pious Muslim family of Saint-Louis, and M'Bengue remembers how Fall's mother would come to visit her son in Dakar and would be observed in prayer throughout the day. People felt she was praying for the soul of her son. Interview with M'Baye M'Bengue and Doudou N'Dao, Dakar, 4 May 1973.
22. High Commissioner of FWA to all Lieutenant Governors, 20 October 1953, ANS, 17G312 143.
23. Renseignements politiques, 18 October 1946, ANS, 2D23 14.
24. Reseignements politiques, 24 June 1947, ANS, 2D23 14. The quotation is from the constitution of the AAAMCC dated 11 May 1954, ANS, 4D172 143.
25. Renseignements politiques, 29 August 1948, origine Bambey, valeur B/2, ANS, 4D178 144.
26. *Réveil*, 14 October 1946, 1.
27. Veterans' lodges are certainly one measure of activity and of political clout. Each territory eventually acquired a lodge in its capital, and Dakar, in addition, rated a lodge as the federal capital. By 1952 Senegal had two regional lodges, one at Ziguinchor and the other at Kaolack; elsewhere, there were five regional lodges in Dahomey, nine in Guinea, eleven in Upper Volta, eighteen in Soudan, and twenty-two in Côte d'Ivoire. Lieutenant Colonel Vatinelle, Head of Military Cabinet and Chairman of the Executive of the *Office Nationale des Anciens Combattants*, report, 21 December 1952, Dakar, ANS, 4D178 144. By

1955, construction on the last of the veterans' lodges, at Boghé in Mauritania, had been undertaken. Benoist, *L'Afrique occidentale française*, 177.
28. Traodec Report, 1 April 1949, Dakar, ANS, 4D178 144.
29. Report of the Saint-Louis local, 1949, ANS, 4D178 144.
30. Liger Report on His Visit to Niger, Dakar, 12 May 1950, ANS, 4D178 144.
31. Morganthau, *Political Parties*, 317–18.
32. Its first number appeared on 1 April 1948 and its last, no. 143, on 2 January 1959. Its first editor was Bakary Camara, who was followed by Armand Angrand.
33. Interview with Al Hajj Doudou Diallo, Dakar, 18 May 1973.
34. See Chapter 6 above.
35. Council Minutes, 18 November 1953 for session held 20–21 October 1953, in ANS, 4D178 144. In his private life, Al Hajj Doudou Diallo could describe a story of success through tenacity. An intelligent and largely self-taught man, he had held the rank of sergeant major in the army. When he was finally granted an honorable discharge by the terms of an Amnesty Law in 1948 at age thirty, he wrote civil service exams on the reserved job program and was successful in qualifying as a clerk in the Post Office. Through perseverance he worked his way up to accountant and after independence became chief of personnel and of finance for the administration of the Cap Vert peninsula (the district of greater Dakar). Interview with Al Hajj Doudou Diallo, Dakar, 18 May 1973.
36. This tradition continued after independence. In the early 1970s in the Casamance region of Senegal, for example, the veterans' lodge at Ziguinchor collected donations of rice for the district of Oussouye, where a food shortage had occurred. Interview with Pierre Diémé, Ziguinchor, 15 April 1973.
37. Traodec report, 7 March 1953, 4D178 144, accuses Reinach and the *ivoirien* veterans of exploiting the widows, but Reinach's reply, enclosed, is a persuasive defense.
38. Director of Political, Social and Administrative Affairs of FWA to the Head of the Military Cabinet, Dakar, 14 January 1946, ANS, 5D31 14.
39. Interview with Jean Ahui, Abidjan, 24 May 1973.
40. Lawler, "Soldiers of Misfortune," vol. 2, 493.
41. Chief of the subdivision of Mali, *cercle* of Labé, Guinea, 22 May 1945, ANS, 4D31 14.
42. Sergeant Moriba Sidibé to Governor-General of FWA, Dakar, 28 December 1944, ANS, 4D31 14.
43. Bouba Traoré to Governor of Soudan, Montpellier, 15 April 1946; Governor-General of FWA to Minister of Colonies, Dakar, 12 July 1946; both in ANS, 4D31 14.
44. High Commissioner of FWA to Minister of Overseas France, Dakar, 16 December 1952, ANS, 4D172 143.
45. Decree of 21 April 1939, ANS, 4D52 28.
46. Interview with Jean Ahui, Abidjan, 24 May 1973.
47. Traodec reported that of all these jobs, the ones most commonly filled by veterans were, first, in the police force, and second, in the postal service (the PTT). Traodec report of 31 December 1948, Dakar, ANS, 4D178 144.
48. Annual Report for Upper Volta in 1948, ANS, 2G48 34.
49. Upper Volta Survey, nos. 4 and 18.
50. Renseignements politiques, Dahomey, October, 1945, ANS, 2D21 14.
51. Renseignements politiques, Dahomey, June, 1946, ANS, 2D21 14.
52. Traodec Report, 31 December 1948, ANS, 4D178 144. This exodus of educated and skilled Dahomeans would end abruptly after independence with the rise of sectarian nationalism and would produce a serious crisis for the new Dahomean state, which was unable to offer alternative employment. See Manning, *Francophone Sub-Saharan Africa*, 160.
53. Upper Volta Survey.

54. All references to veterans in this section are from the Upper Volta Survey unless otherwise indicated.
55. Interview with Léopold Basse, Dakar, 11 April 1973.
56. Interviews with Eba Aoussi, Treichville, Côte d'Ivoire, 25 May 1973; and Joseph Kizerbo, Ouagadougou, 11 June 1973.
57. Doudou Seck to High Commissioner of FWA, Dakar, 31 December 1946, ANS, 4D52 28.
58. For example, veterans in the Kissi region of Guinea were an important modernizing force. Yves Person, "Soixante ans d'évolution en pays Kissi," *Cahiers d'études africaines* 1 (1960), 106–07.
59. Interview with Jean Ahui, Abidjan, 24 May 1973.
60. Lawler, "Soldiers of Misfortune," vol. 2, 519.
61. Interview 80, group of veterans at Dompleu, Prefecture of Man, in Lawler, "Soldiers of Misfortune," vol. 3, 1023.
62. This constitutes good evidence that some officials did not see the benefits of differentiating this intermediary class. Governor-General of FWA to all Lieutenant Governors, 6 November 1945, Dakar, ANS, 2D23 14.
63. See Chapter 9 below.
64. As an example, the basic rate for a retired African soldier with fifteen years of service was 1,060 francs per annum compared to 2,120 for a French counterpart. Acting High Commissioner of FWA Chauvet to Minister of Overseas France, 6 August 1949, Dakar, ANS, 4D178 144.
65. Secretary-General, *Office National*, to Head of Military Cabinet, Dakar, 19 December 1952, ANS, 4D172 143.
66. Director of SICAP to High Commissioner of FWA, 8 April 1954, Dakar, ANS, 4D172 143.
67. High Commissioner of FWA to Minister of Overseas France, Dakar, 30 July 1951, transmitting a dossier of grievance from Fall's association, ANS, 2D95 165.
68. High Commissioner of FWA to Minister of Overseas France, Dakar, 16 December 1952, ANS, 4D172 143.
69. Governor of Niger to Governor-General of FWA Barthes, Niamey, 27 July 1946, ANS, 2D23 14.
70. Liger's Report on Visit to Dahomey, Dakar, 2 May 1950, ANS, 4D178 144. The practice of women but not men engaging in petty commerce was a very old coastal tradition in West Africa. See, for example, Claire C. Robertson, *Sharing the Same Bowl: A Socioeconomic History of Women and Class in Accra, Ghana* (Bloomington, Ind.: University of Indiana Press, 1984).
71. Renseignements politiques, Cotonou, 26 July 1948, ANS, 4D178 144.
72. Interview with Dramane Sarambé, Tougan, Burkina Faso, 15 June 1973.
73. Georges Chaffard, "Un Entretien avec les anciens combattants de Haute-Volta," *Soldats d'outre-mer* (June, 1958): 5.
74. Interview with Aliou Traoré, Koudougou, Burkina Faso, 3 June 1973.
75. Upper Volta Survey, no. 16. Such large sums were not uncommon. Several of the returning veterans who were victims at the Thiaroye disturbances in the fall of 1944 had substantial sums of money on their persons. See Echenberg, *Thiaroye*, 144.
76. Kougsabla's core village had a population of 182, and the extended village, 544, in 1972. Gregory A. Finnegan, "Population Movement, Labor Migration, and Social Structure in a Mossi Village" (Ph.D. dissertation, Brandeis University, Waltham, Mass., 1976), 117.
77. Ibid., 208.
78. Ibid., 209.
79. Interview with Eba Aoussi, Treichville, Côte d'Ivoire, 25 May 1973.
80. This is the view of a thoughtful West African historian and observer, Professor Joseph Kizerbo. Interview, Ouagadougou, 11 June 1973.
81. Interview with Pierre Diémé, Ziguinchor, Senegal, 15 April 1973.

82. Lawler, "Soldiers of Misfortune," vol. 3, 1023-24.
83. Anne Summers and R.W. Johnson have argued that veterans in Guinea after the First World War became a "conspicuous social grouping," but it is doubtful this status could have been sustained very long without the political gains that simply could not be won in the interwar era. See their "World War I Conscription and Social Change in Guinea," *Journal of African History* 19 (1978), 37-38.
84. Georges Chaffard, "Un Entretien," 4.
85. Lawler, "Soldiers of Misfortune," vol. 2, 536-37.
86. See the discussion on military academies in Chapter 7.
87. Benoist, *L'Afrique occidentale française*, 554, table; the same source shows that by 1951 only 170,400 or 6.9 percent of eligible children attended primary school. By 1957 the figures had risen to 356,800, or 13.5 percent.
88. Chaffard, "Un Entretien," 5.
89. The Upper Volta Survey, for example, indicated a strong affirmative response to the idea of a military career for their sons.
90. Interview with Pierre Diémé, Ziguinchor, Senegal, 15 April 1973. Diémé thought his son had made a foolish choice since there was so much unemployment among would-be civil servants but noted that today's youth could not be influenced by their elders.
91. Interview with Paul Vicens, Ouagadougou, 8 June 1973.
92. Renseignements politiques, Dahomey, 3 March 1947, and Renseignements politiques, Porto-Novo, 6 February 1947, both in ANS, 2D23 14.
93. Renseignements politiques, Soudan, 31 May 1947, ANS, 2D23 14.
94. Lawler, "Soldiers of Misfortune," vol. 1, 100-01; and interview with Colonel (retired) Macé, Saint-Raphaël, France, 2 March 1977. A former medical officer in FWA, Colonel Macé was speaking of his experiences while stationed in the Kayes region of Soudan in the 1950s. Rice, most of it imported, has become a common food in francophone West Africa.
95. The extent to which alcoholism was a problem among African ex-servicemen has never been documented. Eyewitnesses of course reported qualitative examples of abuse of alcohol among veterans. Interviews with Dramane Sarambé, Tougan, Burkina Faso, 5 June 1973; and with Eba Aoussi, Treichville, Côte d'Ivoire, 25 May 1973. Aoussi put it well, however, when he observed that there were both positive and negative elements of the westernization which he supported.
96. An excellent discussion of the reintegration into peasantry of ex-servicemen in prerevolutionary Russia may be found in Shanin, *The Awkward Class*.
97. Agricultural innovation among veterans was mentioned in the interviews with Aliou Traoré, Koudougou, Burkina Faso, 3 June 1973, and with Eba Aoussi, Treichville, Côte d'Ivoire, 25 May 1973. Veterans, of course, could afford the odd experiment since, unlike their peasant neighbors, they often were not completely dependent on farming for a livelihood.
98. Lawler, "Soldiers of Misfortune," vol. 3, 889-90.
99. Birago Diop, *Les Contes d'Amadou Koumba* (Paris: Présence Africaine, 1961). The story has been performed in the theater and in 1963 was made into a film by the Senegalese filmmaker Moma Thiam. Guy Hennebelle, ed. *Les Cinémas africains en 1972* (Paris: Société Africaine d'Edition, 1972), 212-14.
100. Interview with Colonel Macé, Saint-Raphaël, France, 2 March 1977.
101. See page 131 and note 27 above for more on veterans' lodges.
102. Constant Flutet, "Historique et legislation de la médaille militaire, 1852-1952" (unpublished pamphlet, ANS, Dakar, 1952).
103. Lunn, "Kandé Kamara Speaks," 48.
104. Veterans were not the only ones to manipulate images of the war to suit their situation. For a similar phenomenon, see Jim Giblin, "The Image of the Loyal African during World

War II and its Postwar Use by the French Communist Party," *Canadian Journal of African Studies* 14 (1980), 319–26.
105. Lawler, "Soldiers of Misfortune," vol. 2, 700.
106. Ibid., vol. 1, 25, quoting *ivoirien* soldier Namble Silue.
107. Chapter 6 above described experiences of *Tirailleurs* in Britain. As for the Soviet Union, African soldiers could have had no illusions about life there based on their own firsthand experience. One group of some thirty African veterans found themselves obliged to spend time there after their liberation by the Red Army from confinement in German POW camps in 1945. These men complained bitterly that the Soviets would not listen to their grievances and made them labor even more unmercifully than had the Germans. Intelligence Report, 28 January 1946, Dakar, ANS, 2D21 14.
108. Renseignements politiques, Dahomey, ANS, 4D31 14.
109. Respondent no. 32, Upper Volta Survey; two others gave similar explanations for discrimination.
110. Doudou Seck to High Commissioner of FWA, 31 December 1946, Dakar, ANS, 4D52 28.
111. Bulletin de renseignements, Guinea, 20 January to 20 February 1945, ANS, 4D31 14.
112. Renseignements politiques, Dahomey, 30 March 1946, ANS, 2D21 14.
113. Interview with Eba Aoussi, Treichville, Côte d'Ivoire, 25 May 1973.
114. Traodec Report, 31 December 1948, ANS, 4D178 144.
115. Resolution by Federal Association President, Seck Douta, in Renseignements politiques, Cotonou, 26 July 1948, ANS, 4D178 144; *Réveil*, 6 September 1948, ran the story with the following front page headline: "Gouverneur de Dahomey Cambon et son Adjoint Comet, insultent et frappent les Anciens Combattants."
116. Renseignements politiques, Cotonou, 26 July 1948, ANS, 4D178 144.
117. Two forceful spokesmen for disabled veterans were Léopold Basse (interview, Dakar, 11 April 1973); and Paul Vicens (interview, Ouagadougou, 8 June 1973). A First World War amputee, Basse became president of the *Association Générale des Amputés et Grands Blessés* of Senegal in 1972, replacing the deceased Momar Baye. Paul Vicens was a *métis*, born in Bobo-Dioulasso in 1916. A French citizen, he saw action in the Free French forces that were part of the Allied invasion of Germany in 1945, where he was severely wounded and subsequently lost his leg. He served as president of the veterans of Upper Volta in the 1960s until "politics" forced him to resign in favor of a Mossi speaker, Nakena Bamba.
118. Olivier Reinach to High Commissioner of FWA, Abidjan, 27 September 1948; Seck Douta to High Commissioner of FWA, Dakar, 12 October 1948; both in ANS, 4D178 144.
119. High Commissioner of FWA Chauvet to Minister of Overseas France, Dakar, 7 October 1949, ANS, 4D178 144.
120. Interview with Al Hajj Lancina Traoré, Ouagadougou, 7 June 1973.
121. This point is made forcefully for soldiers in the Gold Coast by Killingray, "Soldiers, Ex-Servicemen, and Politics," 524.
122. This was clearly the case for the vast majority of Senoufo-speaking *ivoirien* veterans from Korhogo. See Lawler, "Soldiers of Misfortune," passim.
123. Interview with Pierre Diémé, Ziguinchor, Senegal, 15 April 1973.
124. Ibid. Diémé exchanged letters in French with his friend, though in the *Tirailleurs Sénégalais* they had used Bambara as much as French to communicate with each other.
125. Shanin, *The Awkward Class*, 191–92, arrives at a similar conclusion about Russian ex-servicemen after the civil war of the early 1920s in the Soviet Union. Although ex-soldiers gradually adjusted to everyday life in peasant communes and lost their standing as "a special social entity," they continued to have significant impact on rural life by espousing rebellious attitudes towards social relationships and a strong desire for change and innovations generally.
126. Coquery-Vidrovitch, "Vichy et l'industrialisation aux colonies," 69–94.

CHAPTER 9
Veterans and French West African Politics, 1945–1960

1. A. Adu Boahen, *Ghana: Evolution and Change in the Nineteenth and Twentieth Centuries* (London: Longman, 1975), 153–54; and F.K. Buah, *A History of Ghana* (London: Macmillan, 1980), 149.
2. Adrienne M. Israel, "Measuring the War Experience: Ghanaian Soldiers in World War II," *Journal of Modern African Studies* 25 (1987): 159–68; Killingray, "Soldiers, Ex-Servicemen, and Politics"; Richard Rathbone, "Businessmen in Politics: Party Struggle in Ghana, 1949–57," *Journal of Development Studies* 9 (1972–73): 391–401; and Eugene I. A. Schleh, "The Post-War Careers of Ex-Servicemen in Ghana and Uganda," *Journal of Modern African Studies* 6 (1968) : 203–20. A recent article by Wendell P. Holbrook attempts to steer a middle course between the two schools. See Wendell P. Holbrook, "Oral History and the Nascent Historiography for Africa and World War II: A Focus on Ghana," *International Journal of Oral History* 3 (1982) : 148–66.
3. Killingray, "Soldiers, Ex-servicemen, and Politics," 526.
4. Ibid., 533.
5. Thompson and Adloff, *French West Africa*, 229.
6. See below, pp. 160–63.
7. Killingray makes the same point for Gold Coast soldiers, and as a proposition it no doubt applies to most armies in history:

    They [Gold Coast soldiers] came from uncompromising backgrounds and returned to such. The army was in many ways an unprogressive institution, and it attempted, and probably succeeded, in reinforcing conservative values in the majority of the men who served in the ranks. The hierarchical nature of the army, plus battalion and unit loyalty, was more likely to foster identification with, rather than criticism of, the established order. "Soldiers, Ex-servicemen, and Politics," 525.

8. The Russian Revolution is perhaps the most dramatic example of the potential for revolution to be found among disgruntled and disbanded soldiers. See Shanin, *The Awkward Class*, 184–92. Shanin notes that the Soviet Revolution and Civil War added huge numbers of peasant soldiers to the Bolshevik party.
9. Governor-General of FWA, Angoulvant, to Director of Military Services in Paris, 18 March 1919, suggesting modifications for the projected Conscription Law of 1919, ANS, 4D143 100.
10. The major exception was in Dahomey, where, as Chapter 3 indicated, unrest over recruitment had triggered elite opposition on the coast and major rebellions in the interior. One of the leaders of the postwar protest was the Dahomean intellectual Louis Hunkanrin. Despite his basic opposition to military recruitment, he agreed in 1918 to volunteer for military service after he met secretly with Blaise Diagne during the Senegalese deputy's recruiting visit to Porto-Novo. After his discharge in 1920 Hunkanrin resumed his criticism of the militarization of FWA and of colonialism generally. Although French authorities arrested him before the 1923 protests got underway, he was one of the inspirational leaders of the Dahomean national movement. In 1923 these educated elites confronted the colonial administration over its plan to increase taxes, despite two successive years of poor harvests. After strikes in both Porto-Novo and Cotonou and several rural riots, the administration backed down and reduced taxes, but not before punishing the leaders with exile to Mauritania. See Manning, *Dahomey*, 264–66.
11. Charles de Gaulle, *Mémoires de Guerre* (Paris: Plon, 1956), vol. 2, *L'Unité, 1942–1944*, 477–80, contains the general's address at the Brazzaville Conference, in which he declared the war, from France's point of view, to "have been in large part an African war" (p. 478). See also Charles-Robert Agéron, "De Gaulle et la Conférence de Brazzaville," in *"L'Entourage" et de Gaulle*, ed. Gilbert Pilleul (Paris: Plon, 1979), 243–51.

12. Morganthau, *Political Parties*; Benoist, *L'Afrique occidentale française*.
13. Manning, *Francophone Sub-Saharan Africa*, 141.
14. Ibid., 142.
15. Ibid., 143.
16. Jean Suret-Canale, "The French West African Railway Workers' Strike, 1947-48," in *African Labor History*, ed. Robin Cohen, Jean Copans, and Peter Gutkind (Beverly Hills: Sage, 1978), 129-54.
17. Morganthau, *Political Parties*, 27-28.
18. Benoist, *L'Afrique occidentale française*, 514.
19. No records exist indicating how many of an estimated 100,000 veterans voted in these first elections.
20. Thompson and Adloff, *French West Africa*, 227-28; Gutteridge noted that in 1949 in Dahomey, veterans and serving soldiers represented 58 percent of the entire electorate of 54,000 male adults. William Gutteridge, *The Military In African Politics* (London: Methuen, 1969), 5.
21. Opinion of the CIC of FWA, General Magnan, in his letter to Minister of Overseas France, Dakar, 19 June 1946, ANS, 4D90 81.
22. Abiola Irele, ed., *Selected Poems of Léopold Sédar Senghor* (New York: Cambridge University Press, 1977), 9-10; Filipovich, "Léopold Senghor's Approach to Opposition," 2-4.
23. Léopold Sédar Senghor, *Chants d'ombre suivi de Hosties noires: Poèmes* (Paris: Editions du Seuil, 1945).
24. Morganthau, *Political Parties*, 137-38.
25. In retrospect it is equally difficult to picture François Mitterand, for example, as a war veteran turned politician. Yet Mitterand, who was himself a former French POW and Resistance leader, gravitated into French politics at precisely the same time as Senghor, and with a similar base of veterans' popularity. See Azéma, *From Munich to the Liberation*, 237, note 71.
26. Morganthau, *Political Parties*, 138.
27. Ibid., 139-40.
28. Interview with Léopold Basse, Dakar, 11 April 1973.
29. Renseignements politiques, Dakar, 18 October 1946, 2D23 14.
30. Morganthau, *Political Parties*, 154.
31. Mortimer, *France and the Africans*, 136.
32. "Pour les A.C. de l'Afrique noire," *L'AOF*, 1 August 1947, 1.
33. *L'AOF*, 6 April 1948, 1-2.
34. *Condition humaine*, no. 1, 11 February 1948, 1.
35. Ibid., 10 May 1948.
36. Ibid., 5 October 1948.
37. Ibid., 15 February 1949.
38. Ibid., 31 May 1949.
39. While many dossiers relating to the RDA in the archives of former FWA are still closed to researchers, several documents in the private papers of Marius Moutet contain negative comments. See for example, "Transmission de renseignements au sujet de la tournée de conférences entreprise par G. D'Arboussier au Soudan et en Côte d'Ivoire contre le recrutement de tirailleurs," 7 pages, Dakar, 3 June 1947; or Governor Louveau, "Compte rendu de déroulement du Congrès du RDA tenu à Bamako," 18-21 October 1947, Bamako, 8 pages. Both in Papiers Marius Moutet, Papiers Privés, ANSOM, P.A. 28/83, carton 8, Dossier 168, sous-dossier 5, 1947.
40. Haeckel, "Guide à l'usage des officiers de renseignements."
41. Ibid.
42. Oswald Durand, "Transmission au ministre de documents au sujet de la tournée de d'Arboussier et évaluation de l'impact en milieu africain," 26 October 1947, in Papiers

Marius Moutet, Papiers Privés, ANSOM, P.A. 28/83, carton 8, Dossier 168, sous-dossier 5, 1947.
43. When Houphouet-Boigny claimed that by 1950 his party, the RDA/PDCI, had the support of 99 percent of the veterans, French authorities did not dispute him. Lawler, "Soldiers of Misfortune," vol. 2, 490.
44. Lawler, "Soldiers of Misfortune," vol. 2, 501–02.
45. Morgenthau, *Political Parties*, 188–202; Lawler, "Soldiers of Misfortune," vol. 2, 502–03.
46. Manning, *Francophone Africa*, 142.
47. Lawler, "Soldiers of Misfortune," vol. 2, 483.
48. Diori was born in Niger in 1916, was later educated at *Ecole William Ponty*, and had been a schoolteacher and principal before entering politics in 1946 as the RDA standard bearer in Niger. He had no military background himself, but his championing of veterans' issues in the National Assembly clearly contributed to his political power in Niger. Lipschutz and Rasmussen, eds., *African Historical Biography*, 59.
49. See for example, *Journal officiel, assemblée nationale, débates*, second sitting, 30 December 1947, 6532–33.
50. Ibid.
51. Ibid.
52. Reported in *Réveil*, 25 April 1949.
53. Thompson and Adloff, *French West Africa*, 229.
54. *Réveil*, 14 June 1948.
55. See *Réveil*, 25 April 1949, for Coulibaly's report on the National Assembly budget debate of April, 1949; and *Réveil*, 13 June 1949, for an article under his name protesting the use of *Tirailleurs Sénégalais* in Indochina.
56. Its first resolution continued to call for a *statut unique*, that is, full equality of citizenship for overseas subjects. The *Réveil* issue of 24 January 1949 reports fully on the Abidjan congress.
57. Acting High Commissioner of FWA, Chauvet, to Minister of Overseas France, 6 August 1949, Dakar, ANS, 4D178 144.
58. Secretary-General of Office National des Anciens Combattants to Head of Military Cabinet, 19 December 1952, Dakar, ANS, 4D172 143.
59. High Commissioner of FWA to Minister of Overseas France, 30 July 1951, Dakar, ANS, 2D95 165.
60. Thompson and Adloff, *French West Africa*, 228.
61. High Commissioner of FWA to Minister of Overseas France, 5 October 1951, ANS, 4D178 144. The High Commissioner made the point of how useful the *Office* was as a watchdog over the various associations and their leaders in his appeal against a budget cut.
62. Secret Report, Dakar, 4 December 1952, ANS, 4D178 144, contains a copy of a Bulletin.
63. List of council members in 1949, ANS, 4D178 144.
64. Minutes of the Office's Administrative Council, 26 June 1953, ANS, 4D178 144.
65. Minutes of October, 1953 meeting, ANS, 4D178 144.
66. Budget data are from Traodec Report of April, 1949, Dakar, ANS, 4D178 144. Since the federation budget was itself largely a product of locally produced revenues, this meant that the people of FWA were themselves producing a significant part of these funds.
67. Joint letter protesting the 1953 budget cut to the Director of the *Office National*, 10e bureau, Hôtel des Invalides, Paris, 6 November 1952, ANS, 4D178 144.
68. Traodec Report, 31 December 1948, ANS, 4D178 144.
69. Rita Cruise O'Brien, *White Society in Black Africa: The French of Senegal* (London: Faber & Faber, 1972), 108; Morganthau, *Political Parties*, 151.
70. Traodec Report, April, 1949, ANS, 4D10 1.
71. Traodec Report, 7 March 1953, ANS, 4D178 144. Eba Aoussi claims that Yacé used his veterans' support as a springboard to gain election to the executive of the *Parti Démocra-*

tique de la Côte d'Ivoire, Houphouet's regional branch of the RDA. Interview with Eba Aoussi, Treichville, Côte d'Ivoire, 25 May 1973.
72. Traodec Report, no date, 1948, ANS, 4D10 1.
73. Renseignements politiques, Cotonou, 26 July 1948, ANS, 4D178 144.
74. For Durand in Guinea, Traodec Report of 1948, ANS, 4D10 1: for Joubert, Traodec Report, 31 December 1948, ANS, 4D178 144.
75. Traodec Report, 7 March 1953, ANS, 4D178 144. Traodec struck an anti-Semitic note when he stated that Reinach, a Jew, was not a war veteran at all but merely a war victim. His hostility to Reinach was apparently matched by the unidentified French intelligence officer who could write as late as 1948 that Reinach's membership in the French Socialist Party was tantamount to treason. The officer described Reinach as "a convinced and influential socialist, the brother of Léon Blum's secretary, and the member of a revolutionary and international party." Renseignements politiques, 23 September 1948, ANS, 4D178 144.
76. Traodec Report on his trip to Dahomey in November and December of 1948, ANS, 4D178 144.
77. O'Brien, *White Society,* 104 and 195–96; Morganthau, *Political Parties,* 151.
78. Secretary-General of RDA, Denise, to Olivier Reinach, 1 October 1948, enclosed in Traodec Report, 31 December 1948, ANS, 4D178 144.
79. Schmitt to Papa Seck Douta, 20 October 1948, enclosed in Traodec Report, 31 December 1948, ANS, 4D178 144.
80. Renseignements politiques, Kankan, 28 July 1948, ANS, 4D178 144.
81. Renseignements politiques, Guinea, 14 November 1951, ANS, 4D178 144.
82. Liger Mission to Niger, 12 May 1950, Dakar, ANS, 4D178 144.
83. Confidential letter of inquiry from High Commissioner of FWA to Governor of Upper Volta, requesting a security check on these two men, 7 November 1949, ANS, 4D178 144.
84. Traodec Report, 31 December 1948, ANS, 4D178 144.
85. Renseignements politiques, Bobo-Dioulasso, 25 June 1949, ANS, 4D178 144.
86. Renseignements politiques, Bobo-Dioulasso, 9 December 1949; and Veterans of Bobo-Dioulasso, joint letter to Seck Douta, 12 December 1949, both in ANS, 178 144.
87. Traodec to Director of Political Affairs, Dakar, 30 December 1949, ANS, 4D178 144.
88. Renseignements politiques, Bobo-Dioulasso, 4 October 1951, 4D178 144.
89. Report of Liger's Mission to Côte d'Ivoire, 3 September 1949, ANS, 4D178 144.
90. In RDA circular of 13 June 1947, Abidjan, ANS, 2D23 14.
91. Report of Liger's Mission to Côte d'Ivoire, 3 September 1949, ANS, 4D178 144.
92. Ibid.
93. High Commissioner of FWA to all Lieutenant Governors, Dakar, 10 February 1951, ANS, 4D178 144.
94. Rapport politique, Soudan, December, 1946, ANS, 2D23 14.
95. Although a brilliant graduate of *Ecole William Ponty,* Modibo Keita was little known outside Soudan in 1946. He was first elected to the Soudan Territorial Assembly in 1948, and did not win his first seat in the French National Assembly until 1956. He became, of course, leader of the RDA in Soudan and first president of the independent Republic of Mali. Ironically, in view of his early links with the African military, he was overthrown in 1968 in a coup led by Colonel Moussa Traoré, and died in detention in 1977. Lipschutz and Rasmussen, *African Historical Biography,* 103–04; Morganthau, *Political Parties,* 19 and 274.
96. Renseignements politiques, Zinder, Niger, in High Commissioner of FWA to Director of the *Office,* 21 December 1949, ANS, 4D178 144.
97. Liger Report on Mission to Niger, 12 May 1950, ANS, 4D178 144.
98. This is perhaps the moment to insert an important exception to the generalization.

Notes

Mauritania was the one colony in the FWA federation where veterans had little political impact. Since military recruitment for the *Tirailleurs Sénégalais* in this racially divided colony had drawn exclusively on the black African population, colonial officials were reluctant to place emphasis on veterans' affairs for fear of setting the servile black military above the dominant Moors. This policy decision was enunciated in 1946 and repeated a decade later. Report on Mauritania, Saint-Louis, July, 1946, ANS 2D21 14; Report on Veterans in Mauritania, Saint-Louis, August, 1955, ANS 4D172 143.

99. Lamine Guèye to François Mitterand, 28 August 1950; François Mitterand, Minister of Overseas France, to High Commissioner of FWA, 19 September 1950, ANS, 4D178 144.
100. High Commissioner of FWA to Minister of Overseas France, 26 September 1950, ANS, 4D178 144.
101. High Commissioner of FWA to all Lieutenant Governors, Dakar, 20 October 1953, ANS, 17G312 143.
102. *Tropiques*, June, 1957, 23.
103. Final Report of Liger, 13 July 1950, Dakar, ANS, 4D178 144.
104. Ibid.
105. Thompson and Adloff, *French West Africa*, 229.
106. Liger Final Report, 13 July 1950, ANS, 4D178 144.
107. Liger Letter of 16 March 1949, ANS, 4D178 144.
108. Liger Letter, 28 February 1950, ANS, 4D178 144.
109. All of those reports that have survived are in ANS, 4D178 144. Especially helpful is his final report of 13 July 1950.
110. Liger Report of 10 May 1949, ANS, 4D178 144.
111. Vatinelle Report, 19 December 1952, Dakar, ANS, 4D178 144.
112. Manning, *Francophone Sub-Saharan Africa*, 144. One major force in the so-called *ralliement*, or winning over, of Houphouet was then Minister of Overseas France, François Mitterand. See his *Présence française et abandon* (Paris: Plon, 1957), 215–16.
113. *L'Essor*, 10 November 1952.
114. *L'Essor*, 26 May 1953.
115. *Condition humaine*, 7 February 1950. The government complied six months later with its Equality Law of August, 1950.
116. For an interesting overview of Senghor's activities in the French National Assembly, see Janet G. Vaillant, "African Deputies in Paris: The Political Role of Léopold Senghor in the Fourth Republic," in *Double Impact: France and Africa in the Age of Imperialism*, ed. G. Wesley Johnson (Westport, Conn.: Greenwood Press, 1985), 141–52.
117. See Sano's interventions in the *Journal officiel, Assemblée nationale, débats*, sessions of 24 December 1952, 20 January 1953, and 10 March 1955.
118. Thompson and Adloff, *French West Africa*, 229.
119. Major Chailley, "La valorisation des cadres africaines" (unpublished paper, CMIDOM, Versailles, 1955).
120. Ibid.
121. As reported by the Secretary-General, *Office*, to Head of Military Cabinet, 19 December 1952, ANS, 4D172 143.
122. Catherine Coquery-Vidrovitch considers that the *Loi-cadre*, also commonly called the Defferre Law, along with the Treaty of Rome signed in March 1957, after two years of negotiations, marked "the true birthday of independence" in FWA. See Catherine Coquery-Vidrovitch, "The Transfer of Economic Power in French-Speaking West Africa," in *Decolonization and African Independence: The Transfers of Power, 1960–1980*, ed. Prosser Gifford and William Roger Louis (New Haven: Yale University Press, 1988), 113–14; and Manning, *Francophone Sub-Saharan Africa*, 146.
123. Manning, *Francophone Sub-Saharan Africa*, 147–50.

124. Bulletin triestriel from 15 May to 20 August 1957, Dakar, 24 August 1957, ANS, 17G438 152.
125. Bulletin trimestriel du poste de securité de la défense nationale et des forces armées, minister of war, FWA/Togo, period from 15 November 1956 to 15 February 1957, Dakar, ANS, 17G438 152.
126. L'Action, no. 4, February 1956.
127. L'Action, no. 2, October, 1957.
128. It was reported by the French administration that the BDS had "noted with sadness the repression in Algeria" and had therefore decided to support these boycotts. Bulletin trimestriel from 15 May to 20 August 1957, Dakar, 24 August 1957, ANS, 17G438 152.
129. Lawler, "Soldiers of Misfortune," vol. 2, 573.
130. Nancy Lawler has noted that by the 1980s most *ivoiren* veterans of the Second World War were receiving the nonindexed, and therefore paltry, sum of 20,000 CFA, or 400 French francs annually. See "Soldiers of Misfortune," vol. 1, 34, note 31. The same author found that the veterans of Korhogo had been unable to maintain their veterans' lodge. It had fallen into disrepair, and was torn down in the late 1960s to make way for a bank. Ibid., vol. 2, 521.

CHAPTER 10
Epilogue: Collective Memory and the African Military

1. The song was recorded and translated by Nancy Lawler, "Soldiers of Misfortune," vol. 2, 439. The French text is:

> "La France est notre mère,
> C'est elle qui nous nourrit
> Avec les pommes de terre
> Et des fayots pourris!"

2. Diallo, *Force-Bonté*, 31.
3. An obvious New World analogy is the use of old symbols from American plantation slave society such as "Uncle Ben" and "Aunt Jemima" to advertise commercial rice and pancake products.
4. My translation of Léopold Senghor's "Poème liminaire," in his *Poèmes* (Paris: Editions du Seuil, 1964).
5. *L'Action*, December, 1955.
6. References in this paragraph are from Ly, *Mercenaires noirs*, 17-21.
7. Interview with Al Hajj Doudou Diallo, Dakar, 18 May 1973.
8. See also Chapter 4 above.
9. Anonymous Report on the Death of Captain Charles N'Tchoréré (n.d. 1940?); ANSHA, 34N/1081.
10. Ibid.; Bonnet, *Mémorial*, 37; Boisboissel, "Le Centenaire," 22.
11. So incensed over this incident were the French that even under the censorship of Vichy, a toned-down version of N'Tchoréré's tragedy was published. See Bonnet, *Mémorial*, 37.
12. Boisboissel, "Le Centenaire," 22-24; a popular version of the events of Chasselay-Montluzin ran serially in several editions of an African veterans' publication. See "L'Héroique Défense de Chasselay-Montluzin par les Tirailleurs Sénégalais," *Souvenir et Devenir* 7 (1970), passim.
13. Cited in Boisboissel, "Le Centenaire," 24.
14. See pp. 167-68 for the text in translation on these two memorials.
15. Boisboissel, "Le Centenaire," 24; photograph in *Tropiques*, 1946.
16. Echenberg, "Tragedy at Thiaroye," 121-23.
17. Renseignements politiques, no date, ANS, 4D31 14 contains a letter translated from

Arabic and written by one Ibrahima of Dakar to his friend, Abdurahmane Traoré, stationed in a military camp in the Bouches-du-Rhône district.
18. See, for example, *Réveil*, 26 August 1946, 2.
19. Respondent no. 27, Upper Volta Survey. Seventeen of the fifty-five veterans surveyed were familiar with the events of Thiaroye.
20. Interview with Eba Aoussi, Treichville, Côte d'Ivoire, 25 May 1973.
21. *Réveil*, 13 and 27 February and 6 March 1950.
22. Benoist, *L'Afrique occidentale française*, 427.
23. Senghor, *Poèmes*, 90–91.
24. Interview with Al Hajj Doudou Diallo, Dakar, 18 May 1973.
25. It is not often remarked that Sembène had firsthand knowledge of the *Tirailleur* experience. He was called up for his military service in 1942 in his twentieth year (perhaps by the Vichy administration) and later fought for the Free French forces in Europe as an artillery man. He was demobilized in 1946 and participated in the Dakar-Niger rail strike of 1948. While his military service no doubt contributed to his consciousness as a fierce critic of colonialism, it should be noted that even as a teenager he had shown signs of rebellion. He was expelled from school in 1935 for having allegedly come to blows with his school principal. See Françoise Pfaff, *The Cinema of Ousmane Sembène, a Pioneer of African Film* (Westport, Conn.: Greenwood Press, 1984), 181–82.

A NOTE ON SOURCES

This study draws on a wide variety of previously published work. In the colonial era, an unabashedly congratulatory yet functional literature emerged around two themes. First were the epic accounts of the *Tirailleurs Sénégalais* in combat. A sampling of this voluminous literature would include the multivolume *Les Armées françaises d'outre-mer*, published by the Ministry of War for the International Colonial Exhibition of Paris, 1931, of which volume 6 was entitled *Histoire militaire de l'Afrique occidental française* (Paris: Imprimerie Nationale, 1931); and Yves de Boisboissel, *Peaux noirs, coeurs blancs* (Paris: J. Peyronnet, 1954). The second category consisted of practical manuals describing for young French officers the manner in which the colonial army operated. Two such works were Captain Louis-Jules Obissier, "Notice sur les Tirailleurs sénégalais," *Revue des Troupes Coloniales* (July, 1904): 49–69; and Colonel Jean Ferrandi, *L'Officier colonial* (Paris: Larose, 1930). Both approaches sought to inculcate in their largely French audiences a pride in republican France and in its colonial extensions. Both served to legitimize the imperial ideology of domination.

Not all of the popular literature is ponderous, nor is the epic genre the exclusive preserve of French military writers. An African historian, Ibrahim Kaké, has romanticized the exploits of the African soldier through the centuries in a book which he has edited for young readers entitled *Les légions noires* (Paris: Editions ABC, 1976).

In contrast to this military genre stands a body of literature reflecting both liberal and radical perspectives. Among the first to suggest that military service had been a substantial burden for French-speaking West Africans were Raymond Leslie Buell, *The Native Problem in Africa*, vol. 2 (New York: Macmillan, 1928); Abdoulaye Ly, *Mercenaires noirs: Notes sur une forme de l'exploitation des Africaines* (Paris: Présence Africaine, 1957); Jean Suret-Canale, *Afrique noire—l'Ere coloniale, 1900–1945* (Paris: Editions Sociales, 1964); and Michael Crowder, *West Africa under Colonial Rule* (London: Hutchinson, 1968). The only work to devote any attention to French West African veterans is Virginia Thompson and Richard Adloff, *French West Africa* (Stanford: Stanford University Press, 1957).

In his above-mentioned short essay, the Senegalese historian and politician Abdoulaye Ly went so far as to argue that military activities of Africans in French service were unmitigated acts of collaboration with the oppressor. It is perhaps safe to say that Ly's views reflected the bias of most FWA students, intellectuals, and even politicians in the decade of the 1950s. As the possibility of independence approached, the *Tirailleurs Sénégalais* and their history became an embarrassment better left ignored, or denigrated when it could not be.

In recent years, an emerging scholarly literature in French has considerably improved our understanding of the *Tirailleurs* and their place in African history. The definitive study of FWA's total effort in the First World War, and a work which therefore devotes considerable attention to the *Tirailleurs Sénégalais*, is Marc Michel, *L'Appel à l'Afrique: Contributions et réactions à l'effort de guerre en A.O.F., 1914–1919* (Paris: Editions de la Sorbonne, 1982). Other valuable works are: Hélène D'Almeida-Topor, "Les populations dahoméennes et le recrutement militaire pendant la première guerre mondiale," *Revue française d'histoire d'outre-mer* 60 (1973): 196–241; Danielle Domergue-Cloarec, "La Côte d'Ivoire de 1912 à 1920: Influence de la première guerre mondiale sur l'évolution politique, économique et sociale," 2 vol. (thèse de

3ème cycle, L' Université de Toulouse, 1974); and Jean-Loup Salètes, "La Constitution de l'armée noire et le recrutement indigène en AOF pendant la première guerre mondiale (1908–1918)" (mémoire de maîtrise, L'Université d'Aix-en-Provence, 1968).

English-speaking scholars have also contributed to the subject, beginning with a highly original book written over fifty years ago by Shelby C. Davis, *Reservoirs of Men: A History of the Black Troops of French West Africa* (1934; reprint, Westport, Conn.: Negro Universities Press, 1970). A colorful, often personalized work is Charles J. Balesi, *From Adversaries to Comrades-in-Arms: West Africans and the French Military, 1885–1918* (Waltham, Mass.: Crossroads Press, 1979). Two recent American doctoral dissertations have focused on African soldiers in FWA: Nancy E. Lawler, "Soldiers of Misfortune: The *Tirailleurs Sénégalais* of the Côte d'Ivoire in World War Two," 3 vol. (Ph.D. dissertation, Northwestern University, Evanston, Ill., 1988); and J. Malcolm Thompson, "In Dubious Service: The Recruitment and Stabilization of West African Maritime Labor by the French Colonial Military, 1659–1900" (Ph.D. dissertation, University of Minnesota, Minneapolis, 1989. Lastly, Anthony Clayton has reminded readers that the *Tirailleurs Sénégalais* were only part of a larger French military presence in Africa in his *France, Soldiers and Africa* (London: Brassey's, 1988).

Primary sources used in this study are numerous, if widely dispersed. The most important records are to be found in the two series of the *Archives de l'Afrique occidentale française*, which are housed in the *Archives Nationales du Sénégal* (ANS) in Dakar. *Série G* contains political reports of the governor-general of FWA and of the various lieutenant governors of the territories. These frequently comment on political aspects of the annual military levy. More pertinent still is the military section, *Série D*. Here are often richly detailed data for a part of the world where quantitative information is even today difficult to obtain. The colonial army and civilian officials were obligated by the terms of the Conscription Law of 1919 to maintain detailed records for each yearly intake of recruits. In every colony, so-called mobile draft boards maintained these statistics, which formed a part of the Annual Military Report on Recruitment issued under the signature of the *général commandant supérieur* (GCS), or commander in chief (CIC), of the French Colonial Army in FWA. These annual reports, in turn, treated such subjects as the number of absentees from the draft, the physical health of the cohort, and the attitude of the African population toward military conscription. In addition, the chief medical officer of FWA wrote a specialized annual report on the health of the annual cohort.

One unusual feature of the military data is that they emanate from two different administrative authorities. The civilian administrators came to view the military draft as an obstacle to the orderly economic activities of their regions. Their reports stressed the disruptive and burdensome aspects of the annual levy, hoping that it might be reduced if not entirely curtailed in future years. The military, on the other hand, put annual conscription in a positive light, arguing not only that most Africans had come to accept compulsory military service as a fact of colonial life but also that it was beneficial to the men who served and to the colony as a whole.

In France itself, military records dealing with the *Tirailleurs Sénégalais* are widely scattered. The archives of the former Ministry of Colonies, once in the Rue Oudinot, have now been transferred to Aix-en-Provence; they contain the civilian ministry's sometimes jaundiced views of military life in FWA. On the military side, the locale for the archives of the Ministry of War is the Château de Vincennes, just east of Paris. These holdings relate largely to military campaigns of conquest in the late nineteenth century and police operations in the first decade of the twentieth century. *Tirailleurs Sénégalais* activities in the Second World War in particular are difficult to document, as only a handful of regimental diaries seem to have survived what was for France a traumatic defeat. Lastly, because the original colonial army was a branch of the French navy, early records for the period preceding 1900 can be found at the *Centre Militaire d'Information et de Documentation sur l'Outre-mer* (CMIDOM), the Documentation Center of the French Navy, located in Versailles.

A Note on Sources

While the quantity of documentation is impressive, this material is not without its limitations. First, significant gaps exist for the period after the outbreak of the Second World War. The statistically valuable GCS reports from 1920 to 1938 are virtually complete, but reports from 1939 to 1943 are missing, and in the postwar period only the reports of 1946 and 1948 were located. As for the specialized medical reports, only one or two have been found either in Senegal or in France. Finally, much, perhaps most, archival material from the end of the Second World War through to independence, including sensitive information pertaining to the conflicts in Indochina and Algeria, is not yet fully accessible to the public.

A second problem is whether military statistics on African recruitment are as reliable as they would first appear to be. Especially in the early 1920s, most *cercles* or administrative districts relied on African chiefs to present men for the annual draft rather than using systematically prepared census lists of nineteen-year-old males in each jurisdiction. Even when census lists became the rule, colonies such as French Guinea and Senegal tended to have well-prepared data that were internally consistent, while more marginal recruiting regions such as Mauritania, Niger, and parts of Côte d'Ivoire had much less reliable statistics or used shoddy techniques.[1]

A third limitation is more serious. To judge by what has survived in their records, French military officials showed very little interest in how soldiers lived their lives during their leisure hours on base, or while on leave or after discharge. For such important issues as how soldiers ate and entertained themselves, or how they provided for their families, only the briefest traces can occasionally be found.

French military and civil archives, with their weaknesses and with their considerable strengths can only, in the end, provide a ruler's perception of how military service affected Africans. To obtain a perspective from Africans who served in the *Tirailleurs Sénégalais*, it is necessary to combine scarce written records with oral testimonies. Though it must be used cautiously, *Force-bonté* (Paris: F. Rieder, 1926) the autobiography of Bakary Diallo, a First World War veteran from the Upper Senegal valley, is invaluable: it is one of a kind, the only published war memoir by an African in French military service. While Diallo's tone is often obsequious, he does provide muted criticism of the shabby treatment accorded to African veterans of the First World War.

Another African veteran of that war, Captain Abdel-Kader Mademba, wrote *Au Sénégal et au Soudan français: Le Fama Mademba* (Paris: Larose, 1931). This book, a biography of his father, Mademba Sy, included *inter alia* a few references to his own military experiences. In addition, some of Mademba's correspondence with his patron, General Charles Mangin, has also survived and is available in the Mangin Papers at the *Bibliothèque Nationale*.

One recent and valuable contribution that links written and oral mediums is Joe Harris Lunn's presentation of the oral history of Kande Kamara, a Guinean veteran of the First World War. The article is entitled, "Kande Kamara Speaks: An Oral History of the West African Experience in France, 1914-18." in *Africa and the First World War*, ed. Melvin E. Page (London: Macmillan, 1987), 28-53.

Nancy Lawler has provided the most extensive collection to date of oral testimony relating to French military service in West Africa. In 1985 and 1986 she conducted 109 interviews with *Ivoiriens*, mostly from the Korhogo region, who had served in the *Tirailleurs Sénégalais* during the Second World War. These testimonies cover the soldiers' experiences from recruitment and induction in Africa, through combat, and often imprisonment, in France, and end with demobilization and reentry into civilian life in Côte d'Ivoire after 1945. Sixty percent of the interviews were taped, and the cassettes are deposited at the *Université*

---

1. As an illustration, the Annual Military Report on Recruitment for 1932 by the CIC, General Freydenberg, Dakar, 15 September 1932, ANS, 4D70 81.

d'Abidjan, Institut d'Histoire, d'Art et d'Archaéologie Africaines. The entire text of these valuable oral testimonies, covering over 600 pages, are reproduced in Nancy Lawler, "Soldiers of Misfortune."

Mention should also be made here of two sets of oral information I undertook for this study. A list of eighteen interviews that I conducted in Senegal, Upper Volta, and Côte d'Ivoire appears below. In addition, my research assistant, Moussa Traoré, distributed a questionnaire, which I designed, to some fifty-five veterans in Upper Volta in 1973. Respondents were chosen on a random basis in two regions of the country, the Tougan area and the captial city of Ouagadougou. A mix of conscripts and career soldiers, the respondents included one veteran of the First World War, eight of the interwar period, twenty-eight of the Second World War, and eighteen after 1945. The questionnaire is reproduced in the Appendix. The results of this research, which I have named the Upper Volta Survey, proved particularly useful in examining the place of veterans in French West African society, the subject of Chapter 8.

African soldiers have also been the subject of considerable attention from African writers and filmmakers. Birago Diop has written a short story entitled "Sarzan," a mispronunciation of the word for "sergeant," which depicts the alienation of a returning veteran who is driven to insanity. The story appears in Birago Diop, *Les Contes d'Amadou Koumba* (Paris: Présence Africaine, 1961), 173–187. With a different resolution, the same theme is explored by Ousmane Sembène in his film "Niaye."[2] In 1988 Sembène released his feature film "*Le Camp de Thiaroye*," depicting the tragic events unfolding at this Senegalese military barracks in 1944.[3] Much earlier in his career, in 1957, Sembène also treated the theme of a returned Senegalese war veteran in his second novel, *O pays, mon beau peuple* (London: Methuen, 1986). Two French writers, the brothers Jérôme and Jean Tharoud, had much earlier produced the popular work entitled *La Randonnée de Samba Diouf* (Paris: Plon, 1922), and a comic strip with a high degree of ideological content aimed at young soldiers and entitled "Mamadou s'en va-t'en guerre" was produced by the French military. The serialized strip appeared in 1939–1940 in *La Gazette du Tirailleur*, a periodical published for African soldiers by the general staff of the French Colonial Army in French West Africa, Dakar. Both the novel by the Tharouds and the comic strip treated the odyssey of a humble West African peasant in highly romanticized terms, very much from the perspective of the dominant ideology of the metropolis. A similar approach was used by Doumbi-Fakoly for a fictitious West African soldier in the Second World War in his *Morts pour la France* (Paris: Karthala, 1983).

---

2. For a discussion of this film, see Paulin S. Vieyra, *Ousmane Sembène, Cinéaste* (Paris: Présence Africaine, 1971), 47–57.

---

3. See Chapter 6 for a full discussion of these events.

SOURCES

I. Interviews Conducted by the Author

Ahui, Jean. Abidjan, 24 May 1973. Veteran and veterans' official.
Aoussi, Eba. Treichville, Côte d'Ivoire, 25 May 1973. Veteran and veterans' official.
Basse, Léopold. Dakar, 11 April 1973. Veteran and veterans' official.
Diallo, Al Hajj Doudou. Dakar, 18 May 1973. Senior civil servant.
Diémé, Pierre. Ziguinchor, Senegal, 15 April 1973. Veteran.
Dièye, Babacar. Dakar, 11 April 1973. Veteran and veterans' official.
Fall, General Ahmadou. Dakar, 18 May 1973. Former officer in the *Tirailleurs Sénégalais*, and former CIC of the Senegalese National Army.
Doumouya, Soulimane. Dakar, 11 April 1973. Veteran and veterans' official.
Kizerbo, Joseph. Ouagadougou, 11 June 1973. Historian.
Macé, Colonel. Saint-Raphaël, France, 2 March 1977. Former French medical officer.
Mangin, Stanislas. Paris, 16 February 1977. Senator and son of General Charles Mangin.
M'Bengue, M'Baye and Doudou N'Dao. Dakar, 4 May 1973. School principal and teacher, respectively.
Niaré, Cheikh Oumar. Dakar, 14 May 1973. Veteran and veterans' official.
Sané, Mamadou Pop. Dakar, 4 April 1973. Radio technician.
Sarambé, Dramane. Tougan, Burkina Faso, 15 June 1973. Second portion worker in Soudan and farmer.
Traoré, Aliou. Koudougou, Burkina Faso, 3 June 1973. Rural agricultural technician.
Traoré, Al Hajj Lansina. Ouagadougou, 7 June 1973. Veterans' official.
Vicens, Paul. Ouagadougou, 8 June 1973. Veteran and veterans' official.
Zo, Goulli. Tougan, Burkina Faso, 7 April 1967. Veteran.

II. Private Papers Consulted

Mangin MSS, Archives Nationales, Paris: Papers of Charles Mangin.
Moutet MSS, Archives Nationales, Section Outre-Mer, Paris: PA 28/83. Papers of Marius Moutet.

III. Archives Consulted

Archives Nationales du Sénégal, Dakar: Fonds modernes.

    Série G: Politique et Administration Générale.
    Série D: Affaires Militaires.[1]
        Sous-série 1D: Opérations militaires.
        Sous-série 2D: Périodes de Guerre.
        Sous-série 3D: Justice militaire.
        Sous-série 4D: Personnel militaire.

---

1. It is important to note that the codes for Série D, Affaires Militaires, were rearranged after I had used this source. A list of corresponding codes has been prepared at the ANS, and researchers are advised to consult this list if they wish to pursue items cited in my text.

Ministère des Colonies, Archives Nationales (Section Outre-Mer), Rue Oudinot (and now at Aix-en-Provence).
Ministère de la Guerre, Archives Nationales, Service Historique de l'Armée (ANSHA), Vincennes.
Ministère de la Marine, Centre Militaire d'Information et de Documentation sur l'Outre-Mer (CMIDOM), Versailles.

IV. Newspapers and Periodicals

a. Directed by Europeans

Afrique en Guerre. Dakar.
Afrique Nouvelle. White Fathers, Dakar.
L'Afrique française. Bulletin du Comité de l'Afrique Française et du Comité du Maroc.
La Dépêche Coloniale. Paris.
L'Espoir de Nice. Nice.
La France Coloniale. Paris.
Frères d'armes. CMIDOM, Versailles.
Gazette du tirailleur. Etat-major du Général Commandant Supérieur de l'A.O.F., Dakar.
Paris-Dakar. Dakar.
Le Patriote de Nice et du Sud-Est. Nice.
Revue militaire de l'A.O.F. Etat-Major du Général Commandant Supérieur de l'A.O.F., Dakar.
Soldat d'outre-mer. CMIDOM, Versailles.
Souvenir et devenir. Union Française des Anciens Combattants de France, Paris.
Le Temps. Paris.
Tropiques. Revue des Troupes Coloniales; Section d'Études et d'Informations des Troupes Coloniales, Paris.

b. Directed by Africans

L'Action. Mouvement Populaire Sénégalais (MPS)/Rassemblement Démocratique Africain (RDA), Dakar.
L'AOF. Section Française de l'Internationale Ouvrière-Féderation du Sénégal (SFIO), Dakar; Lamine Guèye's newspaper.
Clarté. Parti Socialiste Sénégalais (PSS), Dakar.
La Communauté. Organe des Anciens Militaires Coloniaux de Carrière, Veuves et Orphelins de Guerre, Dakar; Pierre Diagne's newspaper.
La Condition humaine. Bloc Démocratique Sénégalais (BDS), Dakar; Léopold Sédar Senghor's newspaper.
Dakar-étudiant. Association Générale des Etudiants de Dakar, Dakar.
L'Eclair. Rassemblement du Peuple Français (RPF) en A.O.F., Dakar.
L'Essor. Union Soudanaise (US)/RDA, Bamako; Modibo Keita's newspaper.
La Liberté. Parti Démocratique de Guinée (PDG)/RDA, Conakry.
Réveil. RDA, Dakar.
Voix des combattants. Association des Anciens Combattants et Victimes de Guerre de l'Afrique Occidentale Française (AACVGAOF), Dakar.
West African Pilot. London.

V. Government Publications

Journal officiel de la république Française. Paris: Imprimerie Nationale, 1953.
Ministère de la France d'Outre-Mer. Service de la Statistique Générale. *Population de l'A.O.F. en 1950–51 par canton et groupe ethnique (chiffres provisoires)*. Paris: Imprimerie Nationale, 1952.

Ministère de la Guerre. *Annuaire officiel de l'armée française*. Paris: Imprimerie Nationale, 1908, 1911, 1912, and 1914.

_____. *Annuaire officiel des officiers de l'armée active*. Paris: Imprimerie Nationale, 1924.

_____. *Les Armées françaises d'outre-mer. Vol. 6, Histoire militaire de l'Afrique occidentale française*. Paris: Imprimerie Nationale, 1931.

_____. *Code de pensions militaires d'invalidité et des victimes de guerre*.

_____. *Statistique médicale de l'armée métropolitaine et de l'armée coloniale*. 19 vols. Paris: Imprimerie Nationale, 1925-1942.

_____. *Statistique médicale de l'armée métropolitaine et de l'armée coloniale: Statistique du Maroc de 1907 à 1919*. Paris: Imprimerie Nationale, 1922.

Ministère des Colonies. *Annuaire de l'A.O.F., 1917-1921*. Paris: Imprimerie Nationale, 1922.

_____. *Annuaire du territoire du Haut-Sénégal et Moyen Niger, 1900-03*. Paris: Imprimerie Nationale, 1904.

_____. *Service de la statistique généranle: Population de l'AOF par canton et groupe ethnique*. Paris: Imprimerie National, 1952.

VI. Works Cited and Consulted

Abadie, Maurice. *La Défense des colonies*. Paris: Lavauzelle, 1937.

Agéron, Charles-Robert. "De Gaulle et la conférence de Brazzaville." In *"L'Entourage" et de Gaulle*, edited by Gilbert Pilleul, 243–51. Paris: Plon, 1979.

_____. *Les Algériens musulmans et la France*. Vol. 2 Paris: Presses Universitaires de France, 1968.

Allègre, Adrien-Edgar and Amadou N'Diaye Duguay-Clédor. "Déces de Commandant Abd el Kader Mademba." *La France coloniale*, no. 274, 1, September 1932.

Ambler, John S. *The French Army in Politics, 1945-62*. Columbus, Ohio: Ohio State University Press, 1966.

Amin, Samir, ed. *Modern Migration in Western Africa*. London: Oxford University Press, 1974.

Ammi-Oz, Moshe. "Les Interventions extra-militaires des forces armées nationales dans les états d'Afrique noire francophone." Thèse de 3e cycle, Université de Paris I, 1973.

Andrew, C.M., and A.S. Kanya-Forstner. "France, Africa, and the First World War." *Journal of African History* 19 (1978): 11-23.

Annuaire Mémoriale d'EFORTOM, 1956-1965. Saint Raphaël: EFORTOM, n.d.

Archinard, General Louis. "Les Troupes indigènes." In *L'Empire colonial français*. Paris: Plon, 1929.

Asiwaju, A.I. "Migrations as Revolt: The Example of Ivory Coast and the Upper Volta before 1945." *Journal of African History* 17 (1976): 577–94.

_____. *Western Yorubaland under European rule, 1889-1945: A Comparative Analysis of French and British Colonialism*. Atlantic Highlands, N.J.: Humanities Press, 1976.

Azan, Paul. *L'Armée indigène nord-africaine*. Paris: Lavauzelle, 1925.

Azéma, Jean-Pierre. *From Munich to the Liberation, 1938-1944*. 1979. Translated by Janet Lloyd. Cambridge: Cambridge University Press, 1984.

Azevedo, Mario. "Sara Demographic Instability as a Consequence of French Colonial Policy in Chad (1890-1940)." Ph.D. diss., Duke University, Durham, N.C., 1975.

Baier, Stephen, and Paul E. Lovejoy, "The Tuareg of the Central Sudan: Gradations in Servility at the Desert Edge (Niger and Nigeria)." In *Slavery in Africa: Historical and Anthropological Perspectives*, edited by Suzanne Miers and Igor Kopytoff, 391-444. Madison: University of Wisconsin Press, 1977.

Balesi, Charles J. *From Adversaries to Comrades-in-Arms: West Africans and the French Military, 1885-1918*. Waltham, Mass.: Crossroads Press, 1979.

_____. "West African Influence on the French Army of World War I." In *Double Impact: France and Africa in the Age of Imperialism*, edited by G. Wesley Johnson, 93-104. Westport, Conn.: Greenwood Press, 1985.

Barrows, Leland C. "The Impact of Empire on the French Armed Forces, 1830-1920." In *Double Impact: France and Africa in the Age of Imperialism*, edited by G. Wesley Johnson, 53-91. Westport, Conn.: Greenwood Press, 1985.

———. "The Merchants and General Faidherbe: Aspects of French Expansion in Senegal in the 1850's." *Revue française d'histoire d'outre-mer* 61 (1974): 236-83.

Barthes, Roland. *Mythologies*. 1957. Selected and translated by Annette Lavers. London: Granada, 1973.

Bazal, Jean. *Marseille galante*. Marseilles: P. Tacussel, 1980.

Bazin, Jean. "Guerre et servitude à Ségou." In *L'esclavage en Afrique précoloniale*, edited by Claude Meillassoux, 135-181. Paris: Maspero, 1975.

Becker, Charles, Mamadou Diouf, and Mohamed Mbodj. "L'Evolution démographique régionale du Sénégal et du bassin arachidier (Sine-Saloum) au vingtième siècle, 1904-1976." In *African Population and Capitalism: Historical Perspectives*, edited by Dennis D. Cordell and Joel W. Gregory, 76-94. Boulder, Colo.: Westview Press, 1987.

Benoist, Joseph Roger de. *L'Afrique occidentale française de 1944 à 1960*. Dakar: Nouvelles Editions Africaines, 1983.

Bernard-Duquenet, Nicole. *Le Sénégal et le Front populaire*. Paris: L'Harmattan, 1985.

Betts, Raymond F. *Assimilation and Association in French Colonial Theory, 1890-1914*. New York: Columbia University Press, 1961.

Boahen, A. Adu Boahen. *Ghana: Evolution and Change in the Nineteenth and Twentieth Centuries*. London: Longman, 1975.

Boisboissel, Yves de. *Peaux noirs, coeurs blancs*. Paris: J. Peyronnet, 1954.

———. "Contribution à la connaissance de l'histoire militaire de l'Afrique occidentale française: Tirailleurs et spahis sénégalais et soudanais." *Revue internationale d'histoire militaire* 4 (1956): 46-63.

———. "Un siècle d'héroisme au service de la France: Le Centenaire des Tirailleurs Sénégalais," *Tropiques, Revue des troupes coloniales* 55, June (1957), 22-25.

Boisseson, Gilbert de. *Le Recrutement en AOF et AEF*. Paris: CMIDOM, 1956.

Bonnet, Captain G. *Mémorial de l'Empire: A la gloire des troupes coloniales*. Paris: Sequaria, 1941.

Bonnetain, Mme. Paul. *Une Française au Soudan: Sur la route de Tombouctou (du Sénégal au Niger)*. Paris: Librairies Imprimeries réunies, 1984.

Bosquet, Léon, and Ernest Hosten. *Un Fragment de l'epopée sénégalaise: Les tirailleurs sur l'Yser*. Brussels: G. Van Oost, 1918.

Bouche, Denise. "Le retour de l'Afrique occidentale française dans la lutte contre l'ennemi aux côtés des Alliés." *Revue d'histoire de la deuxième guerre mondiale* 29 (1979), 41-68.

———. *Les Villages de liberté en Afrique noire, 1887-1910*. Paris: Mouton, 1968.

Boutillier, J.-L. "Les captifs en A.O.F. (1903-1905)." *Bulletin de l'Institut fondamental de l'Afrique noire*. Sér. B. 30, 2 (1968): 513-35.

Bruge, Roger. *Juin 1940: Le Mois maudit*. Paris: Fayard, 1980.

———. *Les Combattants du 18 Juin*. Vol. 1, *Le Sang versé*. Paris: Fayard, 1982.

Buah, F. K. *A History of Ghana*. London: Macmillan, 1980.

Buell, Raymond Leslie. *The Native Problem in Africa*. Vol. 2. New York: Macmillan, 1928.

Bugnet, Charles. *Mangin*. Paris: Plon, 1934.

Campbell, Bonnie. "Social Change and Class Formation in a French West African State." *Canadian Journal of African Studies* 8 (1974), 285-306.

Careless, J.M.S., and Craig R. Brown., eds. *The Canadians, 1867-1967*. Toronto: Macmillan, 1967.

Catrice, Paul. "L'Emploi des troupes indigènes et leur séjour en France." *Etudes: Revue catholique d'intèrêt général* (20 November 1931): 388-409.

Chaffard, Georges. "Un Entretien avec les anciens combattants de Haute-Volta." *Soldats d'outre-mer* (June, 1958).

_____. "Le Rôle sociale de l'armée en Afrique," *Soldats d'outre-mer* (June, 1958), 14–21.
Chailley, Major. "L'Africanisation des cadres de l'armée." Unpublished paper, CMIDOM, Versailles, 1957.
_____. "La Valorisation des cadres africains." Unpublished paper, CMIDOM, Versailles, 1955.
Challener, Richard D. *The French Theory of the Nation-in-Arms, 1866–1939*. New York: Russell & Russell, 1965.
Clayton, Anthony. *France, Soldiers and Africa*. London: Brassey's, 1988.
Clayton, Anthony, and David Killingray. *Khaki and Blue: Military and Police in British Colonial Africa*. Athens, Ohio: Ohio University Center for International Studies, 1989.
Cohen, Robin, Jean Copans, and Peter Gutkind, eds. *African Labor History*. Beverly Hills: Sage, 1978.
Coleman, James S. *Nigeria, Background to Nationalism*. Berkeley: University of California Press, 1960.
Cooper, Frederick. "From Free Labor to Family Allowances: Labor and African Society in Colonial Discourse." *American Ethnologist* 16 (1989): 745–65.
Coquery-Vidrovitch, Catherine. "The Transfer of Economic Power in French-Speaking West Africa." In *Decolonization and African Independence: The Transfers of Power, 1960–1980*, edited by Prosser Gifford and William Roger Louis, 105–134. New Haven: Yale University Press, 1988.
_____. "Vichy et l'industrialisation aux colonies." *Revue d'histoire de la deuxième guerre mondiale* 29 (April 1979), 69–94.
Coquery-Vidrovitch, Catherine, and Paul E. Lovejoy. "The Workers of Trade in Precolonial Africa." In *The Workers of African Trade*, edited by Catherine Coquery-Vidrovitch and Paul E. Lovejoy, 9–24. Beverly Hills: Sage, 1985.
Corbett, Edward M. *The French Presence in Black Africa*. Washington: Black Orpheus Press, 1972.
Cordell, Dennis D. *Dar al-Kuti and the Last Years of the Trans-Saharan Slave Trade*. Madison: University of Wisconsin Press, 1985.
_____. "Extracting People from Precapitalist Production: French Equatorial Africa from the 1890s to 1930s." In *African Population and Capitalism: Historical Perspectives*, edited by Dennis D. Cordell and Joel W. Gregory, 137–52. Boulder, Colo.: Westview Press, 1987.
Cordell, Dennis D., and Joel W. Gregory. "Labor Reservoirs and Population: French Colonial Strategies in Koudougou, Upper Volta, 1914 to 1939." *Journal of African History* 23 (1982), 205–24.
Cordell, Dennis D., and Joel W. Gregory, eds. *African Population and Capitalism: Historical Perspectives*. Boulder, Colo.: Westview Press, 1987.
Cordell, Dennis D., Joel W. Gregory, and Victor Piché, "African Historical Demography: The Search for a Theoretical Framework." In *African Population and Capitalism: Historical Perspectives*, edited by Dennis D. Cordell and Joel W. Gregory, 14–32. Boulder, Colo.: Westview Press, 1987.
Coulibaly, Sidiki P. "Les Migrations voltaiques: Les origines, les motifs et les perceptions des politiques." Ph.D. diss., Université de Montréal, 1978.
Crowder, Michael. "The 1939–1945 War and West Africa." In *History of West Africa*, edited by J.F.A. Ajayi and M. Crowder, vol. 2, 596–621. London: Longman, 1971.
_____. *West Africa under Colonial Rule*. London: Hutchinson, 1968.
Curtin, Philip D. *Economic Change in Precolonial Africa: Senegambia in the Era of the Slave Trade*. 2 vols. Madison, Wis.: University of Wisconsin Press, 1975.
Cutter, Charles H. "The Genesis of a Nationalist Elite: The Role of the Popular Front in the French Soudan (1936–1939)." In *Double Impact: France and Africa in the Age of Imperialism*, edited by G. Wesley Johnson, 107–39. Westport, Conn.: Greenwood Press, 1985.

D'Aby, F.J. Amon. *La Côte d'Ivoire dans la cité africaine*. Paris: Larose, 1951.
D'Almeida-Topor, Hélène. "Les Populations dahoméennes et le recrutement militaire pendant la première guerre mondiale." *Revue française d'histoire d'outre-mer* 60 (1973): 196–241.
Daget, Serge. "British Repression of the Illegal French Slave Trade: Some considerations." In *The Uncommon Market: Essays in the Economic History of the Atlantic Slave Trade*, edited by Henry A. Gemery and Jan S. Hogendorn, 419–42. New York: Academic Press, 1979.
David, Philippe. *Les Navétanes: Histoire des migrants saisonniers de l'arachide en Sénégambie des origines à nos jours*. Dakar: Nouvelles Editions Africaines, 1980.
Davis, Shelby C. *Reservoirs of Men: A History of the Black Troops of French West Africa*. 1934. Reprint. Westport, Conn.: Negro Universities Press, 1970.
Decalo, Samuel. *Coups and Army Rule in Africa: Studies in Military Style*. New Haven: Yale University Press, 1976.
Delafosse, Maurice. "Les Points sombres de l'horizon en Afrique occidentale." *Bulletin du Comité de l'Afrique française* (June, 1922): 274–77.
Descostes, François. ed. *Au Soudan (1890–1891): Souvenirs d'un tirailleur sénégalais d'après sa correspondance intime*. Paris: Picard, 1893.
Diallo, Bakary. *Force-bonté*. Paris: F. Rieder, 1926.
Digu'en, Abou. *Notre empire africain noir: Ses problèmes politques et militaires*. Paris: Lavauzelle, 1928.
Diop, Birago. *Les Contes d'Amadou Koumba*. Paris: Présence Africaine, 1961.
Domergue-Cloarec, Danielle. "La Côte d'Ivoire de 1912 à 1920: Influence de la Première Guerre Mondiale sur l'évolution politique, économique et sociale." 2 vols. Thèse de 3e cycle, l'université de Toulouse le Mirail, 1974.
Doumbi-Fakoly. *Morts pour la France*. Paris: Karthala, 1983.
Duboc, General A. *Les Sénégalais au service de la France*. Paris: Edgar Malfère, 1939.
Duperray, Anne-Marie. *Les Gourounsi de Haute Volta*. Stuttgart: Franz-Steiner, 1984.
Dutreb, Maurice. *Nos Sénégalais pendant la Grande Guerre*. Metz: Editions des Voix Lorraines, 1922.
Duval, Major. "La législation militaire actuelle en Afrique noire." Unpublished paper, CMIDOM, Versailles, 1956.
Echenberg, Myron. " 'Faire du Nègre': Military Aspects of Population Planning in French West Africa, 1920–1940." In *African Population and Capitalism: Historical Perspectives*, edited by Dennis D. Cordell and Joel Gregory, 95–108. Boulder and London: Westview Press, 1987.
──────. "Jihad and State-building in Late Nineteenth Century Upper Volta: The Rise and Fall of the Marka State of Al-Kari of Boussé." *Canadian Journal of African Studies* 3 (1969): 531–61.
──────. "Les Migrations militaires en Afrique occidentale française, 1900–1945." *Canadian Journal of African Studies* 14 (1980): 429–50.
──────. "Tragedy at Thiaroye: The Senegalese Soldiers' Uprising of 1944." In *African Labor History*, edited by Robin Cohen, Jean Copans, and Peter Gutkind, 109–28. Beverly Hills: Sage, 1978.
Echenberg, Myron, and Jean Filipovich. "African Military Labour and the Building of the Office du Niger Installations, 1925–1950." *Journal of African History* 27 (1986): 533–51.
Enloe, Cynthia H. *Ethnic Soldiers: State Security in Divided Societies*. Harmondsworth: Penguin, 1980.
Faidherbe, General Louis. *Le Sénégal: La France dans l'Afrique occidentale*. Paris: Hachette, 1899.
Fall, Babacar. "Le Travail forcé en Afrique occidentale française, 1900–1946: Cas du Sénégal, de la Guinée et du Soudan." Thèse de 3e cycle, Université de Dakar, 1984.
Fanon, Frantz. *Black Skin, White Masks*. 1952. Translated by Charles Lam Markmann. New York: Grove Press, 1968; original ed. 1952.
Farwell, Byron. *The Great War in Africa (1914–1918)*. New York: Norton, 1986.

Faure, Claude. "La Garrison européenne du Sénégal (1779-1858)." *Revue d'histoire des colonies* 8 (1920): 5-108.
Ferrandi, Jean. *L'Officier colonial*. Paris: Larose, 1930.
Filipovich, Jean. "Léopold Senghor's Approach to Opposition: A Policy of Reconciliation." Master's thesis, McGill University, Montreal, 1981.
Finnegan, Gregory A. "Population Movement, Labor Migration, and Social Structure in a Mossi Village." Ph.D. diss., Brandeis University, Waltham, Mass., 1976.
Flutet, Constant. "Historique et legislation de la médaille militaire, 1852-1952." Unpublished pamphlet, ANS, Dakar, 1952.
Foltz, William J. *From French West Africa to the Mali Federation*. New Haven: Yale University Press, 1965.
Frey, Henri. *Campagne dans le Haut-Sénégal et dans le Haut-Niger (1885-1886)*. Paris: Plon, 1888.
Gardinier, David E. "The French Impact on Education in Africa, 1817-1960." In *Double Impact: France and Africa in the Age of Imperialism*, edited by G. Wesley Johnson, 333-44. Westport, Conn.: Greenwood Press, 1985.
_____. *Historical Dictionary of Gabon*. Metuchen, N.J.: Scarecrow Press, 1981.
_____. "The Second World War in French West Africa and Togo: Recent Research and Writing." *Proceedings of the French Colonial Historical Society* 10 (1985): 261-72.
Gatelet, H. *Histoire de la conquête du Soudan Français, 1878-1899*. Paris: Berger-Leurault, 1901.
Gaulle, Charles de. *Mémoires d'espoir*. Vol. 1, *Le Renouveau, 1958-1962*. Paris: Plon, 1970.
_____. *Mémoires de Guerre*. Vol. 2, *L'Unité, 1942-1944* Paris: Plon, 1956.
_____. *Mémoires de Guerre*. Vol. 3, *Le Salut, 1944-1946*. Paris: Plon, 1959.
_____. *Vers l'armée de métier*. Paris: Berger-Leurault, 1934.
Giblin, Jim. "The Image of the Loyal African during World War II and Its Postwar Use by the French Communist Party." *Canadian Journal of African Studies* 14 (1980): 319-26.
_____. "The Vichy Years in French Africa: A Period of African Resistance to Capitalism." Master's thesis, McGill University, Montreal, 1976.
Gobineau, Hélène de. *Noblesse d'Afrique*. Paris: Fasquelle, 1946.
Gorce, Paul-Marie de la. *The French Army: A Military-Political History*. Translated by Kenneth Douglas. New York: George Braziller, 1963.
Gramsci, Antonio. *Letter from Prison*. Translated by Lynne Lawner. New York: Harper & Row, 1975.
Gregory, Joel W., and Victor Piché, "Mode de production et régime démographique." *Canadian Journal of African Studies* 19 (1985): 73-79.
Guèye, Lamine. *Itinéraire africain*. Paris: Présence Africaine, 1966.
Guez, Nicole. "La Militarisation des organisations de service civique dans les états africains modernes." *Revue française d'histoire d'outre-mer* 68 (1981), 263-67.
Guillaumet, Edouard. *Le Soudan en 1894: La vérité sur Tombouctou, l'esclavage au Soudan*. Paris: Albert Savine, 1895.
Gutteridge, William. *The Military in African Politics*. London: Methuen, 1969.
Haeckel, Lt. Jean. "Guide à l'usage des officiers de renseignements en Afrique." Unpublished paper, 76 pp., CMIDOM, Versailles, 1973.
Hargreaves, J.D., ed. *France and West Africa: An Anthology of Historical Documents*. London: Macmillan, 1969.
Headrick, Rita. "African Soldiers in World War II." *Armed Forces and Society* 4 (1978): 501-26.
Hennebelle, Guy, ed. *Les Cinémas africains en 1972*. Paris: Société Africain d'Edition, 1972.
Historique du 1er Régiment des Tirailleurs sénégalais. Saint-Louis: n.d. (circa 1922).
Hitler, Adolf. *Mein Kampf*. 1925. Translated by Ralph Manheim. Boston: Houghton Mifflin, 1943.
Holbrook, Wendell P. "Oral History and the Nascent Historiography for Africa and World War II: A Focus on Ghana." *International Journal of Oral History* 3 (1982), 148-166.
Horne, Alistair. *A Savage War of Peace: Algeria, 1954-62*. Harmondsworth: Penguin, 1979.
_____. *The Price of Glory*. London: Macmillan, 1962.

---. *To Lose a Battle: France, 1940*. London: Macmillan, 1969.
Houdry, Captain. "Le Recrutement en Afrique Occidentale Française." *Renseignements coloniaux* (1929): 373-77.
Hymans, Jacques L. *Léopold Sédar Senghor: an Intellectual Biography*. Edinburgh: Edinburgh University Press, 1971.
Idowu, H.O. "Assimilation in Nineteenth Century Senegal." *Bulletin de l'Institut fondamental de l'Afrique noire*. Sér. B. 30 (1968): 1422-47.
Igué, Ogunsola John. "Evolution du commerce clandestin entre le Dahomey et le Nigeria depuis la guerre du 'Biafra.' " *Canadian Journal of African Studies* 10 (1976): 235-58.
Irele, Abiola, ed. *Selected Poems of Léopold Sédar Senghor*. New York: Cambridge University Press, 1977.
Irving, David. *The Trail of the Fox*. New York: Avon, 1977.
Isaacman, Allen, and Anton Rosenthal, "Slaves, Soldiers, and Police: Power and Dependency among the Chikunda of Mozambique, ca. 1825-1920." In *The End of Slavery in Africa*, edited by Suzanne Miers and Richard Roberts, 220-53. Madison: University of Wisconsin Press, 1987.
Israel, Adrienne M. "Measuring the War Experience: Ghanaian Soldiers in World War II." *Journal of Modern African Studies* 25 (1987): 159-168.
Johnson, G. Wesley. *The Emergence of Black Politics in Senegal*. Stanford: Stanford University Press, 1974.
---. "William Ponty and Republican Paternalism in French West Africa (1866-1915)." In *African Pro-Consuls: European Governors in Africa*, edited by Lewis Gann and P. Duignan, 127-56. Stanford: Hoover Institute, 1978.
---, ed. *Double Impact: France and Africa in the Age of Imperialism*. Westport, Conn.: Greenwood Press, 1985.
Kaké, Ibrahim Baba, ed. *Les Légions noires*. Paris: Editions ABC, 1976.
Kalck, Pierre. *Central African Republic: A Failure in Decolonization*. Translated by Barbara Thomson. New York: Praeger, 1971.
Kann, Réginald. "Les Troupes indigènes en France." *Le Temps* (26 April 1923).
Keegan, John. *The Face of Battle*. Harmondsworth: Penguin, 1976.
---. *Six Armies in Normandy*. Harmondsworth: Penguin, 1982.
Kelly, George A. *Lost Soldiers: The French Army and Empire in Crisis, 1947-1962*. Cambridge, Mass.: MIT Press, 1965.
Kersaint-Gilly, F. de. "L'Essai sur l'évolution de l'esclavage en Afrique occidentale française: Son dernier stade au Soudan français." *Bulletin du Comité d'études historiques et scientifiques de l'A.O.F.* 9 (1924): 469-78.
Killingray, David. "The Idea of a British Imperial African Army." *Journal of African History* 19 (1979): 421-36.
---. "Soldiers, Ex-Servicemen, and Politics in the Gold Coast, 1939-50." *Journal of Modern African Studies* 21 (1983): 523-34.
Killingray, David, and Richard Rathbone, eds. *Africa and the Second World War*. London: Macmillan, 1986.
Kitching, Gavin. *Class and Economic Change in Kenya: The Making of an African Petite Bourgeoisie, 1905-1970*. New Haven: Yale University Press, 1980.
Klein, Martin, and Paul E. Lovejoy. "Slavery in West Africa." In *The Uncommon Market: Essays in the Economic History of the Atlantic Slave Trade*, edited by H.A. Gemery and J.S. Hogendorn, 181-212. New York: Academic Press, 1979.
Kum'a N'dumbé III, Alexandre. *Hitler voulait l'Afrique: Les Plans secrets pour une Afrique fasciste, 1933-1945*. Paris: L'Harmattan, 1980.
Ladhuie, Paul. "L'Etat d'esprit des troupes noires consécutif à la guerre 1939-1944." Unpublished mémoire d'Ecole Nationale d'Administration de la France d'Outre-Mer (ENFOM), Paris: 1946.

"La France et les cadres africains et malgaches: 1, Les cadres avant l'indépendance." *Frères d'Armes* 10 (1964): 15-23.

Lakroum, Monique. *Le Travail inégal: Paysans et salariés sénégalais face à la crise des années trente.* Paris: L'Harmattan, 1982.

Lassalle-Séré, R. *Le Recrutement de l'armée noire.* Paris: Librairie moderne de droit et jurisprudence, 1929.

Lattre de Tassigny, Jean de. *The History of the French First Army.* Translated by Malcolm Barnes. London: George Allen & Unwin, 1952.

Lawler, Nancy E. "Soldiers of Misfortune: The *Tirailleurs Sénégalais* of the Côte d'Ivoire in World War Two." 3 vols. Ph.D. diss., Northwestern University, Evanston, Ill., 1988.

Lemarchand, René. "The Politics of Sara Ethnicity: A Note on the Origins of the Civil War in Chad." *Cahiers d'Etudes africaines* 20 (1980), 449-71.

"Les Anciens Combattants dans les états de la communauté." *Chroniques de la communauté* (March, 1960): 11-12.

Leys, Colin. "The 'Overdeveloped' Post Colonial State: A Re-evaluation." *Review of African Political Economy* 5 (1976), 39-48.

Lipschutz, Mark R., and R. Kent Rasmussen, eds. *Dictionary of African Historical Biography.* 2d ed. Berkeley: University of California Press, 1986.

Loucou, Jean-Noel. "La deuxième guerre mondiale et ses effets en Côte d'Ivoire." *Annales de l'Université d'Abidjan.* Série 1, Histoire 8 (1980): 183-207.

Lukacs, Georg. *History and Class Consciousness.* Translated by Rodney Livingstone. Cambridge: MIT Press, 1971.

Lunn, Joe Harris. "Kande Kamara Speaks: An Oral History of the West African Experience in France, 1914-18." In *Africa and the First World War,* edited by Melvin E. Page, 28-53. London: Macmillan, 1987.

Ly, Abdoulaye. *Mercenaires noirs: Notes sur une forme de l'exploitation des Africains.* Paris: Présence Africaine, 1957.

Mademba, Abdel-Kader. *Au Sénégal et au Soudan français: Le Fama Mademba.* Paris: Larose, 1931.

———. *Recrutement de Tirailleurs et main d'oeuvre au Soudan français.* Tananarive: n.d., circa 1919.

Mangeot, General P. "Les Troupes coloniales pendant la guerre." *Renseignements coloniaux* 10 (1931): 557-62.

———. "Le Budget de la guerre et de la défense des territoires d'outre-mer." *Renseignements coloniaux* (1932): 259-61.

Mangin, Charles. "Caractères physiques et moraux du soldat nègre." *La Revue anthropologique* 10 (1911), 1-16.

———. *Comment finit la guerre.* Paris: Plon, 1920.

———. "Debate with Torcy." *Le Mois coloniale et maritime* 97 (1911): 87-94.

———. *La Force noire.* Paris: Hachette, 1910.

———. *La Mission des troupes noires, compte-rendu fait devant le Comité de l'Afrique française.* Paris: Comité de l'Afrique Française, 1911.

———. *Lettres de guerre 1914-18.* Paris: Fayard, 1950.

———. *Lettres du Soudan.* Paris: Editions des Portiques, 1930.

———. *Regards sur la France d'Afrique.* Paris: Plon, 1924.

———. *Souvenirs d'Afrique: Lettres et carnets de route.* Paris: Denoël et Steele, 1936.

Manning, Patrick. "L'Affaire Adjovi: La bourgeoisie foncière naissante au Dahomey, face à l'administration." In *Entreprises et Entrepreneurs en Afrique (XIXe et XXe siècles),* vol. 2, edited by Laboratoire "Connaissance du Tiers-Monde," 241-67. Paris: 1983.

———. "The Enslavement of Africans: A Demographic Model." *Canadian Journal of African Studies* 15 (1981), 499-526.

———. *Francophone Sub-Saharan Africa, 1880-1985.* New York: Cambridge University Press, 1988.

———. *Slavery, Colonialism and Economic Growth in Dahomey, 1640–1960*. Cambridge: Cambridge University Press, 1982.
Marceau, Hippolyte-Victor. *Le Tirailleur soudanais*. Paris: Berger-Levrault, 1911.
Markovitz, Irving L. *Léopold Sédar Senghor and the Politics of Negritude*. New York: Atheneum, 1969.
Marseille, Jacques. *Empire colonial et capitalisme français: Histoire d'un divorce*. Paris: Albin Michel, 1984.
McDougall, E. Anne. "A Topsy-Turvy World: Slaves and Freed Slaves in the Mauritanian Adrar, 1910–1950." In *The End of Slavery in Africa*, edited by Suzanne Miers and Richard Roberts, 362–88. Madison: University of Wisconsin Press, 1987.
McLane, Margaret O. "Commercial Rivalries and French Policy on the Senegal River, 1831–58." *African Economic History* 15 (1987): 39–67.
McWilliams, James L., and R. James Steel. *The Suicide Battalion*. Edmonton: Hurtig, 1978.
Memmi, Albert. *The Colonizer and the Colonized*. Translated by Howard Greenfeld. Boston: Beacon Press, 1965.
Michel, Marc. "Citoyenneté et service militaire dans les quatres communes du Sénégal au cours de la première guerre mondiale." In *Perspectives nouvelles sur le passé de L'Afrique noire et de Madagascar: Mélanges offerts à Hubert Deschamps*, 299–314. Paris: Editions de la Sorbonne, 1974.
———. "La Genèse du recrutement de 1918 en Afrique noire française." *Revue française d'histoire d'outre-mer* 58 (1971): 433–50.
———. *La Mission Marchand, 1895–1899*. The Hague: Mouton, 1972.
———. *L'Appel à l'Afrique: Contributions et réactions à l'effort de guerre en A.O.F., 1914–1949*. Paris: Editions de la Sorbonne, 1982.
———. "Un Mythe: La 'Force Noire' avant 1914." *Relations intérnationales* 2 (1974): 83–90.
Miers, Suzanne, and Richard Roberts, eds. *The End of Slavery in Africa*. Madison: University of Wisconsin Press, 1988.
Milcent, Ernest, and Monique Sordet. *Léopold Sédar Senghor et la naissance de l'Afrique moderne*. Paris: Seghers, 1969.
Mitterand, François. *Présence française et abandon*. Paris: Plon, 1957.
Morganthau, Ruth Schachter. *Political Parties in French-speaking West Africa*. Oxford: Clarendon Press, 1964.
Mortimer, Edward. *France and the Africans, 1944–1960: A Political History*. London: Faber & Faber, 1969.
Navarre, General. "Les données de la défense de l'Indochine." *Revue de défense nationale* (March 1956), 271–79.
Nelson, Keith L. "The 'Black Horror on the Rhine': Race as a Factor in Post–World War I Diplomacy." *Journal of Modern History* 42 (1970): 606–27.
Nobécourt, Jacques. *Une Histoire politique de l'armée: De Pétain à Pétain, 1919–1942*. Paris: Editions de Seuil, 1967.
Nyo, General. "L'Evolution des cadres militaires africains de 1914 à 1956." *Tropiques* (June 1956): 33–40.
Obissier, Captain Louis-Jules. "Notice sur les Tirailleurs Sénégalais." *Revue des Troupes Coloniales* (July 1904): 49–69.
O'Brien, Rita Cruise. *White Society in Black Africa: The French of Senegal*. London: Faber & Faber, 1972.
Ollandet, Jérôme. *Brazzaville, capitale de la France Libre: Histoire de la résistance française en Afrique (1940–1944)*. Brazzaville: Editions de la Savane, 1983.
Olusanya, Gabriel O. *The Second World War and Politics in Nigeria, 1939–1953*. London: Evans Bros., 1973.
Page, Melvin E., ed. *Africa and the First World War*. London: Macmillan, 1987.

Painlevé, Paul. "La Politique de guerre de 1917." *Revue de Paris* (15 March 1922): 287–320.
Paxton, Robert O. *Parades and Politicians at Vichy*. Princeton: Princeton University Press, 1966.
Pedroncini, Guy. *Les Mutineries de 1917*. Paris: Presses Universitaires de France, 1967.
Person, Yves. "Soixante ans d'évolution en pays Kissi." *Cahiers d'études africaines* 1 (1960), 86–112.
Pfaff, Françoise. *The Cinema of Ousmane Sembène, a Pioneer of African Film*. Westport, Conn.: Greenwood Press, 1984.
Poujoulat, Fernand. "Evolution de la mentalité des tirailleurs sénégalais au cours de la guerre 1939–1945." Mémoire de l'Ecole Nationale de la France d'Outre-Mer, Paris, 1946.
Poulantzas, Nicos. "The New Petty Bourgeoisie." In *Classes and Class Structure*, edited by A. Hunt, 113–24. London: Lawrence & Wishart, 1977.
Prévaudeau, A. *Joost Van Vollenhoven (1877–1918)*. Paris: Larose, 1953.
Rathbone, Richard. "Businessmen in Politics: Party Struggle in Ghana, 1949–57." *Journal of Development Studies* 9 (1972–73): 391–401.
Renault, François. "L'abolition de l'esclavage au Sénégal: L'attitude de l'administration française (1848–1905)." *Revue française d'histoire d'outre-mer* 58 (1971): 5–80.
Retel-Laurentin, A. "Fécondité et syphilis dans la région de la Volta noire," *Population* 28 (1973): 793–815.
Roberts, Richard and Martin Klein. "The Banamba Slave Exodus and the Decline of Slavery in the Western Sudan." *Journal of African History* 21 (1981): 375–394.
Robertson, Claire C. *Sharing the Same Bowl: A Socioeconomic History of Women and Class in Accra, Ghana*. Bloomington, Ind.: University of Indiana Press, 1984.
Rudé, George. *Ideology and Popular Protest*. New York: Pantheon, 1980.
Sabatier, Peggy. "Did Africans Really Learn to Be French? The Francophone Elite of the Ecole William Ponty." In *Double Impact: France and Africa in the Age of Imperialism*, edited by G. Wesley Johnson, 179–90. Westport, Conn.: Greenwood Press, 1985.
Salan, Raoul. *Mémoires: Fin d'un empire*. Vol. 1, *Le Sens d'un engagement, Juin 1899–Septembre 1946*. Paris: Presses de la Cité, 1970.
Salètes, Jean-Loup. "La Constitution de l'armée noire et le recrutement indigène en AOF pendant la première guerre mondiale (1908–1918)." Mémoire de maitrise, l'Université d'Aix-en-Provence, 1968.
Sarraut, Albert. *La Mise en valeur des colonies françaises*. Paris: Payot, 1923.
Schleh, Eugene I.A. "The Post-War Careers of Ex-Servicemen in Ghana and Uganda." *Journal of Modern African Studies* 6 (1968): 203–20.
Séché, Alphonse. *Les Noirs d'après des documents officiels*. Paris: Payot, 1919.
Sembène, Ousmane. *O pays, mon beau peuple*. London: Methuen, 1986.
Senghor, Léopold Sédar. *Chants d'ombre suivi de Hosties noires: Poèmes*. Paris: Editions de Seuil, 1945.
_____. *Poèmes*. Paris: Editions de Seuil, 1964.
_____. *Senghor: Prose and Poetry*. 1965. Selected and translated by John Reed and Clive Wake. London: Heinemann, 1976; original ed. 1965.
Shanin, Teodor. *The Awkward Class: Political Sociology of Peasantry in a Developing Society: Russia, 1910–1925*. Oxford: Clarendon Press, 1972.
Shiroya, Okete J. E. "The impact of World War II on Kenya: The Role of Ex-Servicemen in Kenyan Nationalism," Ph.D. diss., Michigan State University, East Lansing, Mich., 1968.
Skurnik, W.A.E. "The Military and Politics: Dahomey and Upper Volta." In *Soldier and State in Africa*, edited by Claude E. Welch, Jr., 62–121. Evanston: Northwestern University Press, 1970.
Smith, Carol A. "The Petty Bourgeoisie as a 'Fundamental' Revolutionary Class in Nicaragua," *Labour, Capital and Society* 19 (1986): 8–35.

Spiegler, James S. "Aspects of Nationalist Thought among French-speaking West Africans, 1921-1939." Ph.D. diss., Nuffield College, Oxford, 1968.
Stanley, George F.G. "The Fighting Forces." In *The Canadians, 1867-1967*, edited by J.M.S. Careless and R. Craig Brown, 764-84. Toronto: Macmillan, 1967.
Summers, Anne, and R.W. Johnson, "World War I Conscription and Social Change in Guinea." *Journal of African History* 19 (1978), 25-38.
Suret-Canale, Jean. *Afrique noire—L'Ere coloniale, 1900-1945*. Paris: Editions Sociales, 1964.
———. *Afrique noire occidentale et centrale*: III, 1. *De la colonisation aux indépendances (1945-1960)*. Paris: Editions Sociales, 1972.
———. "The French West African Railway Workers' Strike, 1947-48." In *African Labor History*, edited by Robin Cohen, Jean Copans, and Peter Gutkind, 129-54. Beverly Hills: Sage, 1978.
Tharaud, Jérôme, and Jean Tharaud. *La Randonnée de Samba Diouf*. Paris: Plon, 1922.
Thomas, Roger. "Military Recruitment in the Gold Coast during the First World War." *Cahiers d'études africaines* 15 (1975): 57-83.
Thompson, J. Malcolm. "In Dubious Service: The Recruitment and Stabilization of West African Maritime Labor by the French Colonial Military, 1659-1900." Ph.D. diss., University of Minnesota, Minneapolis, 1989.
Thompson, Virginia, and Richard Adloff. *French West Africa*. Stanford: Stanford University Press, 1957.
Torcy, General de. "Debate with Mangin." *Le Mois coloniale et maritime* 97 (1911): 78-86.
Tronchon, Jacques. *L'Insurrection Malgache de 1947*. Paris: Maspero, 1974.
Van Doorn, Jacques. *The Soldier and Social Change*. Beverly Hills: Sage, 1975.
Vaillant, Janet G. "African Deputies in Paris: The Political Role of Léopold Senghor in the Fourth Republic." In *Double Impact: France and Africa in the Age of Imperialism*, edited by G. Wesley Johnson, 141-52. Westport, Conn.: Greenwood Press, 1985.
Vieyra, Paulin Soumanou. *Sembène Ousmane, Cinéaste*. Paris: Présence Africaine, 1972.
Vigné D'Octon, Paul. *Au pays des fétiches*. Paris: Alphonse Lemerre, 1890.
Voguë, Eugène Melchior de. *Les Morts qui parlent*. Paris: Plon, 1899.
Wallerstein, Immanuel. "The Colonial Era in Africa: Changes in Social Structure." In *Colonialism in Africa, 1870-1960*, vol. 2, edited by Lewis Gann and P. Duignan, 399-421. Cambridge: Cambridge University Press, 1970.
Weber, Eugen. *Peasants into Frenchmen: The Modernization of Rural France, 1876-1914*. Stanford: Stanford University Press, 1976.
Welch, Claude E., ed. *Soldier and State in Africa*. Evanston: Northwestern University Press, 1970.
Weiskel, Timothy C. *French Colonial Rule and the Baulé People: Resistance and Collaboration, 1889-1911*. Oxford: Clarendon Press, 1980.
———. "Labor in the Emergent Periphery: From Slavery to Migrant Labor among the Baoule Peoples, c. 1880-1925." In *The World System of Capitalism: Past and Present*, edited by Walter L. Goldfrank, 207-33. Beverly Hills: Sage, 1979.
White, Luise. "Prostitution, Identity, and Class Consciousness in Nairobi During World War II." *Signs* 11 (1986), 255-73.
———. "A Colonial State and an African Petty Bourgeoisie: Prostitution, Property and Class Struggle in Nairobi, 1936-1940." In *Struggle for the City: Migrant Labor, Capital, and the State in Urban Africa*, edited by Frederick Cooper, 167-94. Beverly Hills: Sage, 1983.
Wolff, Leon. *In Flanders Fields: the 1917 Campaign*. New York: Ballantine Books, 1958.
Woolman, David S. *Rebels in the Rif: Abd el Krim and the Rif Rebellion*. Stanford: Stanford University Press, 1968.
Zuccarelli, F. "Le Régime des engagés à temps au Sénégal (1817-1848)." *Cahiers d'études africaines* 2 (1962): 420-61.

INDEX

Abidjan, 76, 110, 112, 113, 132, 153, 156
Adekampi-Thompson, Benjamin, 20
Adloff, Richard, 128, 147
Africa. *See* French West Africa; North Africa; West Africa
African military, collective memory of, 164–70
   contradictory experiences of African military under the French, 164–66, 169
   ingratitude to, by the Colonial State, 170
   reaction of African military to French colonial policy, 165
   Tirailleurs Sénégalais and their exploits as part of the, 166–69
   Thiaroye events as part of, 166, 169–70
African military ideology after Second World War, 141–45, 164
African officers
   benefits of serving as, 19, 20, 121, 126
   in the early *Tirailleurs Sénégalais*, 3, 4, 19–21, 38–42
   grievances of, 20, 65, 67
   in the interwar era, 64–69
   after 1945, 105, 112, 115, 117–22, 126
   number of, 117, 120–22, 126
   and promotion, 20, 117, 119, 120
   reasons for creating, 64–65
   routes to becoming, 65, 66, 122–26
   schools for, 118–23, 125
Afrique Nouvelle, 110
Airaines, 68, 166, 168
Alcoholism, and African soldiers, 140, 203 n. 95
Algeria, 35, 39, 68, 77, 105, 106, 109, 116, 162, 163
Anglo-French operation at the Suez Canal (1956), 5
Archinard, Louis, 19, 28, 40, 69
Armistice of 1918, 31, 35, 42
Armistice of 1944, 100
Assimilation
   contradictions in the ideal of, 147, 164–66
   of West African colonials to France, 3, 6, 19, 102, 104, 105, 117, 147, 165, 166, 188 n. 95
Association Amicale des Anciens Militaires Coloniaux de Carrière (AAAMCC), 129, 131
Association des Anciens Combattants et Victimes de Guerre de l'Afrique Occidentale Française (AACVGAOF), 129–32, 143, 144, 154–58, 163
Association Générale des Amputés et Grands Blessés de L'Afrique Occidentale Française (AGAGBAOF), 130
Association Quatrième République, 131
Atlantic Charter, 104, 147
Azan, Paul, 35

Bakary, Djibo, 132
Bamako, 15, 23, 79, 106, 110, 112, 135, 148, 155, 163
Bambara states, 14, 15, 18, 28, 32, 44, 63, 115, 140
Bastille Day, 2, 143–44, 163, 169
Bataillon d'Afrique, 8
Baye, Samuel, 129, 130, 155
Baye, Paye Dame, 129, 130
Béchard, High Commissioner, 150
Belgium, 4, 32, 33
Benoit, General, 67
Bingerville, 66, 76
Bloc Démocratique Sénégalais (BDS), 150, 151, 156, 163
Bodo, Biaka, 151
Boisseson, Gilbert de, 112, 121
Bonuses (or premiums) for enlistment, 8, 10, 12, 18, 27, 164
Booty, as incentive for military service, 11, 13, 18
Bordeaux, 39
Bouet-Willaùmez, Governor, 14
Brazzaville, 104, 117, 142, 147–49, 165
Brevet d'etudes du premier cycle (BEPC), 118, 120, 122, 123
British West Africa, 44, 49, 63, 64, 73, 84, 127, 137
Bruge, Roger, 87, 94, 95

229

Index

Cambon, Governor, 143–44
Cameroun, 75
Canada, and the First World War, 34, 37
Cap Vert peninsula, 58, 132
Casablanca, 2, 68, 89, 100
Casamance, 10, 48, 52, 64, 68, 75, 116, 138
Catroux, General, 152
Chad, 75, 76, 98, 157
Chamber of Deputies, 44
Chailley, Major, 105
Chasselay-Montluzin, massacre of *Tirailleurs Sénégalais* at, 166–69
Chemin-des-Dames, 30, 34, 36, 37
Cissé, Souleyman, 157
Clemenceau, Georges, 31, 41, 43–46
Clozet, Governor, 50
Colonies, Ministry of, 32, 48, 79
Communists, 130, 148, 151, 161, 163, 169
Composition of the *Tirailleurs Sénégalais*, 1, 112
   African ethnic and racial groups in, 14–15, 53–56, 59–60, 63, 64, 112
   people largely of low social origin, 14, 18, 24, 63, 112, 125
   porters, 12, 13
   prisoners of war, 12, 18, 27
   skilled soldiers, 112–17
   slaves, 8–10, 12–14, 18, 20, 46
   sons of families of African notables, 15, 18, 20, 24, 45–46, 62, 63, 65, 125
   volunteers, 9, 10, 12–14, 18, 42–46, 58, 74–76, 85, 86, 99, 107–10, 112, 113, 130, 134, 166
Conakry, 76, 112, 156
Condition humaine, 150, 161
Conscription in French West Africa, 5, 7–32, 47–69, 85, 88, 166
   debate regarding, 32, 36, 42–46, 83–86, 88
   and demographic considerations, 47–51, 69
   and mobile draft boards in operation, 51–64
   unpopularity in general of, 41, 45, 46, 63, 69, 74, 83–84, 86
Conscription Law of 1912, 29–30, 42, 84, 89
Conscription Law of 1919, 5, 15, 26, 47, 51, 73, 84, 89
Côte d'Ivoire, 13, 15, 20, 26, 48, 57, 61, 66, 71, 72, 74, 75, 77, 92, 93, 95, 101, 106, 108, 110, 111, 117, 127
   and veterans, 131, 133, 135, 138, 141, 143, 144, 151, 152, 155–58, 160, 161, 163, 169
Cotonu, 75, 79, 137, 143, 144, 156
Counterinsurgency operations. *See* Indochina; North Africa
Cournarie, Governor-General, 103

Dahomey, 20, 22, 44–46, 57, 74, 75, 79, 89, 108, 110–12
   and veterans, 135, 137, 140, 142–44, 155, 156, 160
Dakar, 26–27, 44, 52, 57, 61, 64, 73, 76, 80, 82, 89, 100–02, 104, 106, 110, 112, 113, 115, 116, 122, 162
   and veterans, 129–32, 135–37, 153–55, 169
d'Arboussier, Gabriel, 148, 151, 161
D-Day, 104
de Gaulle, Charles, 68, 88, 93, 98–99, 102, 104, 106, 130, 141, 148, 152, 154, 156, 157, 165, 170
de Larminat, General, 107, 115
de Lattre, Jean, 88, 98, 100, 104
de Vogüé, Eugène Melchior, 28
Death rates
   for European troops in Africa, 8, 10, 14, 21, 80, 81
   for *Tirailleurs Sénégalais*, 21, 77, 80–82, 84–85, 88, 96, 97
   from combat, 21, 22, 46, 88, 96
   from disease, 21, 77, 84–85, 97
   from drunkenness, 21
   in prison camps, 88, 96, 97
Defferre, Gaston, 162
Delmas, Robert, 154–56
Dia, Mamadou, 150, 151
Diagne, Blaise, 40–42, 44–46, 68, 102, 130
Diagne, Pierre, 129, 131
Diagne Laws of 1915 and 1916, 45
Diakité, Yoro, 125
Diallo, Al Hajj Doudou, 132, 147
Diallo, Bakary, 38, 89
Diallo, Hassana, 133
Diarra, Mamadou Otto, 129, 131
Diémé, Pierre, 64, 68, 69
Dien Bien Phu, 5, 110
Diop, Birago, 140–41, 216
Diop, Ogo, 131
Diori, Hamani, 132, 152, 158
Diouf, Galandou, 44, 102, 117
Diouf, Mustapha, 64, 68, 69, 117
Diouf, Sogui, 154
Disease environment in Africa, 8, 10, 14, 21, 39, 189 n. 34, 190 n. 57
   See also Health
Dodds, General, 20
Douta, Papa Seck, 102, 129–30, 132, 154, 155, 157, 158
Duguay-Clédor, Amadou N'Diaye, 41

Ecoles d'application, 122
Ecole de Formation des Officiers du Régime Transitoire des Territoires d'Outre-Mer (EFORTOM), 122–23, 125, 126
Ecole d'Elèves-officiers, 118
Ecole des Enfants de Troupe (EETs), 66, 112, 118, 139, 199 n. 102

Index

Ecoles Militaire Préparatoires Africaines (EMPAs), 118–20, 123, 125, 199 n. 102
Ecole Spéciale des Sous-officiers Indigènes, 66, 68
Ecoles Spéciales Militaires Interarmes (ESMIAs), 122
Emancipation Act, 9
Equality Law of 1950, 136, 147, 153, 161, 162
Essor, L', 112, 161, 163
Europe, 34, 35, 50, 68, 71, 88, 91, 101, 106, 170

Faidherbe, Louis, 2, 7, 8, 10, 12, 19–22, 66, 163
Fall, Ahmadu, 64, 66–69, 118, 121–22, 200 n. 21
Fall, Papa Guèye, 129–31
Fanon, Frantz, 2
Féderation Nationale des Prisonniers de Guerre (FNPG), 129–31
Finnegan, Greg, 137–38
First World War, 8, 25, 30, 94–96, 108, 144, 150, 159
  black African troops in, 2, 5, 15, 18–21, 25, 26, 30–31, 33–39, 66, 68, 77, 86, 88, 91, 99, 130, 137, 166
  claims of veterans of, 160
  effects of, on France, 43, 46
  effects of, on West Africa, 46, 64–65, 87–88, 137–38, 147
  ideology in, 89
  and resistance to conscription, 71
Force Noire, La, 25, 28–34, 50, 65, 85
Forced labor (*indigénât*), 74, 105, 188 n. 1
  abolition of, in colonies, 105, 106, 117, 151
Fourth Republic, 103, 106, 148, 149, 162, 163, 165
France, 42, 79, 80, 89, 94, 99, 104, 154
  attitude of, toward black Africans, 2, 101, 102, 147
  demographic fears of, 29, 32, 43, 48–50, 69
  failure of, to honor its commitments to African veterans, 147, 159
  honoring of its commitments to African veterans by, 161
  loyalty of *Tirailleurs Sénégalais* to, 155, 163
  military schools in, 20, 65, 66, 68
  Second World War and, 87, 88, 94, 96, 97, 103
  use of colonial subjects militarily by, 2, 4–5, 25, 26, 30–32, 68–70, 75, 80, 83–104, 132, 133, 141
Free French Army, 68, 69, 87, 88, 98, 99, 104, 117
  "whitening" of, 98, 104
Fréjus, 35, 66–68, 122, 123
French citizenship, extension of, to West Africans, 44–46, 57, 66, 89, 99, 102, 107, 116, 117, 129, 130, 149, 165

French empire, 70, 77, 80, 106, 108, 165
French Equatorial Africa, 15, 67, 75
French military academies for officers (Saint-Cyr, Ecole Polytechnique, Lyons, Bordeaux, Saint-Maixent, Saumur, Fontainebleau, Versailles, Vincennes), 65
French military records, 1
  omissions in, 14, 58, 87, 88, 108, 110, 113
French Revolution, 19, 165
French West Africa (FWA), 4, 21, 40, 41, 69, 79, 87, 88, 97, 98, 101, 115–18, 125, 159
  birth of African electoral politics in, after Second World War, 146
  civilian authority in, after Second World War, 107, 136, 147, 153–58, 162
  decentralization and regionalism, 162, 163
  demographics of, 1, 29, 31, 32, 43, 47–51
  education in, 105, 110–11, 139, 145
  emerging African elite in, 15, 18, 20, 24, 45–46, 62, 63, 65–69, 112, 118, 126, 165
  ethnic and racial groups in, 14–17, 53–56
  French-African relations in, after Second World War, 105–07, 112, 136, 138, 139, 143, 161
  French militarization of, 4, 7, 13, 31, 77, 84
  growth of, after Second World War, 105, 145
  impact of First World War on, 6, 61, 89, 102–03
  impact of Second World War on, 103–45, 152, 164–66
  independence for, 5–6
  labor shortage in, 29, 32, 46, 48, 50, 82
  nonmilitary groups in, 1, 4, 20, 127
  revolts against French rule in, 41, 45, 46, 106
  social formation in, 3–6
  tensions in, after Second World War, 106, 112, 142, 148, 151, 169
  See also Tirailleurs Sénégalais regiments
Frey, Henri, 22
Fulbe, 48, 72, 133
Futa Jalon, 29, 50

Gabon, 10, 64, 91, 168
Gadel, General, 65
Gamelin, Maurice-Gustave, 93
Gassouin, General, 32, 48, 50
Gazette du Tirailleur, La, 89, 92
German East Africa, 32
Germany, 5, 28, 33, 43, 50, 89, 91–96, 132, 142
  attitude of, toward French colonial troops in First and Second World Wars, 2, 33, 35, 94–97, 102, 148, 167, 168
Ghana, 146
Gobineau, Hélène de, 97
Gold Coast, 72, 73, 138, 142, 146, 162
Gomis, Henri, 41

Gouraud, General, 15, 89
Great Britain, 9, 10, 105, 126, 137, 142, 146
    attitude of, toward French colonial troops in First and Second World Wars, 2, 34, 35, 100
    colonial units of, 2, 4, 5, 7, 126
    and Second World War, 88, 89, 104
Gregory, Joel W., 85
Guderian, Heinz Wilhelm, 94
Guèye, Lamine, 102, 129, 131, 148–50, 158
Guignard, General, 67
Guinea, 9, 15, 18, 20, 22, 26, 29, 48, 50, 57, 72, 74–76, 79, 82, 106, 108, 110, 111
    and veterans, 133, 141, 143, 155–57, 159–61

Haig, Alexander, 34
Hausa, 63, 112, 132
Health, 21, 36, 51, 110–16, 120–23, 165
    demand for health benefits by African veterans, 144, 145, 153
    See also Disease environment in Africa
Hitler, Adolf, 2, 94, 193 n. 36
Hivernage, 35, 72, 76, 99
Ho Chi Minh, 105
Horne, Alistair, 34, 36
Houphouet-Boigny, Félix, 136, 148, 151, 152, 157, 158, 161, 162
Houphouet-Boigny Law, 106, 152

Indépendants d'Outre-Mer (IOM), 148, 150–52, 161
Indochina, 22, 68, 69, 75, 105, 109–10, 121, 154, 165
    colonial counterinsurgency operations in, 2, 104, 109–10, 138, 142, 153, 196 n. 26
Indochina amendment of 1951, 108–09
Interministerial commission's survey and partial census (1922–1923), 48–51
Irving, David, 96–97
Israel, Adrienne, 146
Italy, 69, 88, 98, 142

Jacobinism, 3, 19, 37, 89, 119, 165
Jaurès, Jean, 28

Kamara, Kandé, 78, 141
Kankan, 15, 18, 156, 157
Kann, Reginald, 36
Keegan, John, 36, 38
Keita, Maliké, 156
Kindia, 26, 76, 159
Kita, 15, 133
Koenig, General, 104
Kondé, Laqui, 92, 141
Korhogo region, 61, 92, 93, 163
Koulibaly, Baba, 32–33
Koum-Kelé, Battle of, 40
Krim, Abdel, 130

Labor camps of Germay in Second World War (*Fronts-Stalags*), 87, 96, 132
Labor reserve in West Africe, 1, 32, 46, 48, 50
Lambaye (*cercle*), 131
Lawler, Nancy, 92, 127
League of Nations, 75
Leclerc, General, 104
Levant, 35, 80, 94
Liba, Dibo, 157
Libreville, 67
Liger, Henri, 158–60
Liger campaign: veterans accommodated (1950–1956), 159–62
Loi-cadre (Framework Law), 162, 163, 210 n. 122
Louis XIV, 19
Ly, Abdoulaye, 166
Lyons, 42, 166, 167

McWilliams, James L., 37
Madagascar, 2, 9, 10, 18, 21, 22, 75, 105
Mademba, Abdel Kader, 21, 38–42
Maginot Line, 92
Magnan, General, 107, 117–18
Mali, 69, 125
Man, 133, 136, 138
Mande states, 13, 15
Mangin, Charles, 15, 25, 28–32, 34–36, 40–43, 47–48, 50, 64–66, 77, 85, 89, 94
Marbeau affair, 9
Marceau, Hippolyte-Victor, 7, 14–15
Marne, 42
Marraines (godmothers), 77, 78
Marseilles, 44, 78, 89
Marshall, S. L. A., 38
Marx, Karl, 4, 176 n. 15
Mauritania, 20, 46, 57, 69, 108, 110, 111, 131
M'Baye, Macodou, 20
M'Bengue, M'Baye, 116, 147
Médaille militaire, 141
Memmi, Albert, 3
Messmer, High Commissioner, 69
Migration in French West Africa, 75–80, 137–38
    absentee rates of men on draft lists (1923–1946), 70–75
    methods of, 71-72, 75, 85
    to other parts of Africa, 75–77
    overseas, 75, 77
    reasons for, 1, 49, 71–74
    as resistance to conscription, 71–75, 85
    risks of, 71, 73
    seasonal, 72
    and substitution, 73, 75, 76, 83, 85
    and women and families, 77–80
Mitterand, François, 158, 206 n. 25
Molabé, Karmoko Alfa, 133
Monshire Camp (at Huyton, England), 100

Index

Moors, 48, 64
Morganthau, Ruth, 149–50
Mori, Sarankeni, 15
Morocco, 21, 25, 27, 30, 34, 35, 38, 42, 67, 69, 77, 80, 89, 115, 136, 162
Morts qui parlent, Les (de Vogüé), 28
Moscow, 151
Mossi, 15, 29, 44, 63, 137, 169
Muller, Major, 155
Muslims, 61, 75, 89, 101–02, 106, 169, 200 n. 21
   vs. Christians, 18, 21, 80, 130, 140

Namoro, Bahouna, 157
Napoleon III, 10, 51
National Assembly, 68, 106, 109, 148, 151, 153, 159, 161, 166
NATO, 106
Navétanes (seasonal migrants), 72
N'Dao, Doudou, 116, 147
Nehring, Oberst, 94
N'Guetta, Edmond, 93
Niger, 18, 29, 38, 46, 48, 57, 63, 74–76, 106, 108, 110–12, 119, 125
   and veterans, 131, 132, 135, 137, 151, 152, 155–58, 160
Nigeria, 63, 64, 75, 76
Nivelle, General, 36
North Africa, 35, 78, 80, 88, 98, 117, 154, 165
   counterinsurgency operations in, 2, 18, 105, 109
N'Tchórérè, Charles, 64, 67–69, 87, 91, 166–68
N'Tchórérè, Jean-Baptiste, 87, 166
Nyo, General, 119

Office National des Anciens Combattants, 130, 132, 153–61
   budget of, 154–55
   criticism of, 158
   policy of, to try to control veterans' affairs, 156–61
Oran, 150
Ottomans, 13
Ouagadougou, 134–35, 138, 139
Ouattara, Namongo, 152

Painlevé, Paul, 34–35
Panzer divisions, 92, 167, 168
Paris, 46, 93, 97, 148–51, 153, 154, 158, 159, 161–63
Parti Africain de l'Indépendance, 170
Partisans, 99, 103
Petit nègre vocabulary of the *Tirailleurs Sénégalais*, 115–17, 140, 164
Piché, Victor, 85
Poison gas, 34, 36, 37
Ponty, William, 25, 29, 30, 50

Porto-Novo, 20, 75, 110, 111, 137, 144
Portugal, 10
Portuguese Guinea, 75
Poulantzas, Nicos, 4
Prison and transit camps, 87, 94, 96–99
   composition of, 96–97
   hardships of, 96–99
Prisoners of war (POWs), 93–97, 100, 101, 103, 110, 128, 129, 131, 132, 142, 149, 158, 161, 169, 170
Provence, 87, 92, 98, 99, 104
Puypeyroux, General, 65

Rachat (or repurchase) system, 8–12, 27
Racial bias in French West Africa and the African army, 78, 80, 94, 100, 117, 121, 142–43, 164, 175 n. 5, 193 n. 36
Reassemblement Démocratique Africain (RDA), 132, 138, 143, 144, 147, 148, 151–53, 157, 158, 161, 163, 169
Reassemblement du Peuple Français (RPF), 157
Rathbone, Richard, 146
Recruitment for the *Tirailleurs Sénégalais*, 40–42, 46, 132
   during the Conquest era (1886–1905), 11–15, 18–19
   early methods of, 8–11
   incentives for, 8–11, 13–14, 18
   incorporation of prisoners-of-war and porters, 12–13
   rachat system, 8–12
   rates of, 80, 82
   and re-enlistment, 82, 91
   resistance to, 18–20
   skill levels of soldiers of, 113–17
   volunteering, 9, 10, 12–14, 18, 30, 42–43, 45–46, 58, 60, 85, 99
   after Second World War, 107–13, 121
Régiments d'Infanterie Coloniale Mixtes Sénégalais (RICMS), 91
Reinach, Olivier, 155, 156
Resistance, the, 97, 104
Réveil, 131, 144, 166, 169
Rhineland, 15, 35, 67, 94
Rommel, Erwin, 2, 96
Roosevelt, Franklin Delano, 99

Sahara, 5, 14, 48
Sahel, 48, 64
Saint-Cyr, 65–67, 122, 198 n. 89
Saint-Louis, 12, 14, 20, 22, 38, 64, 66, 68, 69, 76, 80, 91, 115, 120, 129–31, 163
Saint-Maixent, 65, 66, 69, 123
Sala, Sidibé, 157
Salan, Raoul, 98
Sané, Mamadou Pop, 116, 117
Sansanding, 38, 42
Sarambé, Dramane, 61

Sarraut, Albert, 50
Saumur, 65, 66, 123
Schleh, Eugene, 146
Schools for children of African notables to ensure their cooperation with France, listing of, 20
Seck, Doudou, 135-36
Second Constituent Assembly in Paris, 148, 149
Second World War, 5, 38, 45, 50, 78
  African consciousness of themselves after, 103, 104, 106, 112, 118, 136, 141-45, 152, 164-66
  birth of African electoral politics after, 146
  contribution of *Tirailleurs Sénégalais* in, 92, 94-95, 98, 103, 104, 166-67
  impact of, on Africans, 87-88, 100-05, 147
  impact of, on France, 104-05
  military contribution of French West Africa, in, 88, 104
  See also Second World War and the *Tirailleurs Sénégalais*
Second World War and the *Tirailleurs Sénégalais*, 87-104, 159, 160
  change in attitude toward France after, 102-04
  combat, 92-96, 127
  ideology and mobilization, 89-92, 166
  integration of African regiments with French ones, 91-92
  prison and transit camps, 94, 96-99
  and repatriation to French West Africa, 97-99
  Thiaroye and other soldiers' protests, 100-04
Section Française de l'International Ouvrière (SFIO), 148, 150, 151
Ségou, 18, 38, 40, 79, 140
Sembène, Ousmane, 170, 216
Senegal, 2, 3, 7-10, 14, 15, 19-22, 45, 46, 48-49, 52, 66, 68, 69, 72, 74-76, 80, 82, 89, 91, 100, 102, 106, 108, 110, 111, 113, 115, 116, 119
  privileged position of, regarding conscription, 57-58
  and veterans (1945-1960), 129-32, 135, 136, 138, 143, 145, 148-50, 154, 155, 163, 166, 169, 170
Senegalese National Army, 68, 69
Senegambia, 7, 8, 18, 38
Senghor, Léopold Sédar, 2, 3, 97, 102, 130-32, 148-51, 156, 161-63, 165, 170
Senoufo speakers, 139
Shanin, Teodor, 4
Sidibé, Moriba, 133
Sikasso, 12, 15, 22
Silue, Dossigutta, 101

Slavery
  and racial ideologies, 64
  and Western Europe, 9, 11-12
  winding down of, in Africa, 9, 11-12, 177 n. 25
Slaves into soldiers, 7-24, 46
  African officers in the early *Tirailleurs Sénégalais*, 19-21
  beginnings of, 7-8
  conscription vs. volunteers, 9, 10, 18, 42-46
  daily life, 21-24
  motivations for entering military service, 8-11, 13-14, 18, 62-63
  purpose of, 164
Sodegbé, Louis, 144, 155
Sofa (slave soldiers), 13
Soldiers' skills, 112-17
  and competence in French, 113-15, 117
  and recruitment, 113
  types of, 113, 115
Somme, 2, 37, 42, 68, 69, 92, 94, 96, 166, 168
Songhay speakers, 132
Soudan, 15, 22, 28, 32, 38, 41, 42, 49, 57, 61, 66, 69, 72, 74-76, 79, 82, 85, 108, 110-12, 119, 168
  and veterans, 113, 134, 140, 144, 148, 151, 155, 158, 161, 163
Soumaré, Lt. Colonel, 121-22, 159
Soviet Union, 99, 142
Spahis (cavalry corps), 64
Stanley, G. F. G., 34, 37
Steel, R. James, 37
Sub-Saharan Africa, 64
Sy, Mademba, 38-40, 42
Sy, Mahmadou Racine, 21, 38
Sy, M'Baye, 38
Syria, 42, 80

Tal, Al Hajj Umar, 38, 40
Tall, Seydou Nourou, 101-02
Tanganyika, 5
Thiaroye and other soldiers' protests, 100-04
  aftermath of, 101-05, 132, 147, 166, 169, 170
Third Republic, 28, 83
Thompson, Virginia, 127-28, 147
Tijani empire, 38, 40
Timbuktu, 15
Tirailleurs Sénégalais regiments
  benefits of service in, 3-4, 10, 19, 23-24, 69, 74, 105-13
  burdens of service in, 40-42, 71-74, 84, 109-10
  combat experiences of, 2, 5, 9, 20, 21, 25-27, 32-38, 130
  as a conquest army (1857-1905), 5, 7-24

Index

as a conscript army (1919-1945), 5, 46–70, 76
counterinsurgency operations, 2, 5, 18
diet of, 21, 22
evaluation of, 1, 3–4, 24, 33, 69, 89, 117–18, 121, 125–26
family life of, 1, 21–23, 77–79
founding of, 2, 7, 10, 163
growth of (1820-1914), 7–8, 10, 11, 25, 26, 28, 32, 47, 48
ideology of, 3, 10, 24, 38, 42, 45, 69, 79, 87, 88, 141–45, 164, 175 n. 10
images of, abroad and in French West Africa, 2, 18, 19, 32–38, 94–95
incentives to join, 10–11, 62–63, 88, 108–13, 164
as mercenaries, 12, 18, 24, 37, 42, 153
as a mirror of contradictions of colonialism, 3–4, 125–26, 164–66, 169
as an occupation army (1905-1919), 5, 24–26
phaseout of, 2, 78
professionalization of (1946-1960), 5–6, 10, 104–26
purpose of, 4–5, 8–10, 21, 46–51, 85, 105–06, 191 nn. 6, 11
quotes for, 51–57, 70–75, 112
size of, 8, 10, 11, 14, 25–27, 32, 48, 50, 88, 106, 108–09, 112, 195 n. 25
skill levels of, 113–17
as strikebreakers, 106, 153
tensions in, after Second World War, 105
upward mobility through, 68, 69, 84, 116, 138, 145
wages of, 21, 23, 67, 73, 75, 86, 112, 113, 117

Tirailleur Soudanais, Le, 7
Togoland, 32
Torodbe Fulbe clan, 38, 40
Touré, Samori, 13–15, 18
Traodec, René, 130, 154–60
  criticism of, 158, 160
Traoré, Bouba, 134
Trotsky, Leon, 4, 176 n. 13
Tuaregs, 18, 48, 64
Tukolor soldiers, 14–15
Tunisia, 3, 162
Tuo, Lielourou, 95–96

United States, 35, 96, 99, 142
  black American troops of, 2
Upper Volta, 15, 49, 57, 61, 69, 72, 74–76, 83, 85, 108, 110, 111, 115, 119, 125, 127
  veterans and, 134–35, 137, 139, 142, 145, 151, 157, 158, 160, 169
Upper-Senegal-and-Niger, 14, 26, 44, 48

Urbanization of French West Africa, 82, 83, 112, 113, 145

Van Vollenhoven, Joost, 41, 43–47
Verdun, 30, 34, 36, 41
Versailles, 100
Veterans' Affairs, Ministry of, 153–54, 168, 169
Veterans and French West Africa (1945-1960), 127, 129–31, 145–63
  and African Press, 148, 150, 161, 166
  and birth of African electoral politics after Second World War, 146, 148, 162
  definition of, 128
  discrimination against, 142–44, 147, 165
  diversity of, 144
  economic benefits for, 104, 116, 133–39, 144
  education for, 139, 165, 203 n. 87
  evaluation of, 133, 134, 138–39
  handicapped and disabled, 128, 135, 144, 150, 153, 161
  importance of, 128, 134, 147, 165
  and intermarriage, 140
  Liger campaign: veterans accommodated (1950-1956), 159–62
  lobbies of, 129–33, 145
  lodges of, as centers for a helping hand, 132, 201 n. 27
  loyalty of, to the French, 147, 162–66, 168, 169
  military careers for sons of, 123, 125, 139–40, 203 n. 89
  new African ideology, 141–45, 164–65
  number of, 128
  Office National des Anciens Combattants, 153–61
  and political parties, 145, 149–58, 161–62, 163
  preferential access to reserved jobs in the private and public sectors for, 134–35, 138, 145, 156, 159, 165
  questionnaire for, 171–73
  regionalism as a danger to, 162, 163
  sources on, 127–28
  as a special-interest, differentiated group, 127, 139–41, 144, 145
  struggle for justice by, 100, 104, 105, 152
  veteran vs. ex-soldier, 128
Veterans' return migration, 80–83
  distribution of veterans by residence (1952), 83
  and urbanization, 82, 83
Vichy interlude, 69, 88, 97, 102, 106, 129, 149
Villages de liberté, 27
Villard, Lt. Colonel, 120–21
Vimy Ridge, 2, 37
Von Lettow-Vorbeck, Paul, 5

Wells, H. G., 34
West Africa, 78, 100, 136, 153
   demographic data about, 48–51, 80–82, 85, 89, 103, 164
   distribution of the *Tirailleurs Sénégalais* in, 76
   French community in, 44, 106–07, 142
   group consciousness of Africans in, after World War II, 103, 104, 141–45
Wiltord, Governor, 150

Wolff, Leon, 34
Wolofs, 8, 9, 14

Yameogo, Maurice, 139
Yoruba speakers, 75
Ypres, 34, 36, 37
Yacé Philippe, 99, 155

Zerbo, Seye, 125
Zinder, 106, 107